# Bacterial Signal Transduction: Networks and Drug Targets

# ADVANCES IN EXPERIMENTAL MEDICINE AND BIOLOGY



---

Recent Volumes in this Series

Volume 623
ALTERNATIVE SPLICING IN THE POSTGENOMIC ERA
Edited by Benjamin J. Blencowe and Brenton R. Graveley

Volume 624
SUNLIGHT, VITAMIN D AND SKIN CANCER
Edited by Jörg Reichrath

Volume 625
DRUG TARGETS IN KINETOPLASTID PARASITES
Edited by Hemanta K. Majumder

Volume 626
GENOMIC IMPRINTING
Edited by Jon F. Wilkins

Volume 627
TRANSGENESIS AND THE MANAGEMENT OF VECTOR-BORNE DISEASE
Edited by Serap Aksoy

Volume 628
BRAIN DEVELOPMENT IN DROSOPHILA MELANOGASTER
Edited by Gerhard M. Technau

Volume 629
PROGRESS IN MOTOR CONTROL
Edited by Dagmar Sternad

Volume 630
INNOVATIVE ENDOCRINOLOGY OF CANCER
Edited by Lev M. Berstein and Richard J. Santen

Volume 631
BACTERIAL SIGNAL TRANSDUCTION
Edited by Ryutaro Utsumi

---

# Bacterial Signal Transduction: Networks and Drug Targets

Edited by
Ryutaro Utsumi, PhD
*Department of Bioscience, Graduate School of Agriculture, Kinki University, Nakamachi, Nara, Japan*

**Springer Science+Business Media, LLC**
**Landes Bioscience**

Springer Science+Business Media, LLC
Landes Bioscience



Printed in the USA

Springer Science+Business Media, LLC, 233 Spring Street, New York, New York 10013, USA
http://www.springer.com

Please address all inquiries to the publishers:
Landes Bioscience, 1002 West Avenue, 2nd Floor, Austin, Texas 78701, USA
Phone: 512/ 637 6050; FAX: 512/ 637 6079
http://www.landesbioscience.com

*Bacterial Signal Transduction: Networks and Drug Targets* edited by Ryutaro Utsumi, Landes Bioscience / Springer Science+Business Media, LLC dual imprint / Springer series: Advances in Experimental Medicine and Biology.

ISBN: 978-0-387-78884-5



**Library of Congress Cataloging-in-Publication Data**

A C.I.P. catalog record for this book is available from the Library of Congress.

# PREFACE

A decade has passed since Drs. Hoch and Silhavy edited their comprehensive work entitled *Two-Component Signal Transduction*. This fascinating book encouraged many microbiologists and students to enter the new world of signal transduction in microbiology. In 2003, Dr. Inouye edited *Histidine Kinase in Signal Transduction*, which focused on histidine kinases and presented the wealth of information accumulated on this protein family.

Bacteria usually possess a number of Two-Component Systems (TCSs), ranging from a few to over 100. In *E. coli*, 29 histidine kinases, 32 response regulators, and 1 histidine-containing phosphor transmitter (HPt) domain have been found by analyses of the K-12 genome. Several examples of in vitro and in vivo cross-talks and signal transduction cascades between TCSs in *E. coli* have been reported, which suggests the existence of a TCS network (Chapter 1).

Interactions among different TCSs enable one system to respond to multiple signals, which is important for bacteria to minutely adjust themselves to complex environmental changes. Such interactions are found or predicted in various bacteria in this book. Many of these interactions might be connected by small proteins such as B1500 (Chapter 1) and PmrD (Chapter 2). More examples of such proteins should be identified in the near future in order to fill-in the missing parts of the bacterial signal transduction network, a new paradigm that is increasingly recognized as the signal transduction pathway in bacterial cells. For drug discovery, this pathway is considered as important as the signal transduction pathway in animal cells (Chapters 15 and 16).

Over the past decade, a vast amount of exciting new information on the signal transduction pathway in bacteria has been brought to light. Reports on these developments have been put together in this book, *Bacterial Signal Transduction: Networks and Drug Targets*. My aim is to provide an incentive for graduate students, academic scientists, and researchers in the pharmaceutical industry to further elucidate the TCS networks and apply them in the search for novel drugs.

I express my sincere gratitude to all of the contributors for their excellent chapters, and I am also grateful to the staff of Landes Bioscience: Ronald G. Landes, Cynthia Conomos and Megan Klein for production of this book.

*Ryutaro Utsumi, PhD*

# ABOUT THE EDITOR...

RYUTARO UTSUMI, PhD, was born and raised in Osaka, Japan, and is a professor in the Department of Bioscience of the Graduate School of Agriculture, Kinki University, Japan. His research interests in bacterial signal transduction include the molecular mechanisms of the small protein B1500 connecting TCSs. He has a particular interest in drug discovery targeting bacterial signal transduction. Before coming to Kinki University, Dr. Utsumi received the PhD in Agricultural Chemistry from Kyoto University, where he studied under Prof. Tohru Komano. He studied bacterial signal transduction via hybrid receptor Tazl for 1 year (1988-1989) in the Department of Biochemistry (Masayori, Inouye), Robert Wood Johnson Medical School (Piscataway, NJ). Now, in order to develop a new class of inhibitors of bacterial signal transduction, including histidine kinases and response regulators, he is promoting the national project on the "Research and Development Program for New Bio-industry Initiatives (2006-2010) of Bio-oriented Technology Research Advancement Institution (BRAIN), Japan".

# PARTICIPANTS

Carl E. Bauer
Department of Biology
Indiana University
Bloomington, Indiana
USA

Dagmar Beier
Lehrstuhl für Mikrobiologie
Biozentrum
University of Würzburg
Würzburg
Germany

Andrew N. Binns
Plant Science Institute
Department of Biology
University of Pennsylvania
Philadelphia, Pennsylvania
USA

Cara C. Boutte
Department of Biochemistry
and Molecular Biology
The University of Chicago
Gordon Center for Integrative Science
Chicago, Illinois
USA

Daelynn R. Buelow
Department of Biological Sciences
University of Alberta
Edmonton, Alberta
Canada

Mark J. Buttner
Department of Molecular Microbiology
John Innes Centre
Norwich Research Park
Colney, Norwich
UK

Sean Crosson
Department of Biochemistry
and Molecular Biology
*and*
The Committee on Microbiology
The University of Chicago
Gordon Center for Integrative Science
Chicago, Illinois
USA

Dennis G. Cvitkovich
Oral Microbiology
University of Toronto
Toronto, Ontario
Canada

Sarah Dubrac
Biology of Gram-Positive Pathogens
Department of Microbiology
Institut Pasteur
Paris
France

Yoko Eguchi
Department of Bioscience
Graduate School of Agriculture
Kinki University
Nakamachi, Nara
Japan

Rong Gao
Center for Fundamental and Applied
Molecular Evolution
Departments of Chemistry and Biology
Emory University
Atlanta, Georgia
USA

Yasuhiro Gotoh
Department of Bioscience
Graduate School of Agriculture
Kinki University
Nakamachi, Nara
Japan

Eduardo A. Groisman
Department of Molecular Microbiology
Howard Hughes Medical Institute
Washington University School of Medicine
St. Louis, Missouri
USA

Roy Gross
Lehrstuhl für Mikrobiologie
Biozentrum
University of Würzburg
Würzburg
Germany

Dieter Haas
Département de Microbiologie Fondamentale
Université de Lausanne
Lausanne
Switzerland

Regine Hengge
Institut für Biologie-Mikrobiologie, FB Biologie, Chemie und Pharmazie Freie
Universität Berlin
Berlin
Germany

Hee-Jeon Hong
Department of Biochemistry
University of Cambridge
Cambridge
UK

Matthew I. Hutchings
School of Biological Sciences and School of Medicine
Health Policy and Practice
University of East Anglia
Norwich
UK

Sumiko Inouye
Department of Biochemistry
Robert Wood Johnson Medical School
Piscataway, New Jersey
USA

Akinori Kato
Department of Bioscience
Graduate School of Agriculture
Kinki University
Nakamachi, Nara
Japan

Yi-Han Lin
Center for Fundamental and Applied Molecular Evolution
Departments of Chemistry and Biology
Emory University
Atlanta, Georgia
USA

David G. Lynn
Center for Fundamental and Applied Molecular Evolution
Departments of Chemistry and Biology
Emory University
Atlanta, Georgia
USA

Dawn M. MacRitchie
Department of Biological Sciences
University of Alberta
Edmonton, Alberta
Canada

Tarek Msadek
Biology of Gram-Positive Pathogens
Department of Microbiology
Institut Pasteur
Paris
France

Hirofumi Nariya
Department of Biochemistry
Robert Wood Johnson Medical School
Piscataway, New Jersey
USA

Ario Okada
Department of Bioscience
Graduate School of Agriculture
Kinki University
Nakamachi, Nara
Japan

Nancy L. Price
Department of Biological Sciences
University of Alberta
Edmonton, Alberta
Canada

Erin B. Purcell
Department of Biochemistry
and Molecular Biology
The University of Chicago
Gordon Center for Integrative Science
Chicago, Illinois
USA

Tracy L. Raivio
Department of Biological Sciences
University of Alberta
Edmonton, Alberta
Canada

Dilani Senadheera
Oral Microbiology
University of Toronto
Toronto, Ontario
Canada

Yoshitsugu Shiro
Biometal Science Laboratory
RIKEN SPring-8 Center
Harima Institute
Kouto, Sayo
Hyogo
Japan

Toru Tobe
Division of Applied Bacteriology
Graduate School of Medicine
Osaka University
Osaka
Japan

Ryutaro Utsumi
Department of Bioscience
Graduate School of Agriculture
Kinki University
Nakamachi, Nara
Japan

Claudio Valverde
Programa Interacciones Biológicas
Departamento de Ciencia y Tecnología
Universidad Nacional de Quilmes
Bernal
Argentina

Takafumi Watanabe
Department of Bioscience
Graduate School of Agriculture
Kinki University
Nakamachi, Nara
Japan

Jiang Wu
Department of Biology
Indiana University
Bloomington, Indiana
USA

Seiji Yamada
Biometal Science Laboratory
RIKEN SPring-8 Center
Harima Institute
Kouto, Sayo
Hyogo
Japan

# CONTENTS

## 7. DUAL REGULATION WITH SER/THR KINASE CASCADE AND A HIS/ASP TCS IN *MYXOCOCCUS XANTHUS* ..................111

Sumiko Inouye and Hirofumi Nariya



## 8. TWO-COMPONENT SIGNALING SYSTEMS AND CELL CYCLE CONTROL IN *CAULOBACTER CRESCENTUS* .................. 122

Erin B. Purcell, Cara C. Boutte and Sean Crosson



## 9. REGB/REGA, A GLOBAL REDOX-RESPONDING TWO-COMPONENT SYSTEM .................. 131

Jiang Wu and Carl E. Bauer

# CHAPTER 1

# Introduction to Bacterial Signal Transduction Networks

**Yoko Eguchi and Ryutaro Utsumi***

## Abstract

Transcriptional analysis using a DNA microarray is an extremely efficient method for analyzing two-component signal transduction networks. Here we introduce three such networks in *Escherichia coli* that were clarified using a DNA microarray: a PhoQ/PhoP system that senses extracellular $Mg^{2+}$ and controls the gene expression for adaptation to environmental $Mg^{2+}$ deprivation ($Mg^{2+}$ stimulon); an EvgS/EvgA signal transduction system that activates expression of multiple drug efflux pumps and acid resistance genes; and a signal transduction cascade between EvgS/EvgA and PhoQ/PhoP.

## Introduction

Bacteria uses the two-component system (TCS) to sense environmental stresses and transduce the information inside the cells for adaptation. This system is basically composed of a histidine kinase (HK, sensor) residing in the inner membrane and a cognate response regulator (RR) in the cytoplasm. In *E. coli,* 29 HKs, 32 RRs and 1 HPt (histidine-containing phosphotransmitter) domains have been found by analyzing the *E. coli* K-12 genome (Fig. 1).[1] In addition to each sensor responding to individual environmental stresses, a complex regulatory network between the TCSs is assumed to exist, enabling *E. coli* to cope with a variety of environmental conditions.

In order to analyze such TCS networks of *E. coli,* Oshima et al systematically examined the mRNA profiles of 36 two-component deletion mutants, which included all of the TCSs (http:ecoli.aist-nara.ac.jp/xp_analysis/2_components).[2] Their results proposed the existence of a network of functional interactions such as cross talks and cascades of signal transductions between different TCSs. This complex regulatory network contributes to the fine-tuning of the adaptive ability of the cell, which leads to better survival. This chapter focuses on two TCSs, the PhoQ/PhoP and EvgS/EvgA systems and describes a signal transduction cascade from EvgS/EvgA to PhoQ/PhoP TCSs as one example of complex regulatory networks.

## $Mg^{2+}$ Stimulon (PhoQ/PhoP TCS)[3]

The PhoQ/PhoP TCS was first recognized in *Salmonella enterica* serovar *Typhimurium* as a regulatory system that monitors the availability of extracellular $Mg^{2+}$ (see Chapter 2). The PhoQ protein functions as an $Mg^{2+}$ sensor[4] and, in the presence of micromolar concentrations of $Mg^{2+}$, phosphorylates the PhoP regulator. Phosphorylated PhoP activates the transcription of some 30 different genes.[5,6] The PhoQ/PhoP system is present in many nonpathogenic, gram-negative bacteria, suggesting that it plays a fundamental physiological role in response to $Mg^{2+}$ starvation.[7] In

*Corresponding Author: Ryutaro Utsumi—Department of Bioscience, Graduate School of Agriculture, Kinki University, 3327 204 Nakamachi, Nara 631-8505, Japan.
Email: utsumi@nara.kindai.ac.jp

*Bacterial Signal Transduction: Networks and Drug Targets,* edited by Ryutaro Utsumi.

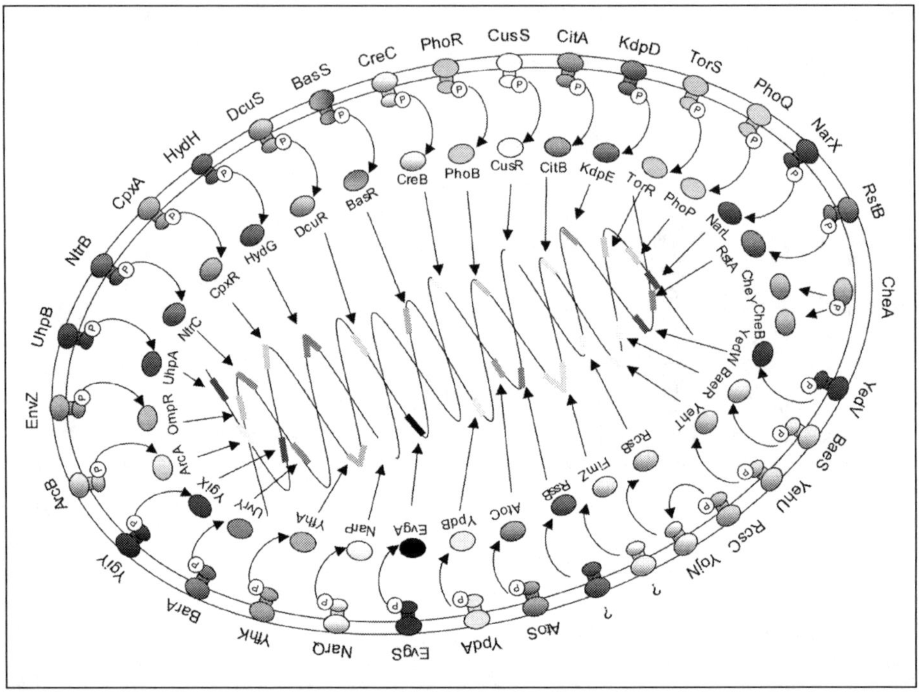


Figure 1. Schematic diagram of the TCSs of *E. coli*. Histidine kinase sensors reside in the inner membrane and cognate response regulators in the cytoplasm. Each sensor responds to individual environmental stresses to cope with the fluctuating environmental conditions that *E. coli* faces. In addition to signal transduction from a sensor to its cognate regulator, a network of functional interactions is proposed, including crosstalks and signal transduction cascades between different TCSs. Used with permission from Eguchi Y, Utsumi R. Signal Transduction Network for Environmental Adaptation in Echerichia coli. From survival and death in bacteria. In: Yamada M, ed. Survival and Death in Bacteria. Research Signpost, 2005:70.

order to identify the member genes of the $Mg^{2+}$ stimulon in *E. coli*, we carried out a genome-wide transcription profile analysis in the presence or absence of $MgCl_2$ by using a DNA microarray. The mRNA levels of W3110 (wild type) in the presence of 30 mM $MgCl_2$, WP3022 (*phoP* defective) and WQ3007 (*phoQ* defective) were compared with those of W3110 in the absence of $MgCl_2$. The expression ratios, for a total of 232 genes, were <0.75 in all three strains (http://www.nara.kindai.ac.jp/nogei/seiken/array.html), suggesting that the PhoQ/PhoP system is involved directly or indirectly in the transcription of these genes. Of those, 26 contained the PhoP box-like sequences with direct repeats of (T/G)GTTTA within 500 bp upstream of the initiation codon. Furthermore, S1 nuclease assays of 26 promoters were performed to verify nine $Mg^{2+}$ stimulon genes: *vboR, hemL, nagA, rstAB, slyB, yrbL, phoPQ, mgrB* and *mgtA*. Using gel shift and DNase I footprinting assays, all of these genes were found to be regulated directly by PhoP. Accordingly, we concluded that *phoPQ, mgrB, mgtA, hemL, nagA, rstAB, slyB, vboR* and *yrbL* genes constitute the $Mg^{2+}$ stimulon in *E. coli* (Fig. 2).

## EvgS/EvgA TCS

The EvgS/EvgA TCS[8] in *E. coli* is highly homologous to the virulence-related BvgS/BvgA system of *Bordetella pertussis* (see Chapter 10). A BLAST search indicates the presence of homologous systems in *Klebsiella pneumoniae, Vibrio cholerae* and *Pseudomonas aeruginosa*. EvgS is a histidine kinase hybrid sensor and it is composed of an N-terminal periplasmic region and

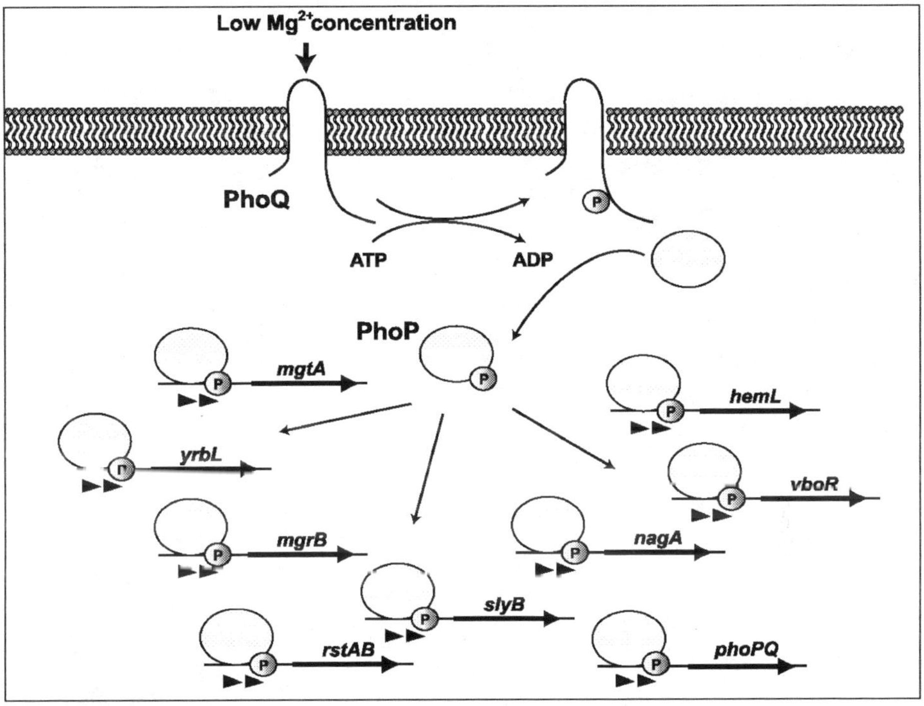


Figure 2. Proposed $Mg^{2+}$ stimulon of *E. coli*.

a C-terminal cytoplasmic region that is divided into four domains: linker, transmitter, receiver and output (Hpt).[9] The periplasmic domain of EvgS is involved in signal recognition, ultimately transducing the signal into a transcriptional regulation network via a cascade of phosphorylation.[8,10] Autophosphorylation of the cytoplasmic region of EvgS was reported to be inhibited in vitro by an oxidized ubiquinone-0,[11] as is also the case with the anaerobic sensor ArcB.[12]

The EvgS/EvgA system is not activated in mid-exponential-phase cells grown in a rich medium. The environmental signal to which the EvgS sensor responds remains unidentified, but studies using a mutant with a constitutively active EvgS (*evgS1* mutant)[13] or using overexpression of the response regulator EvgA have demonstrated that the EvgS/EvgA system confers multidrug resistance to a drug-hypersusceptible strain that lacks constitutive multidrug efflux genes *acrAB*[14,15] as well as acid resistance to exponentially growing cells.[16,17]

To examine transcriptional levels of multidrug efflux pump genes in an *E. coli* mutant expressing the constitutive mutant EvgS1, an *E. coli* DNA microarray analysis was performed (http://www.nara.kindai.ac.jp/nogei/seiken/array2.html).[18] As a result, most of the putative multidrug efflux-related genes were below the detection level under the steady-state of the exponential growth phase, but at least five genes or operons, *emrKY, yhiUV, acrA, mdfA* and *tolC*, showed increased expression in the presence of the constitutive mutant EvgS1.[18] Among these five genes or operons, only *emrKY* is directly regulated by EvgA; the others require additional transcriptional regulators. Thus, activation of the sensor EvgS initiated a transcriptional network, leading to multidrug resistance by a coordinated expression of multidrug efflux pump genes.[18]

In addition to the up-regulation of the multidrug efflux pump genes, activation of EvgS also up-regulated the acid-resistance-related genes, *gadA, gadBC, hdeAB, ydeP, ydeO* and *gadE*, as in the EvgA-overexpressed strain.[16,17] Acid resistance was also conferred to an exponentially grown *evgS1* mutant (to be submitted). Interestingly, *yhiUV*, which are multidrug efflux pump genes

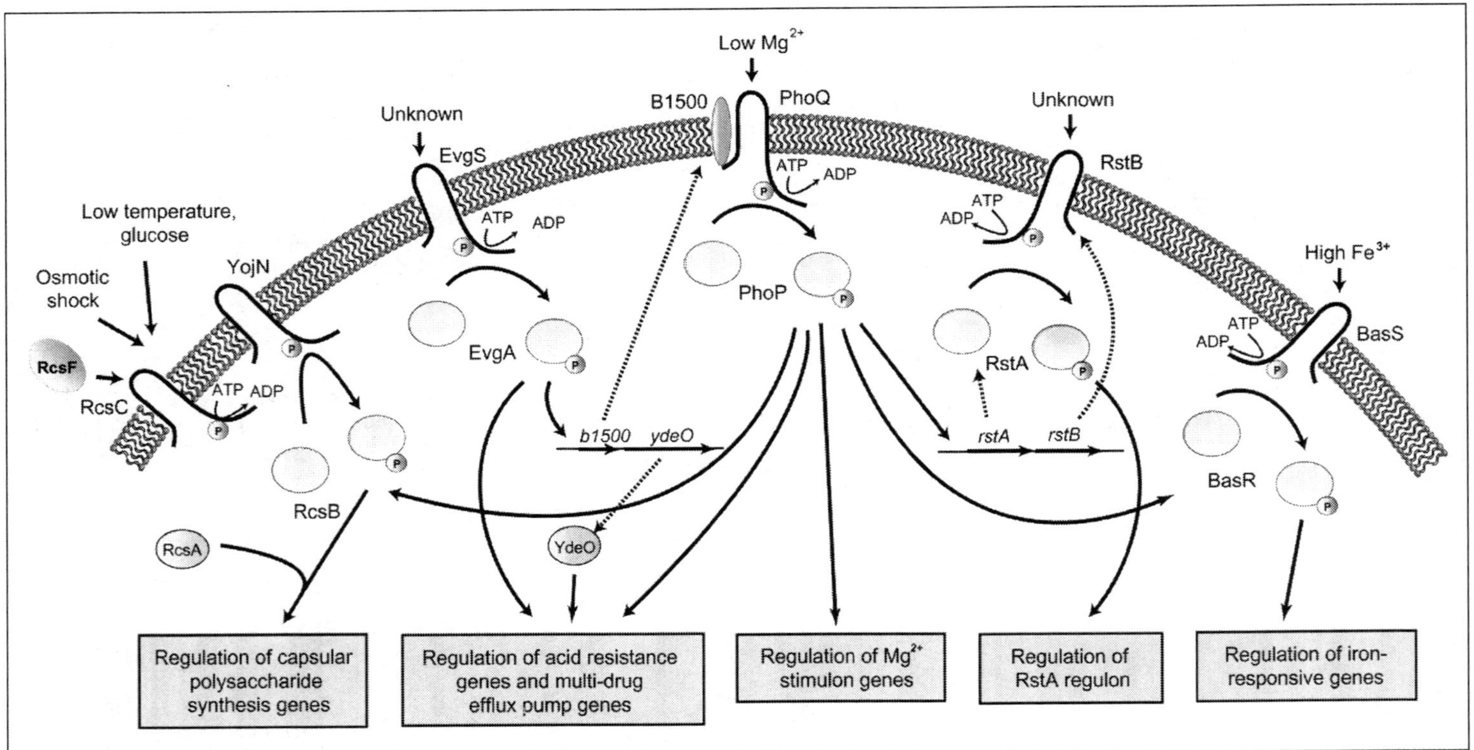


Figure 3. Signal transduction cascade in *E. coli*. Multiple signal transduction cascades have been found in *E. coli* using microarray analyses. The EvgS/EvgA system activates the PhoQ/PhoP system with B1500, whose gene expression is regulated by EvgS/EvgA.[19,21] The PhoQ/PhoP system activates the BasS/BasR system[22] in addition to RcsC/YojN/RcsB.[22] The phosphorylated PhoP also directly promotes expression of the RstB/RstA system.[3] Regulons of these TCSs have been reported.[3,18,19,22-24]

regulated by EvgS/EvgA, constitute an operon with *gadE*, encoding a critical regulator involved in expression of some acid resistance genes. These results indicate that by activation of the EvgS/EvgA system, the MDR phenotype occurs concurrently with the acid resistance phenotype through a regulation network of multiple genes.

## Signal Transduction Cascade between EvgS/EvgA and PhoQ/PhoP TCSs

To search for TCSs interacting with EvgS/EvgA, we first selected up-regulated genes of other TCSs by picking genes that were down-regulated by deletion of each individual TCS (http://ecoli.aist-nara.ac.jp/xp_analysis/2_components).[2] We then matched them with the up-regulated genes by the activation of the EvgS sensor (http://www.nara.kindai.ac.jp/nogei/seiken/array2.html).[18] As a result, the expression of 13 (*crcA, ompT, proP, hemL, mgtA, phoP, phoQ, rstA, rstB, slyB, ybjG, yrbL* and *mgrB*) of 27 PhoQ/PhoP up-regulated genes was enhanced in the *evgS1* mutant, clearly indicating interaction between the EvgS/EvgA and PhoQ/PhoP systems. Of these 13 genes, 9 (*hemL, mgtA, phoP, phoQ, rstA, rstB, slyB, yrbL* and *mgrB*) were members of the PhoQ/PhoP-dependent $Mg^{2+}$ stimulon.[3] These genes are regulated directly by PhoP via phosphorylation from the PhoQ sensor.

To further analyze the promoter regions of EvgS/EvgA-enhanced PhoP regulons, S1 mapping was performed for the 13 genes. Enhanced expression in the *evgS1* mutant for these genes was confirmed, validating the microarray data.[19] This regulatory network between the two systems also occurred as a result of overexpression of the EvgA regulator; however, enhanced transcription of the *phoPQ* genes did not further activate expression of the PhoQ/PhoP-regulated genes. These results demonstrated signal transduction from the EvgS/EvgA system to the PhoQ/PhoP system in *E. coli*. Similar signal transduction between two TCSs also exists in *Salmonella* and Kato et al proved that a small protein, PmrD, connected the two *Salmonella* TCSs' PhoQ/PhoP and PmrB/PmrA.[20] Since there was a possibility that a protein similar to PmrD might also connect the EvgS/EvgA and PhoQ/PhoP in *E. coli*, a shotgun screening was performed and B1500, a small protein of only 65 amino acids, was identified as a factor connecting this signal transduction cascade.[21] B1500 is localized in the inner membrane and bacterial two-hybrid data showed that B1500 formed a complex with the sensor PhoQ. These results indicate that the small membrane protein, B1500, connected the signal transduction between EvgS/EvgA and PhoQ/PhoP systems by directly interacting with PhoQ, thus activating the PhoQ/PhoP system (Fig. 3). Hagiwara et al[22] found a cascade from the PhoQ/PhoP system to the BasS/BasR system (a *Salmonella* PmrB/PmrA homolog) and one from the PhoQ/PhoP system to the RcsC/YojN/RcsB system (HK, RcsC; RR, RcsB; Hpt domain, YojN),[23] but the factors connecting these systems are still not identified. The PhoQ/PhoP system is also connected to the RstB/RstA system.[3,24] In this case, the phosphorylated PhoP directly up-regulates the transcription of the *rstAB* operon by binding to its promoter. The signal transduction networks surrounding the EvgS/EvgA system in *E. coli* is illustrated in Figure 3.

## Future Perspectives

Interactions among different TCSs enable one system to respond to multiple signals, which is important to allow bacteria to minutely adjust themselves against complex environmental changes. Many of these interactions might be connected by small proteins such as PmrD and B1500. More examples of such proteins should be identified in the near future, which would help to fill in the yet unknown parts of the bacterial signal transduction network. It seems that a new paradigm, here termed bacterial signal transduction networks, is emerging.

### *Acknowledgements*

This work was supported by the Research and Development Program for New Bio-industry Initiatives (2006-2010) of the Bio-oriented Technology Research Advanced Institution (BRAIN), Japan and the Academic Frontier Project for Private Universities matching fund subsidy from Ministry of Education, Culture, Sports, Science and Technology (2004-2008), Japan.

## References


1. Mizuno T. Compilation of all genes encoding two-component phosphotransfer signal transducers in the genome of Escherichia coli. DNA Res 1997; 4:161-168.
2. Oshima T, Aiba H, Masuda Y et al. Transcriptome analysis of all two-component regulatory system mutants of Escherichia coli K-12. Mol Microbiol 2002; 46:281-291.
3. Minagawa S, Ogasawara H, Kato A et al. Identification and molecular characterization of the $Mg^{2+}$-stimulon of Escherichia coli. J Bacteriol 2003; 185:3696-3702.
4. Vescovi EG, Soncini FC, Groisman EA. $Mg^{2+}$ as an extracellular signal: environmental regulation of Salmonella virulence. Cell 1996; 84:165-174.
5. Soncini FC, Groisman EA. Transcriptional autoregulation of the Salmonella typhimurium phoPQ operon. J Bacteriol 1996; 177:4364-4371.
6. Soncini FC, Vescovi EG, Groisman EA. Two-component regulatory systems can interact to process multiple environmental signals. J Bacteriol 1996; 178:6796-6801.
7. Groisman EA. The pleiotropic two-component regulatory system PhoP-PhoQ. J Bacteriol 2001; 183:1835-1842.
8. Utsumi R, Katayama S, Taniguchi M et al. Newly identified genes involved in the signal transduction of Escherichia coli K-12. Gene 1994; 140:73-77.
9. Perraud AL, Weiss V, Gross R. Signaling pathways in two-component phosphorelay systems. Trends Microbiol 1999; 7:115-120.
10. Utsumi R. Bacterial signal transduction for environmental adaptation and their inhibitors. Microbes Enviroment 2002; 17:1-5.
11. Bock A, Gross R. The unorthodox histidine kinases BvgS and EvgS are responsive to the oxidation status of a quinone electron carrier. Eur J Biochem 2002; 269:3479-3484.
12. Georgellis D, Kwon O, Lin EC. Quinones as the redox signal for the arc two-component system of bacteria. Science 2001; 292:2314-2316.
13. Kato A, Ohnishi H, Yamamoto K et al. Transcription of emrKY is regulated by the EvgA-EvgS two-component system in Escherichia coli K-12. Biosci Biotechnol Biochem 2000; 64:1203-1209.
14. Nishino K, Yamaguchi A. Overexpression of the response regulator evgA of the two-component signal transduction system modulates multidrug resistance conferred by multidrug-resistance transporters. J Bacteriol 2001; 183:1455-1458.
15. Nishino K, Yamaguchi A. Analysis of a complete library of putative drug transporter genes in Escherichia coli. J Bacteriol 2001; 183:5803-5812.
16. Masuda N, Church GM. Escherichia coli gene expression responsive to levels of the response regulator EvgA. J Bacteriol 2002; 184:6225-6234.
17. Masuda N, Church GM. Regulatory network of acid resistance genes in Escherichia coli. Mol Microbiol 2003; 48:699-712.
18. Eguchi Y, Oshima T, Mori H et al. Transcriptional regulation of drug efflux genes by EvgAS, a two-component system in Escherichia coli. Microbiology 2003; 149:2819-2828.
19. Eguchi Y, Okada T, Minagawa S et al. Signal transduction cascade between EvgA/EvgS and PhoP/PhoQ two-component systems of Escherichia coli. J Bacteriol 2004; 186:3006-3014.
20. Kato A, Groisman EA. Connecting two-component regulatory systems by a protein that protects a response regulator from dephosphorylation by its cognate sensor. Genes Dev 2004; 18:2302-2313.
21. Eguchi Y, Itou J, Yamane M et al. B1500, a small membrane protein, connecting EvgS/EvgA and PhoQ/PhoP two component systems of Escherichia coli. Proc Natl Acad Sci USA; 2007; 104:18712-18717.
22. Hagiwara D, Yamashino T, Mizuno T. A genome-wide view of the Escherichia coli BasS-BasR two-component system implicated in iron-response. Biosci Biotechnol Biochem 2004; 68:1758-1767.
23. Hagiwara D, Sugiura M, Oshima T et al. Genome-wide analyses revealing a signal network of the RcsC-YojN-RcsB phosphorelay system in E. coli. J Bacteriol 2003; 185:5735-5746.
24. Ogasawara H, Hasegawa A, Kanda E et al. Genomic SELEX search for target promoters under the control of the PhoQP-RstBA signal relay cascade. J Bacteriol 2007; 189:4791-4799.

# Chapter 2

# The PhoQ/PhoP Regulatory Network of *Salmonella enterica*

**Akinori Kato and Eduardo A. Groisman***

## Abstract

The PhoQ/PhoP two-component regulatory system is a major regulator of virulence in the enteric pathogen *Salmonella enterica* serovar Typhimurium. It also controls the adaptation to low $Mg^{2+}$ environments by governing the expression and/or activity of $Mg^{2+}$ transporters and of enzymes modifying the $Mg^{2+}$-binding sites on the bacterial cell surface. The regulator PhoP modifies expression of ~3% of the *Salmonella* genes in response to the periplasmic $Mg^{2+}$ concentration detected by the PhoQ protein. Genes that are directly controlled by the PhoP protein often differ in their promoter structures, resulting in distinct expression levels and kinetics in response to the low $Mg^{2+}$ inducing signal. PhoP regulates a large number of genes indirectly: via other transcription factors and two-component systems that form a panoply of regulatory architectures including transcriptional cascades, feedforward loops and the use of connector proteins that modify the activity of response regulators. These architectures confer distinct expression properties that may be important contributors to *Salmonella*'s lifestyle.

## Introduction

A pathogen's ability to cause disease is multifactorial, not only requiring factors that mediate adherence to, entry into and/or survival within host cells but also regulatory proteins and RNAs that ensure that these virulence factors are produced at the right time, at the appropriate place and in the correct amounts. Indeed, many of the genes identified as being required for virulence often encode regulatory proteins. These regulators could be specific to a pathogen or be also present in related nonpathogenic species that do not normally associate with a eukaryotic host. The latter regulators typically play roles in bacterial physiology and in a microbe's ability to gain access to and prosper inside animals or plants.

As discussed in this book, two-component regulatory systems constitute the most prevalent form of signal transduction used by bacteria to respond to changes in its surroundings. Not surprisingly, two-component systems are required for virulence in numerous bacterial species and it is often the case that several two-component systems participate in the control of pathogenic functions in a particular species. Importantly, virulence attenuation may result not only from the lack of a functional two-component system but also from its constitutive activation.

In this chapter, we discuss the mechanisms utilized by the PhoQ/PhoP two-component regulatory system of the Gram-negative pathogen *Salmonella enterica* serovar Typhimurium to control gene expression. We begin with an introduction to the role that the PhoQ/PhoP system

*Corresponding Author: Eduardo A. Groisman—Department of Molecular Microbiology, Howard Hughes Medical Institute, Washington University School of Medicine, Campus Box s8230, 660 S. Euclid Ave, St. Louis, MO 63110, USA. Email: groisman@borcim.wustl.edu

*Bacterial Signal Transduction: Networks and Drug Targets*, edited by Ryutaro Utsumi.

plays in controlling *Salmonella* virulence properties. We then describe how differential expression of PhoP-regulated genes is accomplished by analyzing the promoter features of genes that are directly controlled by the PhoP protein and by discussing the particular regulatory architectures in which PhoP participates when controlling gene expression indirectly. This chapter focuses on the *Salmonella* PhoQ/PhoP system; however, it is likely that many of the discussed general principles be applicable to orthologous systems in other species and/or paralogous systems in *Salmonella*.

## The *Salmonella phoP* Gene: from Phosphatase Regulator to Virulence Controller

The *phoP* gene was first identified in a screening for mutants affected in the production of a nonspecific acid phosphatase (NSAP) in *S. enterica* serovar Typhimurium. This screening identified two loci—designated *phoP* and *phoN*—encoding a regulatory and a structural gene for NSAP.[1] The moniker "*pho*" to designate *phoP* originated in this report and does not reflect *phoP*'s role as it is presently understood. We now know that the *phoP* locus is a two-gene operon—*phoPQ*—that encodes a two-component regulatory system that controls the expression of >120 genes including *phoN*, which encodes the NSAP. The *phoN* gene is *Salmonella*-specific,[2] which is in contrast to the wide phylogenetic distribution of the *phoP* gene among enteric bacteria.[3] Although "*pho*" is usually utilized to designate genes involved in phosphate metabolism (the transcription of which is regulated by phosphate levels), neither *phoP* nor *phoN* is affected by changes in the phosphate levels in the growth media. Thus, the *Salmonella* PhoQ/PhoP system does not govern the response to phosphate limitation, which is carried out by the PhoR/PhoB two-component system; and the PhoP protein should not be confused with the *Bacillus subtilis* PhoP protein, despite both of them being response regulators.

Ten years after its description as a regulator of NSAP, *phoP* was found to be required for virulence in mice [4,5] and survival inside phagocytic cells[4,5] and a *Salmonella phoP* mutant was reported to have potential as a live vaccine strain.[6] Inactivation of the *phoP* gene resulted in an increase in the median lethal dose ($LD_{50}$) for mice inoculated intraperitoneally of five orders of magnitude, putting the *phoP* strain among the most attenuated *Salmonella* mutants described to date. This is likely due to the PhoQ/PhoP system regulating *Salmonella*'s abilities to invade epithelial cells and to survive within phagocytic cells. Moreover, *phoP* and *phoQ* mutants exhibit hypersensitivity to antimicrobial peptides,[4,7] hydrogen peroxide[8] and acid pH,[9] and are defective for growth in low $Mg^{2+}$.[10,11] Interestingly, inactivation of the *phoP* gene attenuates virulence also in several other bacterial pathogens including *Shigella flexneri*,[12] *Yersinia pestis*,[13] *Photorhabdus luminescens*[14] and *Erwinia carotovora*,[15] which cause different diseases in a variety of animals and plants.

## Structural and Functional Properties of the PhoP and PhoQ Proteins

The *Salmonella* PhoP protein is a member of the OmpR family of response regulators. It consists of a conserved N-terminal domain that includes the essential aspartate residue (Asp 52) that constitutes the site of phosphorylation and a DNA binding helix-turn-helix motif at the C-terminus.[16] The PhoQ protein is a prototypical bi-functional sensor that has kinase (i.e., autokinase and phosphotransferase) and phosphatase activities. It is an inner membrane protein that harbors two transmembrane regions that define a 146 amino acid periplasmic region and cytoplasmic domain that includes the subdomain harboring the conserved histidine residue (His277) that is the site of PhoQ phosphorylation and required for activity, as well as the subdomain involved in phosphatase activity and ATP binding.[17,18] Low $Mg^{2+}$ is the signal that, when detected by the periplasmic region of the sensor PhoQ protein,[19-22] triggers phosphorylation of the PhoP protein,[16,18,20,23] allowing binding of the PhoP-P protein to its target promoters and resulting in activation or repression of gene transcription.[16,23] It has been proposed that the *Salmonella* PhoQ protein also senses antimicrobial peptides[24] and acid pH[25] via its periplasmic region, though this notion has been questioned.[26]

Null mutations in either the *phoP* or *phoQ* genes, which mimics the system in non-inducing conditions by eliminating either the active form of the PhoP protein (i.e., phospho-PhoP) or the phospho-donor of PhoP (i.e., PhoQ), respectively, abolish expression of PhoP-activated genes. *phoQ* mutants are as defective as *phoP* mutants, arguing that PhoQ is the sole phosphodonor for the PhoP protein.

Two types of mutants render expression of PhoP-regulated genes partially or completely blind to the presence of signals.[19] On the one hand, there is a *phoQ* allele that is regularly referred to as $PhoP^C$ even though the mutation maps to the *phoQ* gene (it was originally designated *pho24* and the mutant *phoQ* encodes a PhoQ protein with a single amino acid substitution—T48I—in its periplasmic domain[27]) and the strain with this allele still responds to $Mg^{2+}$.[19] A *Salmonella pho24* mutant is as attenuated[28] as a strain a *phoQ* null allele. This indicates that for *Salmonella* to cause a lethal infection in mice, expression of PhoP-regulated genes must be tightly controlled. There is also a *phoP* allele—designated PhoP* and with a single amino acid substitution in the N-terminal domain of the PhoP protein—that can promote gene transcription in the absence of the *phoQ* gene.[17] Activation of the PhoP* protein is still dependent on phosphorylation on the conserved Aspartate. The PhoP* protein gets phosphorylated from acetyl phosphate in vivo and exhibits an increased ability to autophosphorylate from this chemical in vitro.[17]

## Defining the PhoP Regulon

It is estimated that the PhoQ/PhoP system controls the expression of ~3% of the *Salmonella* genome.[28] PhoP-regulated genes have been identified by a variety of approaches including transcriptome[29] and proteome[30] analyses, in vivo expression technology (IVET) in mice and inside macrophages,[31] differential fluorescence induction (DFI) inside macrophages[32] as well as classical genetic screenings looking for strains harboring PhoP-regulated *lac*-transcriptional fusions or identifying the PhoP-regulated genes whose defective expression is responsible for particular phenotypes exhibited by *phoP* mutant *Salmonella*.[5,10,11,33-41] PhoP-regulated genes can be classified using a variety of criteria including their phylogenetic distribution, the time at which a gene is expressed after induction of the PhoQ/PhoP system, and whether there is a requirement for additional regulatory proteins for expression to name just a few. However, the improved understanding of the sequences that are recognized by the PhoP protein allows classification of PhoP-regulated targets into those under direct and indirect (Fig. 1) control of the PhoP protein. This understanding is beginning to suggest interesting correlations between the various classifications.

## Direct Transcriptional Control by the PhoP Protein

Examination of PhoP-dependent promoters has revealed that many of them harbor a motif—(T/G)GTTTA-5 bp-(T/G)GTTTA (Fig. 1)—designated PhoP box[8,29,42-49] that was originally found in certain PhoP-regulated promoters in *Salmonella*[50] and *Escherichia coli*.[51,52] In the first set of analyzed promoters, this motif was found 11-13 bp upstream of the -10 region for RNA polymerase at a position normally occupied by the -35 region. Nonetheless, the search for PhoP-regulated promoters has not been easy because the PhoP box can be found at various distances from the -10 region, in both possible orientations and accompanied or not by a -35 region. This is further complicated by variation in the PhoP box motif itself. Not surprisingly, this has led several investigators to propose that certain PhoP-regulated genes were regulated by the PhoP protein indirectly because their corresponding promoters did not resemble prototypical PhoP-activated promoters in terms of PhoP box sequence, distance and orientation.[29]

We have successfully used a novel procedure developed in our laboratory—termed gene promoter scan (GPS)[47]—to recognize PhoP-regulated promoters in bacterial genomes. This procedure, which provides increased sensitivity while maintaining specificity,[29,47] groups promoters that share different features including the PhoP box submotif, the distance of the PhoP box to the -10 region, the orientation of the PhoP box, the presence of a -35 region, the existence of predicted binding sites for other regulatory proteins. Many PhoP-regulated promoters identified using GPS have been confirmed to be directly regulated by the PhoP protein using DNase I footprinting and chromatin immunoprecipitation (ChIP) analysis.[29]

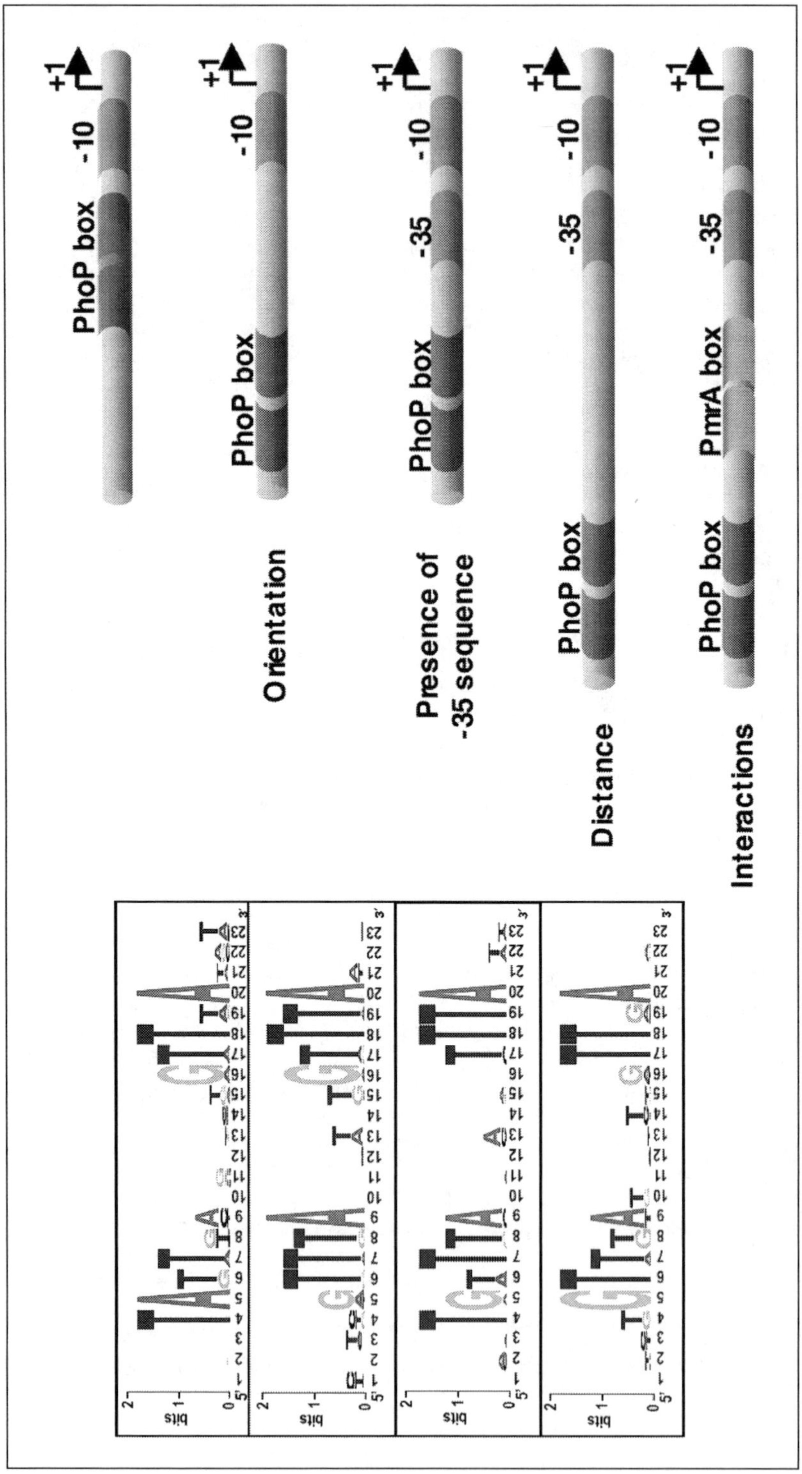


Figure 1. Regulatory submotifs constituting the PhoP box and schematic of promoter structures regulated by the PhoP protein. This figure is adapted from a figure reported in ref. 29.

The prototypical PhoP-activated promoter harbors the PhoP box 11~13 bp upstream of the consensus -10 region for RNA polymerase.[8,29,44,45,51,52] This includes the promoters for the *phoPQ, pmrD, slyB, pagP, mgrB* and *rstA* genes, as well as that corresponding to the *mgtA* gene. It has been demonstrated that the latter promoter can be efficiently transcribed in vitro using purified PhoP protein and RNA polymerase from *E. coli*.[52] It is not surprising that the *phoPQ* and *mgtA* promoters, which display the best matches to the PhoP box consensus, exhibit the highest affinity for the *E. coli* PhoP protein in vitro.[44] This scenario may also play in other enteric bacterial species given the widespread distribution of the *mgtA* gene and that the *phoPQ* operon is positively autoregulated in *Salmonella* and *E. coli*.[50,51,53] Like *mgtA*, the genes listed above, which harbor a "prototypical" PhoP-activated promoter, are found in several organisms in addition to *Salmonella*.

PhoP regulates transcription of many genes with an unusual GC context and a limited phylogenic distribution, suggesting they were acquired by horizontal gene transfer. Different features characterize the promoters of horizontally acquired genes in *Salmonella*,[54] such as those corresponding to the *ugtL, mig-14, pagC* and *mgtC* genes.[29,46,55] In these promoters, the PhoP box is the opposite strand relative to the prototypical orientation exemplified by the *mgtA* promoter. Moreover, the PhoP box is located further upstream of the -10 region and these promoters often have a recognizable -35 region, which is in contrast to the overlap of the PhoP box and -35 region that characterizes the prototypical PhoP-activated promoters. Transcription of some of these promoters is also dependent on the DNA binding protein SlyA,[46] which, as described below, is regulated by the PhoP protein. Thus, there appears to be a correlation between the structure of PhoP-regulated promoters and the phylogenetic distribution of the corresponding genes.

## The How and Why of PhoQ/PhoP Positive Autoregulation

Like many other two-component systems, PhoQ/PhoP positively regulates its own transcription (Fig. 2).[50] The *phoPQ* operon is transcribed from two promoters: the constitutive p1 promoter that provides the basal level of expression of the PhoP and PhoQ proteins so *Salmonella* can detect the

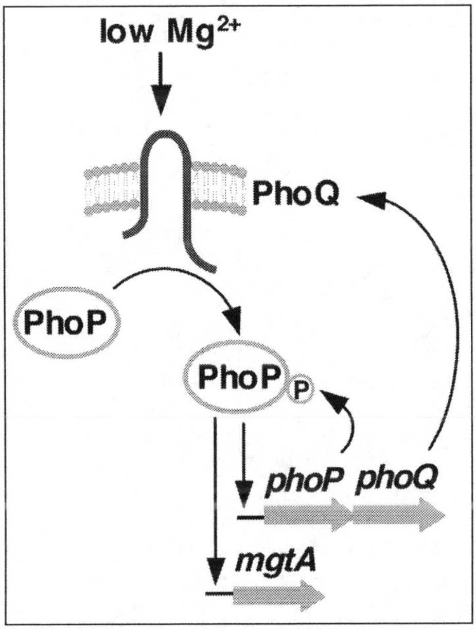


Figure 2. Model illustrating direct transcriptional control where low $Mg^{2+}$ detected by the PhoQ protein promotes phosphorylation of the PhoP protein, which binds to the promoters of the *mgtA* gene and the *phoPphoQ* operon to stimulate gene transcription.

presence of an inducing signal, and the PhoP-activated p2 promoter that increases the amount of both the PhoP and PhoQ proteins in response to the low $Mg^{2+}$ inducing signal.[19] The biological significance of the *phoPQ* autoregulation was not clear until recently when it was demonstrated that autoregulation of the *phoPQ* operon is critical for the initial surge of PhoP activity that *Salmonella* exhibits when it experiences the low $Mg^{2+}$ inducing signal. An engineered *Salmonella* strain expressing constitutively high levels of the PhoP and PhoQ proteins from a PhoP-independent promoter reached the same steady state levels of transcription of PhoP-activated genes as the one with the wild-type autoregulated promoter, but did so monotonically (i.e., without the surge). The engineered strain was attenuated for virulence in mice, highlighting the critical role played by *phoPQ* autoregulation. It has been proposed that the autoregulation-mediated surge jump-starts the PhoQ/PhoP system, helping establish a new phenotypic state[23,56] and that the expression levels taking place after the surge are responsible for maintaining the newly-established phenotypic state.

## Indirect and Nontraditional Transcriptional Control by the PhoP Protein

Genetic and genomic analyses have implicated PhoP in controlling gene expression in nontraditional ways. On the one hand, there are genes whose expression is PhoP-dependent but that lack a PhoP box in their promoters. On the other hand, the promoter of certain DNA-binding regulatory proteins harbors a PhoP box but also binding sites for other regulatory proteins. Over the last few years, there has been considerable progress in determining the nature of the connections existing among PhoP, various regulatory proteins and regulated structural genes. These studies have demonstrated that the PhoP protein participates of a variety of regulatory architectures to control gene expression including typical and atypical transcriptional cascades, feedforward loops and the use of connector proteins to control the activity of other response regulators.

## Typical and Atypical Transcriptional Cascades

The PhoP protein participates in classical transcriptional cascades by acting as a direct transcriptional regulator of genes encoding DNA binding regulatory proteins. For example, the RstA/RstB two-component system is upregulated by the PhoP protein at a transcriptional level (Fig. 3).[29,44,57] Because response regulators bind to DNA in vivo only when phosphorylated, one would expect the genes that are directly controlled by the response regulator RstA to affect gene expression only

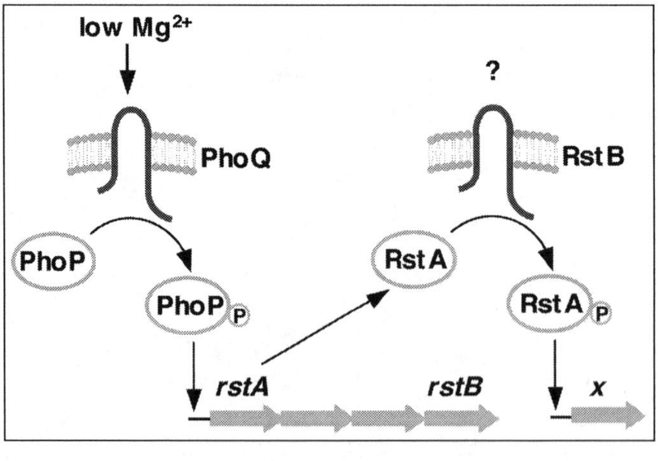


Figure 3. Model illustrating a transcriptional cascade in which the PhoP protein binds to the promoter of the *rstArstB* operon to promote RstA and RstB expression. The RstB responds to a yet unidentified signal to promote the phosphorylation state of the RstA protein, which would then bind to the promoter of a target gene *x* (yet undefined in *Salmonella*) and stimulate its transcription.

when the organism experiences the simultaneous presence of the signals that activate the sensors PhoQ and RstB: low $Mg^{2+}$ to generate PhoP-P protein for binding to the *rstA* promoter and the still unknown signal activating the RstB to generate RstA-P. This suggests that this cascade is likely to function as a Boolean AND gate. Yet, one may conceive a scenario in which a promoter has binding sites for both the PhoP and RstA proteins, with one protein acting as a activator and the other as a repressor. In this scenario, the cascade would function as a Boolean NAND gate so that expression of the target gene would occur only when the activating signal for the activator is present and the activating signal for the repressor is absent.

The PhoQ/PhoP system has opposite effects on the expression of genes involved in *Salmonella* invasion of and survival within mammalian cells, being a repressor of the former[58] and an activator of the latter.[59] PhoP exerts this control primarily by regulating the expression of proteins that are the direct regulators of genes coding for two different type III secretion systems, which are protein delivery machines that inject bacterial proteins into a eukaryotic cell cytosol.[60] Activation of the PhoQ/PhoP system represses expression of the HilA protein, a transcriptional activator of the InvA/Spa type III secretion system mediating invasion of host cells. The molecular mechanism by which this occurs remains unclear and appears to involve an AraC-like regulator named HilD that directly controls expression of the *hilA* gene.[61,62]

SpiR/SsrB is a *Salmonella* specific two component system that is essential for the production of the Spi/Ssa type III secretion system and its secreted effector proteins mediating *Salmonella* survival within phagocytic cells.[63] The SpiR/SsrB system is, in turn, regulated by both the PhoQ/PhoP and the EnvZ/OmpR two-component systems. Whereas the regulator OmpR controls the initiation of transcription from the separate *ssrB* and *spiR* promoters, PhoP governs expression of the regulator SsrB at a transcriptional level by binding to the *ssrB* promoter, and of the sensor SpiR at a posttranscriptional level (Fig. 4).[64] The PhoP protein appears to control translation of the *spiR* message, which includes a long 5′ untranslated region (UTR) with the potential to adopt a secondary structure occluding the ribosome-binding site. Consistent with this notion, deletion of a segment of the *spiR* ‵5 UTR restored SpiR expression to a *phoP* mutant.[64]

What is the physiological reason for the differential regulation of the SsrB and SpiR proteins by the PhoQ/PhoP and EnvZ/OmpR systems? One possibility is that activation via PhoQ/PhoP and EnvZ/OmpR generate different amounts of phosphorylated SsrB protein thereby affecting

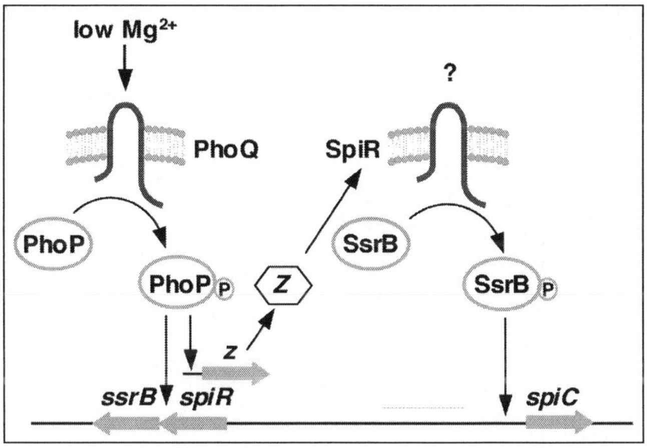


Figure 4. Model illustrating the atypical cascade in which the PhoP protein binds to the *ssrB* promoter to activate *ssrB* transcription and to a yet unidentified gene *z* that promotes SpiR expression. In response to a yet to be discovered signal, the SpiR protein promotes phosphorylation of the SsrB protein, which binds to its target promoters such as that corresponding to the *spiC* gene.

binding to and expression from different SsrB-dependent promoters inside macrophage.[64] Indeed, SpiR expression requires *ompR*, but not *phoP*, two hours after the internalization.[65] By contrast, the situation is virtually the opposite, six hours after the internalization.[65,66] Although this regulatory architecture resembles a Boolean OR gate, further experiments are required to determine whether the activation pathways mediated by the PhoQ/PhoP and OmpR/EnvZ systems produce the same transcription output of SsrB-dependent mRNAs.

## A Feedforward Loop Regulating Expression of Horizontally-Acquired Genes

The feedforward loop is the most often encountered network motif in microorganisms.[67] The PhoP protein utilizes this architecture to promote expression of *ugtL* and possibly other PhoP-activated genes. This entails PhoP activating transcription[46] and/or activity[55] of the DNA-binding protein SlyA and then both the PhoP and SlyA proteins binding to the *ugtL* promoter to stimulate its expression (Fig. 5).[46] The behavior of feedforward loops can be categorized by connectivity (i.e., Boolean logic AND gate or OR gate) and types of each interaction (i.e., activation or repression).[68] The PhoP- and SlyA-dependent control of the *ugtL* gene appears to be a Coherent type 1 AND feedforward loop, which is anticipated to shorten the period of time it takes for the *ugtL* gene to be expressed. It is noteworthy that *ugtL*, like other PhoP- and SlyA-regulated genes, is *Salmonella*-specific and/or exhibits a very limited phylogenetic distribution, suggesting that they were incorporated into the *Salmonella* genome by horizontal gene transfer. These and other horizontally-acquired genes are bound by the histone-like nucleoid-structuring (H-NS) protein[69] in vivo, which represses their expression. The requirement for both the PhoP and SlyA proteins in transcription of certain PhoP-activated promoters may obey the need to overcome repression by H-NS.

## PhoP as a Co-Activator Protein

The *Salmonella ugd* promoter harbors binding sites for three different response regulators: PhoP, PmrA and RcsB.[70] The PmrA protein stimulates *ugd* transcription in strains deleted for both the *phoP* and *rcsB* genes from one promoter. By contrast, a different promoter is used in the RcsB-promoted *ugd* expression, which requires different co-activators depending on the inducing

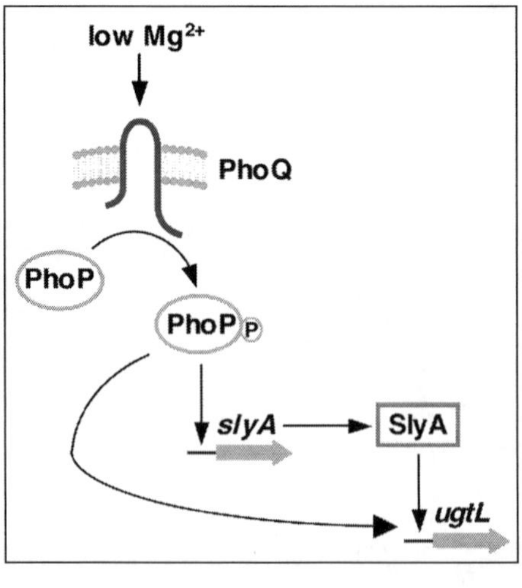


Figure 5. Model illustrating the feedforward loop by which *ugtL* transcription is activated with the PhoP protein binding to the promoter of the *slyA* gene to stimulate its expression and both the PhoP and SlyA proteins binding to the *ugtL* promoter.

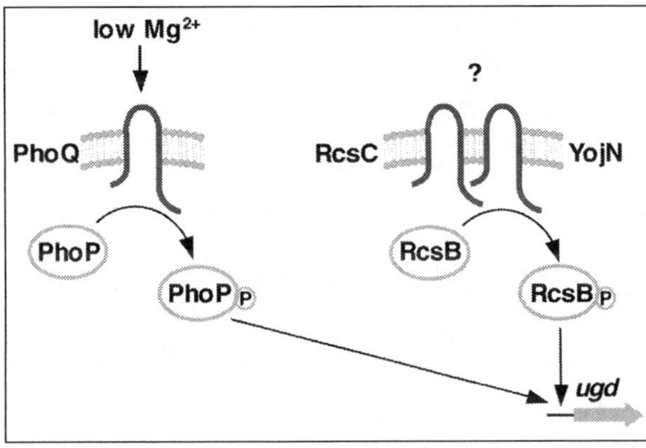


Figure 6. Model illustrating PhoP as an RcsB co-activator of *ugd* gene transcription, which takes place when the activating signals for both the PhoQ and RcsC/YojN proteins are present.

condition.[71] In a strain harboring a mutation in the *tolB* gene, which is a condition that activates the RcsC/YojN/RcsB phosphorelay system, *ugd* expression requires both RcsB and RcsB's co-activator RcsA but neither PhoP nor PmrA. However, the *ugd* gene is expressed in a *pmrA* mutant strain experiencing low $Mg^{2+}$ and $Fe^{3+}$ in a process that is RcsA-independent but dependent on both RcsB and PhoP (Fig. 6). Although the PhoP protein has been shown to footprint the *ugd* promoter,[71] its specific biochemical role in promoting *ugd* transcription together with the RcsB protein remains undefined. This is an example of a Boolean AND gate because the signals activating both RcsB and PhoP must be present for *ugd* transcription to take place.

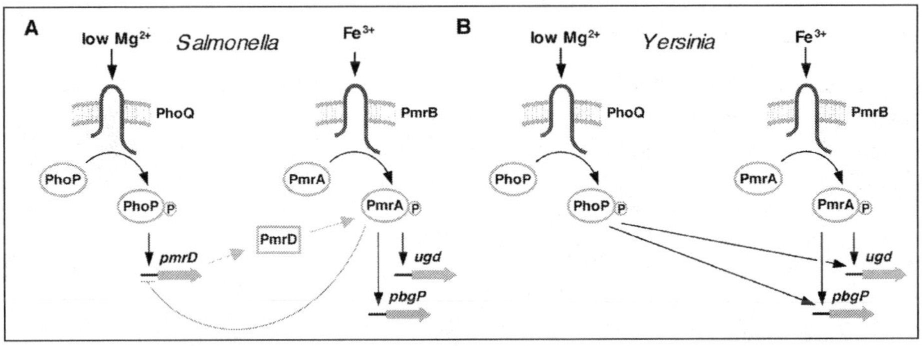


Figure 7. A) Model illustrating the connector-mediated pathways by which genes regulated directly by the PmrA protein are induced in response to the signal activating the PhoQ protein. Phosphorylation of the PhoP protein promotes its binding to the *pmrD* promoter resulting in expression of the PmrD protein. PmrD binds to the phosphorylated form of the PmrA protein (green line), resulting in binding to the *pbgP* and *ugd* promoters and transcription of the *pbgP* and *ugd* genes. The phosphorylated PmrA protein represses transcription of the *pmrD* gene (red line). $Fe^{3+}$ activates the PmrA/PmrB system directly. B) Model illustrating direct regulation of the *pbgP* and *ugd* genes where the PhoQ/PhoP system promotes transcription of the *pbgP* and *ugd* genes directly in response to low $Mg^{2+}$ in *Yersinia*. $Fe^{3+}$ activates the PmrB/PmrA system directly. A color version of this figure is availble at www.eurekah.com.

## Connector Proteins that Regulate Response Regulators

Recent reports have shown that small proteins—termed connectors—have the ability to connect two-component systems at a posttranslational level.[8,42,72,73] These novel proteins exhibit low or no sequence similarity in primary structure. However, they appear to function in a similar manner: by binding to response regulators to modulate their activities.

The best-studied connector is the PmrD protein, which integrates the signal activating the PhoQ/PhoP system with actions taken by the PmrA/PmrB two-component system (Figs. 7A and Table 1). The latter system directly controls the expression of genes encoding enzymes that modify the bacterial lipopolysaccharide at different positions.[74-76] These PmrA-dependent changes increases bacterial resistance to serum[77] to the antibiotic polymyxin B[35,42,49,75,78,79] and to $Fe^{3+}$.[78] When *Salmonella* experiences low $Mg^{2+}$, the PhoP protein binds to the *pmrD* promoter and stimulates *pmrD* transcription.[42] The purified PmrD protein binds specifically to the phosphorylated form of the PmrA protein (PmrA-P) and protects it from dephosphorylation by the sensor PmrB.[72] Because PmrA-P exhibits higher affinity for its target promoters than unphosphorylated PmrA, transcription of PmrA-activated genes is stimulated and that of PmrA-repressed genes are inhibited. A variant PmrA protein that promotes transcription in a PmrD-independent fashion was shown to be resistant to dephosphorylation by the PmrB protein, supporting the notion that PmrD's function is to prevent PmrB from dephosphorylating PmrA-P. The activity of the PmrD protein resembles that exhibited by some members of the eukaryotic 14-3-3 family in that both protect the phosphorylated form of regulatory proteins.[72,80]

PhoQ/PhoP's sole role in the PmrD-mediated pathway is to promote *PmrD* expression because transcription of the *pmrD* gene from a heterologous promoter renders production of PmrA-activated mRNAs *phoP*- and *phoQ*-independent.[42] Moreover, the PhoP-regulated PmrD protein functions at a posttranslational level because expression of PmrA-activated genes was still *phoP*- and *pmrD*-dependent when the *pmrA* and *pmrB* genes were expressed using a heterologous promoter and ribosome-binding site.

The 85-amino acid PmrD constitutes the first example of a protein that integrates the signal of one two-component system into the action of a different two-component system. This Boolean OR gate allows *Salmonella* to be resistant to polymyxin B not only in response to the environments denoted by the $Fe^{3+}$ signal that activates the PmrB protein, but also in low $Mg^{2+}$ sensed by the PhoQ protein.[42] When both $Fe^{3+}$ and low $Mg^{2+}$ are present, the PmrA proten binds to the *pmrD* promoter and represses its expression,[81] in this way preventing accumulation of PmrA-P.

The PhoP protein relies on a different connector protein—termed IraP—to enhance the levels of the alternative sigma factor RpoS (Table 1).[8,73] This allows *Salmonella* to express RpoS-dependent

***Table 1. Comparison of the connector proteins PmrD and IraP***

| Connector | PmrD | IraP |
|---|---|---|
| Length | 85 aa | 88 aa |
| Isoelectric point | 8.9 | 5.6 |
| Target | Phospho-PmrA | RssB (MviA) |
| Mechanism of action | Protecting PmrA from dephosphorylation catalyzed by the PmrB protein | Protecting RpoS from RssB-dependent degradation catalyzed by the ClpXP protease |
| Interaction | phosphorylation-dependent protein-protein interaction | protein-protein interaction |
| Cis-element | PhoP box (activation site) PmrA box (repression site) | PhoP box (activation site) |
| Ortholog distribution | *Salmonella, E. coli, Shigella* | *Salmonella, E. coli, Shigella, Enterobacter* |

genes and display RpoS-dependent phenotypes, such as resistance to hydrogen peroxide, in response to the low $Mg^{2+}$ signal sensed by the PhoQ protein. This activation entails binding of the PhoP protein to the *iraP* promoter, which stimulates *iraP* transcription. The synthesized IraP protein competes with RpoS for binding to the response regulator MviA (referred to in *E. coli* as RssB or SprE). The MviA protein, which lacks a DNA binding domain, exerts its regulatory effect by binding to the RpoS protein and delivering it to the ClpXP protease for degradation.[82] Thus, the produced IraP titrates the MviA protein, allowing the free RpoS to reprogram RNA polymerase to transcribe RpoS-dependent promoters.

The 86-amino acid IraP protein from *Salmonella* is highly conserved with the *E. coli* one suggesting that these orthologs function in a similar fashion. It has been demonstrated that the *E. coli* IraP protein binds to the RssB protein in vitro and that this binding is not dependent on RssB being phosphorylated. This distinguishes IraP from PmrD in that the latter interacts only with the phosphorylated form of the response regulator PmrA (Table 1). The IraP-mediated connection between PhoP and RpoS suggests that the strong virulence attenuation resulting from inactivation of the *phoP* gene might due, in part, to the inability of a *phoP* mutant to produce the needed levels of the RpoS protein as *rpoS* mutants are also attenuated for virulence.[4,83]

It is presently unclear whether the connector-mediated interactions between PhoP and other regulatory proteins identified in *Salmonella* can be used to infer analogous interactions in other bacterial species. This is because: first, connector proteins may be conserved but their regulation may not. For example, the conserved IraP protein is not under transcriptional control of the PhoP protein in *E. coli*.[8] Second, orthologous connectors may lose their recognized function but retain similar regulation. This is the case of the *E. coli* PmrD protein, which is only 55.3% identical to the *Salmonella* PmrD and unable to activate the PmrA/PmrB system in *E. coli*; yet, the two *pmrD* genes are similarly induced in low $Mg^{2+}$ in a PhoP-dependent fashion.[79] And third, non-orthologous replacement can provide similar connectivities despite the absence of previously identified regulators.

## The Biological Consequences of Different Network Designs

As discussed above, the PhoP protein uses a variety of regulatory architectures to control expression of its target genes. Qualitatively, these architectures achieve the same goal: to promote/repress gene expression in response to the low $Mg^{2+}$ signal that activates the PhoQ/PhoP system. However, it is becoming increasingly clear that different regulatory architectures may display important quantitative differences in gene expression. In this section, we focus on the quantitative differences that exist between direct regulation and the connector-mediated pathway because these two architectures result in distinct expression behaviors not only when contrasting different PhoP-regulated genes in *Salmonella* but also when comparing orthologous genes between the related enteric species *Salmonella* and *Yersinia pestis* (Fig. 7).

In *Salmonella*, PhoP-activated genes regulated via the PmrD protein are induced to higher levels than those that are directly controlled by the PhoP protein.[57] Signal amplification is also detected when one compares expression of the same gene(s) in two isogenic strains: one harboring the PmrD-mediated pathway operating in wild-type *Salmonella* and one lacking the *pmrD* gene and harboring a PhoP box in place of the PmrA box at the target promoter(s).[57] Mathematical modeling of the two architectures suggests that the rapid formation of the PmrD-PmrA-P complex and the low production rate of PmrA-P are essential for the PmrD-mediated signal amplification.[57] As predicted for the presence of an extra step in transcriptional activation, the PmrD-mediated pathway displays slower induction kinetics than direct activation.

We contrasted the polymxyin B resistance exhibited by a *Salmonella pmrD* mutant where PhoP controls expression of the polymyxin B resistance *pbgP* and *ugd* genes directly to that of a *Salmonella* with the PmrD-mediated pathway. The latter strain was more resistant to polymyxin B than the former when bacteria were grown under inducing concentrations of $Mg^{2+}$.[57] The resistance enhancement was significant around 0.75-1 mM $Mg^{2+}$, which is the physiological $Mg^{2+}$ concentration in mammalian tissues.[84]

The PmrD-mediated pathway exhibits expression persistence: that is, gene transcription continues for some time after disappearance of the signal activating the PhoQ/PhoP system.[57] This is because the PmrD protein that was synthesized before the PhoQ/PhoP system was shut off continues to bind to PmrA-P, allowing PmrA-dependent gene transcription. Consistent with this notion, increasing the stability of the PmrD protein resulted in longer expression persistence.[57]

We would like to propose that the particular regulatory design used to control gene expression reflects the spectrum of niches that particular bacterial species occupy. This is in line with a recent report suggesting coherence of regulatory architectures and organisms' lifestyles.[85] For example, *S. enterica* can infect a broad range of animal species and survive in soil and water for an extended time.[86] By contrast, *Y. pestis* lives in rodents, spreads via a flea vector without an environmental cycle.[87] These differences in the natural histories for these two organisms could be due, in part, to the particular designs controlling expression of these lipopolysaccharide-modifying genes—connector-mediate for *Salmonella* and direct for *Yersinia* (Fig. 7). This is because the 4-aminoarabinose modification of the LPS is required not only for resistance to polymyxin B[35,74,88] but also to metal ions present in soil.[89] Moreover, the 4-aminoarabinose modification genes have been implicated in *S. enterica* infection of chicken macrophages[90] and in *Y. pestis* survival within murine macrophages.[91]

## Conclusions

The *Salmonella* PhoQ/PhoP two-component system controls the expression of a large number of genes both directly and via a several regulatory proteins that PhoP controls at the transcriptional, posttranscriptional and posttranslational levels. Thus, PhoP is part of a variety of regulatory architectures that demonstrate qualitative and quantitative differences. It is possible that additional connectors and/or other types of regulatory proteins may be discovered experimentally rather than bioinformatically as small proteins are often not annotated, exhibit limited conservation and display discrete phylogenetic distributions.

### *Acknowledgements*

Our research on the PhoQ/PhoP and PmrB/PmrA two-component systems is supported by grants from the National Institutes of Health (USA) and from the Howard Hughes Medical Institute.

### References

1. Kier LD, Weppelman RM, Ames BN. Regulation of nonspecific acid phosphatase in Salmonella: phoN and phoP genes. J Bacteriol 1979; 138:155-161.
2. Groisman EA, Saier MH Jr, Ochman H. Horizontal transfer of a phosphatase gene as evidence for mosaic structure of the Salmonella genome. EMBO J 1992; 11:1309-1316.
3. Groisman EA, Chiao E, Lipps CJ et al. Salmonella typhimurium phoP virulence gene is a transcriptional regulator. Proc Natl Acad Sci USA 1989; 86:7077-7081.
4. Fields PI, Groisman EA, Heffron F. A Salmonella locus that controls resistance to microbicidal proteins from phagocytic cells. Science 1989; 243:1059-1062.
5. Miller SI, Kukral AM, Mekalanos JJ. A two-component regulatory system (phoP phoQ) controls Salmonella typhimurium virulence. Proc Natl Acad Sci USA 1989; 86:5054-5058.
6. Galan JE, Curtiss R 3rd. Virulence and vaccine potential of phoP mutants of Salmonella typhimurium. Microb Pathog 1989; 6:433-443.
7. Groisman EA, Parra CA, Salcedo M et al. Resistance to host antimicrobial peptides is necessary for Salmonella virulence. Proc Natl Acad Sci USA 1992; 89:11939–11943.
8. Tu X, Latifi T, Bougdour A et al. The PhoP/PhoQ two-component system stabilizes the alternative sigma factor RpoS in Salmonella enterica. Proc Natl Acad Sci USA 2006; 103:13503-13508.
9. Bearson BL, Wilson L, Foster JW. A low pH-inducible, phoPQ-dependent acid tolerance response protects Salmonella typhimurium against inorganic acid stress. J Bacteriol 1998; 180:2409-2417.
10. Soncini FC, García Véscovi E, Solomon F et al. Molecular basis of the magnesium deprivation response in Salmonella typhimurium: identification of PhoP-regulated genes. J Bacteriol 1996; 178:5092-5099.
11. Blanc-Potard A-B, Groisman EA. The Salmonella selC locus contains a pathogenicity island mediating intramacrophage survival. EMBO J 1997; 16:5376-5385.

12. Moss JE, Fisher PE, Vick B et al. The regulatory protein PhoP controls susceptibility to the host inflammatory response in Shigella flexneri. Cell Microbiol 2000; 2:443-452.
13. Oyston PC, Dorrell N, Williams K et al. The response regulator PhoP is important for survival under conditions of macrophage-induced stress and virulence in Yersinia pestis. Infect Immun 2000; 68:3419-3425.
14. Derzelle S, Turlin E, Duchaud E et al. The PhoP-PhoQ two-component regulatory system of Photorhabdus luminescens is essential for virulence in insects. J Bacteriol 2004; 186:1270-1279.
15. Flego D, Marits R, Eriksson AR et al. A two-component regulatory system, pehR-pehS, controls endopolygalacturonase production and virulence in the plant pathogen Erwinia carotovora subsp carotovora. Mol Plant Microbe Interact 2000; 13:447-455.
16. Shin D, Groisman EA. Signal-dependent binding of the response regulators PhoP and PmrA to their target promoters in vivo. J Biol Chem 2005; 280:4089-4094.
17. Chamnongpol S, Groisman EA. Acetyl phosphate-dependent activation of a mutant PhoP response regulator that functions independently of its cognate sensor kinase. J Mol Biol 2000; 300:291-305.
18. Castelli ME, Garcia Vescovi E, Soncini FC. The phosphatase activity is the target for $Mg^{2+}$ regulation of the sensor protein PhoQ in Salmonella. J Biol Chem 2000; 275:22948-22954.
19. García Véscovi E, Soncini FC, Groisman EA. $Mg^{2+}$ as an extracellular signal: environmental regulation of Salmonella virulence. Cell 1996; 84:165-174.
20. Chamnongpol S, Cromie M, Groisman EA. $Mg^{2+}$ sensing by the $Mg^{2+}$ sensor PhoQ of Salmonella enterica. J Mol Biol 2003; 325:795-807.
21. García Véscovi E, Ayala M, Di Cera E et al. Characterization of the bacterial sensor protein PhoQ. Evidence for distinct binding sites for $Mg^{2+}$ and $Ca^{2+}$. J Mol Biol 1997; 272:1440-1443.
22. Cho US, Bader MW, Amaya MF et al. Metal bridges between the PhoQ sensor domain and the membrane regulate transmembrane signaling. J Mol Biol 2006; 356:1193-1206.
23. Shin D, Lee EJ, Huang H et al. A positive feedback loop promotes transcription surge that jump-starts Salmonella virulence circuit. Science 2006; 314:1607-1609.
24. Bader MW, Sanowar S, Daley ME et al. Recognition of antimicrobial peptides by a bacterial sensor kinase. Cell 2005; 122:461-472.
25. Prost LR, Daley ME, Le Sage V et al. Activation of the bacterial sensor kinase PhoQ by acidic pH. Mol Cell 2007; 26:165-174.
26. Groisman EA, Mouslim C. Sensing by bacterial regulatory systems in host and nonhost environments. Nat Rev Microbiol 2006; 4:705-709.
27. Gunn JS, Hohmann EL, Miller SI. Transcriptional regulation of Salmonella virulence: a PhoQ periplasmic domain mutation results in increased net phosphotransfer to PhoP. J Bacteriol 1996; 178:6369-6373.
28. Miller SI, Mekalanos JJ. Constitutive expression of the phoP regulon attenuates Salmonella virulence and survival within macrophages. J Bacteriol 1990; 172:2485-2490.
29. Zwir I, Shin D, Kato A et al. Dissecting the PhoP regulatory network of Escherichia coli and Salmonella enterica. Proc Natl Acad Sci USA 2005; 102:2862-2867.
30. Guina T, Yi EC, Wang H et al. A PhoP-regulated outer membrane protease of Salmonella enterica serovar Typhimurium promotes resistance to alpha-helical antimicrobial peptides. J Bacteriol 2000; 182:4077-4086.
31. Heithoff DM, Conner CP, Hanna PC et al. Bacterial infection as assessed by in vivo gene expression. Proc Natl Acad Sci USA 1997; 94:934-939.
32. Valdivia RH, Falkow S. Fluorescence-based isolation of bacterial genes expressed within host cells. Science 1997; 277:2007-2011.
33. Behlau I, Miller SI. A PhoP-repressed gene promotes Salmonella typhimurium invasion of epithelial cells. J Bacteriol 1993; 175:4475-4484.
34. Belden WJ, Miller SI. Further characterization of the PhoP regulon: identification of new PhoP-activated virulence loci. Infect Immun 1994; 62:5095-5101.
35. Groisman EA, Kayser J, Soncini FC. Regulation of polymyxin resistance and adaptation to low-$Mg^{2+}$environments. J Bacteriol 1997; 179:7040-7045.
36. Guo L, Lim KB, Poduje CM et al. Lipid A acylation and bacterial resistance against vertebrate antimicrobial peptides. Cell 1998; 95:189-198.
37. Hilbert F, García del Portillo F, Groisman EA. A periplasmic D-alanyl-D-alanine dipeptidase in the gram-negative bacterium Salmonella enterica. J Bacteriol 1999; 181:2158-2165.
38. Gunn JS, Belden WJ, Miller SI. Identification of phoP-phoQ activated genes within a duplicated region of the Salmonella typhimurium chromosome. Microb Pathog 1998; 25:77-90.
39. Groisman EA. The pleiotropic two-component regulatory system PhoP-PhoQ. J Bacteriol 2001; 183:1835-1842.
40. Mouslim C, Hilbert F, Huang H et al. Conflicting needs for a Salmonella hypervirulence gene in host and nonhost environments. Mol Microbiol 2002; 54:1019-1027.

41. Shi Y, Cromie MJ, Hsu FF et al. PhoP-regulated Salmonella resistance to the antimicrobial peptides magainin 2 and polymyxin B. Mol Microbiol 2004; 53:229-241.
42. Kox LF, Wosten MM, Groisman EA. A small protein that mediates the activation of a two-component system by another two-component system. EMBO J 2000; 19:1861-1872.
43. Mouslim C, Groisman EA. Control of the Salmonella ugd gene by three two-component regulatory systems. Mol Microbiol 2003; 47:335-344.
44. Minagawa S, Ogasawara H, Kato A et al. Identification and molecular characterization of the $Mg^{2+}$ stimulon of Escherichia coli. J Bacteriol 2003; 185:3696-3702.
45. Lejona S, Aguirre A, Cabeza ML et al. Molecular characterization of the $Mg^{2+}$-responsive PhoP-PhoQ regulon in Salmonella enterica. J Bacteriol 2003; 185:6287-6294.
46. Shi Y, Latifi T, Cromie MJ et al. Transcriptional control of the antimicrobial peptide resistance ugtL gene by the Salmonella PhoP and SlyA regulatory proteins. J Biol Chem 2004; 279:38618-38625.
47. Zwir I, Huang H, Groisman EA. Analysis of differentially-regulated genes within a regulatory network by GPS genome navigation. Bioinformatics 2005; 21:4073-4083.
48. Monsieurs P, De Keersmaecker S, Navarre WW et al. Comparison of the PhoPQ regulon in Escherichia coli and Salmonella typhimurium. J Mol Evol 2005; 60:462-474.
49. Winfield MD, Latifi T, Groisman EA. Transcriptional regulation of the 4-amino-4-deoxy-L-arabinose biosynthetic genes in Yersinia pestis. J Biol Chem 2005; 280:14765-14772.
50. Soncini FC, García Véscovi E, Groisman EA. Transcriptional autoregulation of the Salmonella typhimurium phoPQ operon. J Bacteriol 1995; 177:4364-4371.
51. Kato A, Tanabe H, Utsumi R. Molecular characterization of the PhoP-PhoQ two-component system in Escherichia coli K-12: identification of extracellular $Mg^{2+}$-responsive promoters. J Bacteriol 1999; 181:5516-5520.
52. Yamamoto K, Ogasawara H, Fujita N et al. Novel mode of transcriptional regulation of divergently overlapping promoters by PhoP, the regulator of two-component system sensing external magnesium availability. Mol Microbiol 2002; 45:423-438.
53. Zhou D, Han Y, Qin L et al. Transcriptome analysis of the $Mg^{2+}$-responsive PhoP regulator in Yersinia pestis. FEMS Microbiol Lett 2005; 250:85-95.
54. Ochman H, Lawrence JG, Groisman EA. Lateral gene transfer and the nature of bacterial innovation. Nature 2000; 405:299-304.
55. Navarre WW, Halsey TA, Walthers D et al. Coregulation of Salmonella enterica genes required for virulence and resistance to antimicrobial peptides by SlyA and PhoP/PhoQ. Mol Microbiol 2005; 56:492-508.
56. Bijlsma JJ, Groisman EA. Making informed decisions: regulatory interactions between two-component systems. Trends Microbiol 2003; 11:359-366.
57. Kato A, Mitrophanov AY, Groisman EA. A connector of two-component regulatory systems promotes signal amplification and persistence of expression. Proc Natl Acad Sci USA 2007; 104:12063-12068.
58. Bajaj V, Hwang C, Lee CA. hilA is a novel ompR/toxR family member that activates the expression of Salmonella typhimurium invasion genes. Mol Microbiol 1995; 18:715-727.
59. Deiwick J, Nikolaus T, Erdogan S et al. Environmental regulation of Salmonella pathogenicity island 2 gene expression. Mol Microbiol 1999; 31:1759-1773.
60. Ly KT, Casanova JE. Mechanisms of Salmonella entry into host cells. Cell Microbiol 2007; 9:2103-2111.
61. Ellermeier CD, Ellermeier JR, Slauch JM. HilD, HilC and RtsA constitute a feed forward loop that controls expression of the SPI1 type three secretion system regulator hilA in Salmonella enterica serovar Typhimurium. Mol Microbiol 2005; 57:691-705.
62. Ellermeier JR, Slauch JM. Adaptation to the host environment: regulation of the SPI1 type III secretion system in Salmonella enterica serovar Typhimurium. Curr Opin Microbiol 2007; 10:24-29.
63. Ochman H, Soncini FC, Solomon F et al. Identification of a pathogenicity island required for Salmonella survival in host cells. Proc Natl Acad Sci USA 1996; 93:7800-7804.
64. Bijlsma JJ, Groisman EA. The PhoP/PhoQ system controls the intramacrophage type three secretion system of Salmonella enterica. Mol Microbiol 2005; 57:85-96.
65. Lee AK, Detweiler CS, Falkow S. OmpR regulates the two-component system SsrA-SsrB in Salmonella pathogenicity island 2. J Bacteriol 2000; 182:771-781.
66. Feng X, Oropeza R, Kenney LJ. Dual regulation by phospho-OmpR of ssrA/B gene expression in Salmonella pathogenicity island 2. Mol Microbiol 2003; 48:1131-1143.
67. Alon U. The feed-forward loop network motif. In: Alon U, ed. An Introduction to Systems Biology: Design Principles of Biological Circuits. Boca Raton: Chapman and Hall/CRC Press, 2006:41-74.
68. Mangan S, Alon U. Structure and function of the feed-forward loop network motif. Proc Natl Acad Sci USA 2003; 100:11980-11985.

69. Navarre WW, Porwollik S, Wang Y et al. Selective silencing of foreign DNA with low GC content by the H-NS protein in Salmonella. Science 2006; 313:236-238.
70. Mouslim C, Groisman EA. Control of the Salmonella ugd gene by three two-component regulatory systems. Mol Microbiol 2003; 47:335-344.
71. Mouslim C, Latifi T, Groisman EA. Signal-dependent requirement for the co-activator protein RcsA in transcription of the RcsB-regulated ugd gene. J Biol Chem 2003; 278:50588-50595.
72. Kato A, Groisman EA. Connecting two-component regulatory systems by a protein that protects a response regulator from dephosphorylation by its cognate sensor. Genes Dev 2004; 18:2302-2313.
73. Bougdour A, Wickner S, Gottesman S. Modulating RssB activity: IraP, a novel regulator of sigma(S) stability in Escherichia coli. Genes Dev 2006; 20:884-897.
74. Gunn JS, Lim KB, Krueger J et al. PmrA-PmrB-regulated genes necessary for 4-aminoarabinose lipid A modification and polymyxin resistance. Mol Microbiol 1998; 27:1171-1182.
75. Lee H, Hsu FF, Turk J et al. The PmrA-regulated pmrC gene mediates phosphoethanolamine modification of lipid A and polymyxin resistance in Salmonella enterica. J Bacteriol 2004; 186:4124-4133.
76. Tamayo R, Choudhury B, Septer A et al. Identification of cptA, a PmrA-regulated locus required for phosphoethanolamine modification of the Salmonella enterica serovar typhimurium lipopolysaccharide core. J Bacteriol 2005; 187:3391-3399.
77. Delgado MA, Mouslim C, Groisman EA. The PmrA/PmrB and RcsC/YojN/RcsB systems control expression of the Salmonella O-antigen chain length determinant. Mol Microbiol 2006; 60:39-50.
78. Wösten MM, Kox LF, Chamnongpol S et al. A signal transduction system that responds to extracellular iron. Cell 2000; 103:113-125.
79. Winfield MD, Groisman EA. Phenotypic differences between Salmonella and Escherichia coli resulting from the disparate regulation of homologous genes. Proc Natl Acad Sci USA 2004; 101:17162-17167.
80. Tzivion G, Avruch J. 14-3-3 proteins: active cofactors in cellular regulation by serine/threonine phosphorylation. J Biol Chem 2002; 277:3061-3064.
81. Kato A, Latify T, Groisman EA. Closing the loop: the PmrA/PmrB two-component system negatively controls expression of its posttranscriptional activator PmrD. Proc Natl Acad Sci USA 2003; 100:4706-4711.
82. Becker G, Klauck E, Hengge-Aronis R. Regulation of RpoS proteolysis in Escherichia coli: the response regulator RssB is a recognition factor that interacts with the turnover element in RpoS. Proc Natl Acad Sci USA 1999; 96:6439-6444.
83. Fang FC, Libby SJ, Buchmeier NA et al. The alternative sigma factor katF (rpoS) regulates Salmonella virulence. Proc Natl Acad Sci USA 1992; 89:11978-11982.
84. Reinhart RA. Magnesium metabolism: a review with special reference to the relationship between intracellular content and serum levels. Arch Intern Med 1988; 148:2415-2420.
85. Madan Babu M, Teichmann SA, Aravind L. Evolutionary dynamics of prokaryotic transcriptional regulatory networks. J Mol Biol 2006; 358:614-633.
86. Winfield MD, Groisman EA. Role of nonhost environments in the lifestyles of Salmonella and Escherichia coli. Appl Environ Microbiol 2003; 69:3687-3694.
87. Perry RD, Fetherston JD. Yersinia pestis—etiologic agent of plague. Clin Microbiol Rev 1997; 10:35-66.
88. Trent MS, Ribeiro AA, Lin S et al. An inner membrane enzyme in Salmonella and Escherichia coli that transfers 4-amino-4-deoxy-L-arabinose to lipid A: induction on polymyxin-resistant mutants and role of a novel lipid-linked donor. J Biol Chem 2001; 276:43122-43131.
89. Nishino K, Hsu FF, Turk J et al. Identification of the lipopolysaccharide modifications controlled by the Salmonella PmrA/PmrB system mediating resistance to Fe(III) and Al(III). Mol Microbiol 2006; 61:645-654.
90. Zhao Y, Jansen R, Gaastra W et al. Identification of genes affecting Salmonella enterica serovar enteritidis infection of chicken macrophages. Infect Immun 2002; 70:5319-5321.
91. Grabenstein JP, Fukuto HS, Palmer LE et al. Characterization of phagosome trafficking and identification of PhoP-regulated genes important for survival of Yersinia pestis in macrophages. Infect Immun 2006; 74:3727-3741.

# Chapter 3

# Structural Basis of the Signal Transduction in the Two-Component System

**Seiji Yamada and Yoshitsugu Shiro***

## Abstract

Two-component system (TCS) consists of two multi-domain proteins, a sensor histidine kinase (HK) and a response regulator (RR). In response to environmental change, the signal is transduced from HK to RR through phosphoryl transfer. At the first stage of structural biology of TCS, crystallographic and NMR analyses of domain blocks revealed the folds and the remarkable regions of sensor, dimerization and catalytic domains of HK and receiver and effecter domains of RR. As the second stage, the advanced researches of their multi-domain form and HK/RR complex showed the inter-domain and inter-molecular interactions and implied that the dynamic conformation changes are required in the signaling process. Thus, this chapter describes what these structural analyses of TCS proteins have contributed in understanding the cell signaling mechanism; signal input → phosphoryl transfer → signal output.

## Introduction

Bacteria, fungi and plants have exploited the "two-component system (TCS)" to respond a sudden change of environments such as nutrients, oxygen, light, osmotic pressure. The system consists of two proteins, a sensor histidine kinase (HK) and a response regulator (RR). HK senses an extra- or an intra-cellular signal and transmits the sensed signal to RR through phosphoryl transfer reactions. HK is a multi-domain protein composed of sensor and histidine kinase domains (Fig. 1A) and the histidine kinase domain is further divided into a dimerization and a catalytic domain. The structure and mechanism of histidine kinase domain of HK is highly conserved among TCS. In response to the sensing of an external stimuli or a ligand in the sensor domain, the special His residue in the dimerization domain is phosphorylated using ATP, which is catalyzed by the catalytic domain (autophosphorylation). HKs function in the dimeric form and the autophosphorylation reaction occur in *trans*-manner; i.e., the catalytic domain of one subunit phosphorylates His in other subunit.[1] The phosphoryl group on the His residue is transferred to the special Asp residue in RR (RR-kinase).

The sequence analysis classified HKs into Class I, hybrid type of Class I and Class II (Fig. 1B) by their domain compositions. Even though the sequence similarity of HK is not high, their amino acid motifs of H, N, G1, F and G2 boxes are identified in the all HKs. Among the Class I HK, the H box (HExxxP) contains a phosphorylated His residue lying on the dimerization domain and the N (NLxxxN), G1 (DxGxG), F (FxPF) and G2 (GxGxGL) boxes create the ATP-binding and the catalytic sites in the catalytic domain (Fig. 1B). In the case of hybrid type HK, the histidine kinase

*Corresponding Author: Yoshitsugu Shiro—Biometal Science Laboratory, RIKEN SPring-8 Center, Harima Institute, 1-1-1 Kouto, Sayo, Hyogo 679-5148, Japan. Email: yshiro@riken.jp

*Bacterial Signal Transduction: Networks and Drug Targets*, edited by Ryutaro Utsumi.

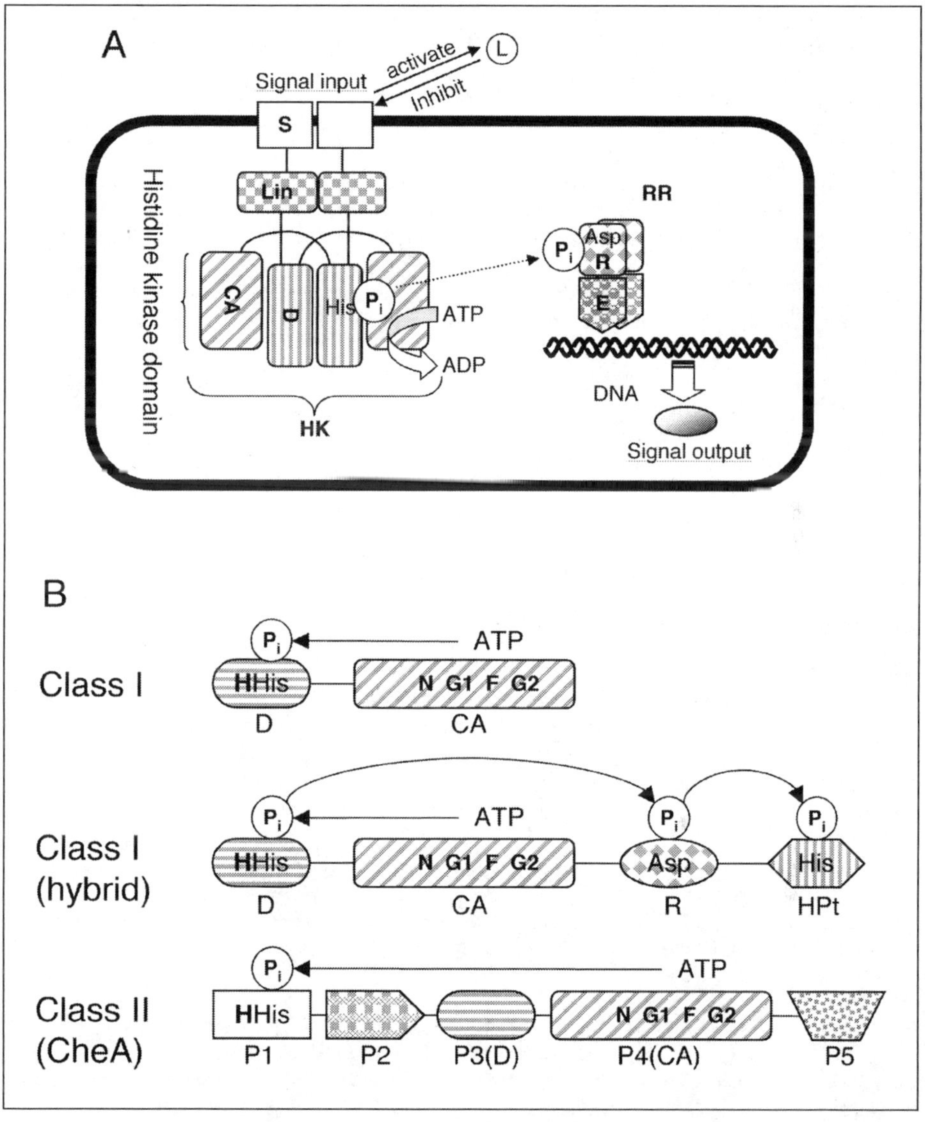


Figure 1. Schematic diagrams of TCS. A) Phosphotransfer in TCS proteins. S; Sensor, Lin; Linker, D; Dimerization, CA; Catalytic, R; Receiver, E; Effecter domains. A phosphoryl group ($P_i$) produced from ATP is transferred to His in dimerization domain and subsequently to Asp in receiver domain. When a specific ligand (L) binds to sensor domain, the kinase activity is inhibited. B) Classification of HKs. H, N, G1, F and G2 are the conserved motifs. HPt; Histidine-containing Phosphotransfer domain. Allows indicate the sequence of phosphoryl transfer.

domain is followed by an Asp-containing receiver domain and a His-containing phosphotransfer domain (e.g., anaerobic sensor ArcB, Fig. 1B).[2,3] Therefore, the phosphoryl group migrates in the His-Asp-His sequence in the case of the hybrid HK (reviewed in ref. 4). On the other hand, Class II HK, chemotaxis sensor CheA, has five domains per monomer (Fig. 1B, reviewed in ref. 5, 6). The

P1 domain contains a phosphorylated His residue,[7] the P2 domain docks RR,[8,9] the P3 domain mediates only dimerization,[10] the P4 domain binds ATP,[10,11] and the P5 domain regulates the kinase activity.[10,12] Therefore, location of the phosphorylated His (P1) in CheA, relative to the catalytic domain (P4), is different from those in Class I HKs (Fig. 1B).

RR is activated upon accepting the phosphoryl group from HK. The phosphorylated RR functions as a transcription factor to promote expression of target genes or an activator for the down stream protein. RR comprises of a receiver and an effecter domain. The receiver domain contains the phosphorylated Asp residue and the structure is highly conserved among all RRs. In contrast, structure of the effecter domains of RR vary depending on type of the signal output (reviewed in ref. 13). Twenty-nine out of 32 RRs in TCS of *Escherichia coli* are transcription factors, which have a DNA binding ability in the effecter domains. On the other hand, one of chemotaxis RR, CheY, has no DNA binding domain and another one, CheB, has metylesterase domain as the effecter domain.[14,15]

Since 1990s, structural analyses of the TCS proteins have revealed characteristics of each domain in detail. In the first half of this chapter, the functional and structural aspects of the domain blocks in the TCS, especially in the Class I HK/RR system, are outlined. The latter half of this chapter focuses on their multi-domains form and HK/RR complex, which have recently been reported. These structures revealed relative orientation of each domain and provided structural basis for discussion on inter-domain and inter-molecular mechanism of the signal transduction in TCS.

## Structures of Each Domain in Two-Component System

### *Histidine Kinase Domain*

#### Dimerization Domain

The functional form of HK is a dimer, which is formed through interaction of the dimerization domain of each subunit. NMR structure of the dimerization domain of EnvZ, osmosensor in *E. coli*, revealed that two long helices (I and II) in one subunit are packed with those of the other subunit to form a four-helix bundle (Fig. 2).[16] The amino acid sequence of the helix I in the N-terminal side is more conserved than those of the helix II, since the helix I has the H box motif, which contains the phosphorylated His. The H-box region is more flexible than the other regions, but the significance has been unknown yet. The side-chain of the phosphorylated His residue is extremely exposed to bulk solvent (Fig. 2). The orientation of the His residue is apparently suitable for accepting the phosphoryl group from the catalytic domain in *trans*-autophosphorylation[1] and for subsequently donating it to its cognate RR.

#### Catalytic Domain

The structures of catalytic domain of HK have been solved for CheA,[10,11] PhoQ (*E. coli* $Mg^{2+}$ sensor),[17] EnvZ,[18] NtrB (*E. coli* nitrate sensor),[19] and PrrB (*Mycobacterium tuberculosis*)[20] (Figs. 3A to 3F). All these structures exhibit $\alpha/\beta$ sandwich fold, in which three helices ($\alpha$1, $\alpha$2 and $\alpha$4) is covered with the $\beta$-sheet (the topology of $\beta$2-$\beta$4-$\beta$5-$\beta$7-$\beta$6). An N-terminal short $\beta$-sheet ($\beta$1 and $\beta$3) seals the $\alpha$1 and $\alpha$2 helices. The loop region between $\alpha$3 and $\alpha$4 helices, termed as the ATP-lid, is flexible and has been often disordered in the crystal structures (Figs. 3E and 3F). The four conserved motifs of HK, N, G1, F and G2 boxes, are located in the cavity for the ATP binding (Figs. 3G and 3H). In addition to the basic fold, CheA and NtrB have extra secondary structures. CheA has two additional helices; i.e., one ($\alpha$I1) is inserted between $\alpha$2 and $\alpha$4 and the other ($\alpha$I2) preceded the ATP-lid (Figs. 3A and 3B).[10,11] NtrB has a $\beta$-hairpin ($\beta$4' and $\beta$5') inserted between $\beta$4 and $\beta$5 for binding of HK inhibitor (PII protein), as was illustrated in Figure 3E.[19]

These structural characteristics of the catalytic domain are distinct from Thr/Ser/Tyr kinases in mammalian system. Instead, the folds of catalytic domain are similar to that of GHL ATPases, GyrB (DNA gyrase),[21] MutL (DNA mismatch repair protein)[22] and Hsp90 (eukaryotic molecular chaperon).[23]

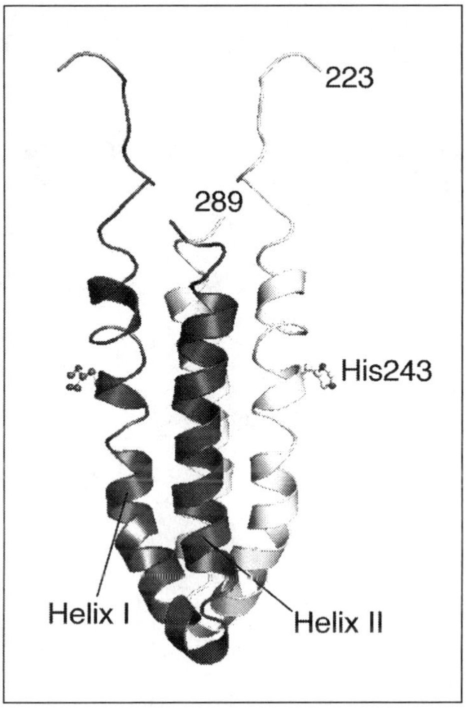


Figure 2. Structure of dimerization domain of EnvZ (1joy.pdb). The phosphorylated His residue (His243) is represented as ball-and-stick. Each subunit formed dimer is colored by white and dark-gray, respectively.

HK requires ATP and divalent cation, $Mg^{2+}$ or $Mn^{2+}$, for the kinase activity. Structures of the complex with nonhydrolysable ATP analogue and divalent cation incorporated in the ATP binding site have been reported for some HK (Figs. 3B, 3C and 3D). The N box in the helix α2 contains the conserved Asn residue (N409 in CheA and N389 in PhoQ), the Oδ atom of which coordinates with the divalent cation (Figs. 3G and 3H). In the G1 box at the end of β5, the Oδ atom of the conserved Asp residue (D449 in CheA and D415 in PhoQ) forms a hydrogen bond with amino group (N6) in the adenine base and the main-chain N and O atoms of the conserved Gly residue (G453 in CheA and G419 in PhoQ) interact with the adenine ring with intervention of water molecules. The F box in the end of α3 and the G2 box in the beginning of α4 are present at the edges of ATP-lid.

In structural comparison between HKs in the ATP analogue-free and -bound forms, the most significant difference was observed in the conformation of the ATP-lid. Upon the nucleotide binding, the ATP-lid is apparently converted from the open conformation to the closed conformation, consequently covering the ATP-binding site.[11] In the case of CheA, the ATP-lid displays a loop conformation in the free form (Fig. 3A), while a loop-helix transition is induced upon the ADPCP-$Mg^{2+}$ binding (Fig. 3B). Lys494 and Ser498 in the induced helix interact with the γ-phosphate of ADPCP (Fig. 3G). All these interactions with the ATP-lid as well as the N and G1 boxes can stabilize the ATP (or ATP analogue) binding and, eventually the γ-phosphate is exposed to bulk solvent region, possibly being accessible to the phosphorylated His.[11] On the other hand, the ATP-lid of PhoQ takes a loop conformation even in the ADPNP- $Mg^{2+}$ bound form, but Arg 434, Arg439 and Gln442 in the ATP-lid interact with the β- or the γ-phosphate of ADPNP (Fig. 3H). In this case, the γ-phosphate faces toward the α2 helix (Fig. 3H) and it would rotate toward the

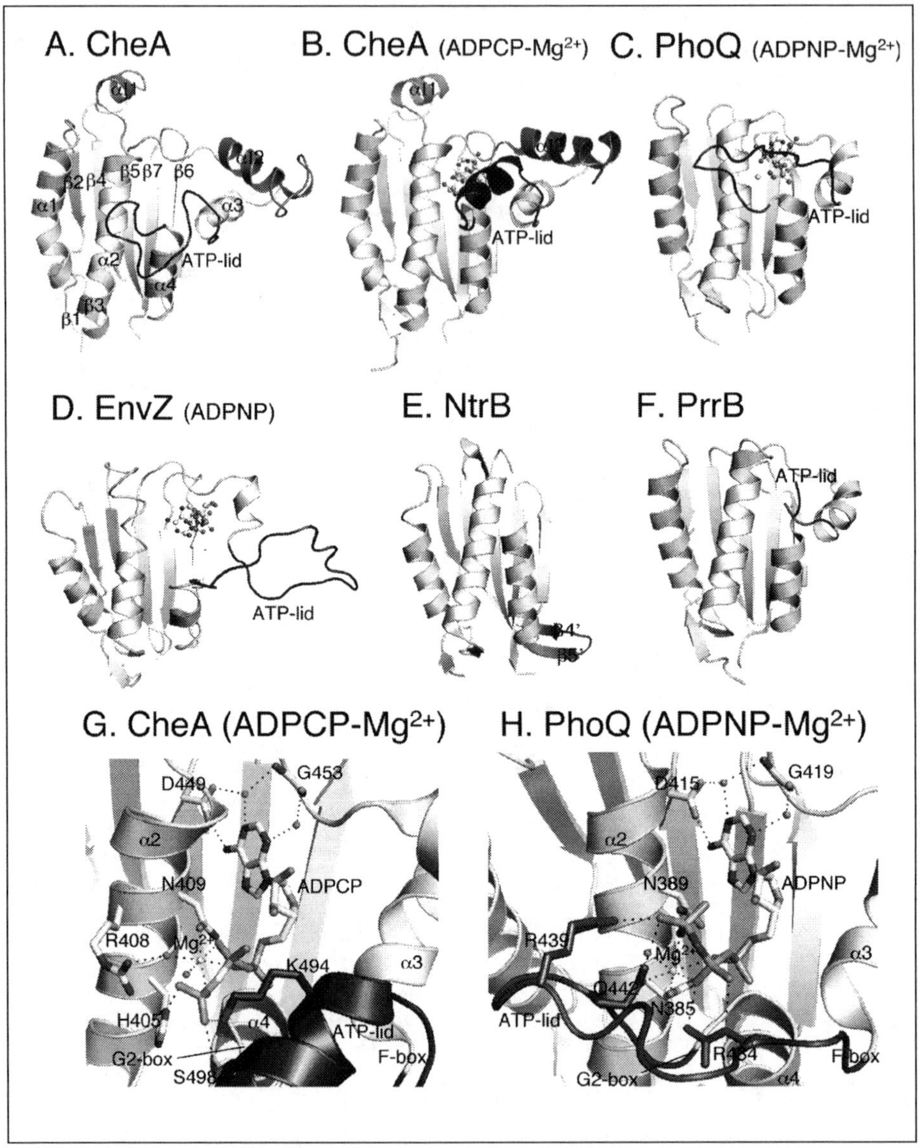


Figure 3. Structures of catalytic domain. (A) CheA (1b3q.pdb), (B) CheA complexed with ADPCP-$Mg^{2+}$ (1i58.pdb), (C) PhoQ complexed with ADPNP-$Mg^{2+}$ (1id0.pdb), (D) EnvZ complexed with ADPNP (1bxd.pdb), (E) NtrB (1r62.pdb) and (F) PrrB (1ysr.pdb). ATP analogues are represented as ball-and-stick. ATP-lid and extra secondary structures (in CheA and NtrB) are colored by dark-gray and light-gray, respectively. (G and H) ATP-binding sites of CheA (1i58.pdb) and PhoQ (1id0.pdb), respectively. ATP analogues and amino acid residues interacted with ATP analog are represented as stick. $Mg^{2+}$ and water molecules are represented as white and gray spheres, respectively. The dashed lines indicate hydrogen bonds. ATP-lid and the involving residues are colored by dark-gray. H405, R408 and N409 in (G) and N385 and N389 in (H) are involved in N-box, respectively. D449 and G453 in (G) and D415 and G419 in (H) are involved in G1 box, respectively. The boundaries between white and dark-gray loops correspond F and G2 boxes, respectively.

solvent region in the autophosphorylation reaction.[17] Thus, the ATP-lid is responsible for nucleotide binding in HKs, although it does not show high sequence similarity excluding F- and G2- boxes. The structural transition of ATP-lid produces new surface nature and it would plays an important role in interaction with phosphorylated His residue for autophosphorylation reaction.[5]

## *Sensor Domain of Histidine Kinase*

In sharp contrast to the histidine kinase domain, the structures of the sensor domain of HK vary depending on signals (ligands). The sensor domains are classified into three types: the periplasmic receptor, the transmembrane receptor and receptor in cytoplasm, as were illustrated in Figure 4. An example of the first category is the chemotaxis receptor, which can commonly recognize methyl-accepting chemotaxis proteins (MCPs); e.g., Tar and Tsr are the MCP of aspartate and serine receptors, respectively.[24-26] The periplasmic region of these receptors is composed of eight-helix bundle in dimer (Fig. 4A). The binding of aspartate and serine to these subunits stabilizes the dimeric form and causes the inhibition of kinase activity of CheA.[27] Another example is quorum sensor, LuxP/LuxQ,[28-30] which will be appeared in the following section.

Ethylene is a ripening hormone for higher plants. The ethylene sensor in higher plants, ETR1, has the transmembrane sensor domain,[31] which contains three membrane spanning helices per subunit and forms dimer (Fig. 4B). A $Cu^{1+}$ ion is flanked by the two subunits of the helices.[32] Cys65 and His69 in the second helix in *Arabidopsis* ETR1 are coordination residues of the $Cu^{1+}$ ion. The $Cu^{1+}$ is a possible binding site of ethylene.

GAF (cyclic GMP binding-Adenyl cyclase-FhlA) is a sensor domain located in cytoplasm, which functions as a red/far-red light sensor in phytochromes (reviewed in ref. 33). A biliverdin molecule ligated via a thioether linkage to a Cys residue is located in the ligand-binding pocket of the GAF domain of bacteriophytochrome (Fig. 4C).[34] The photoisomerization of the biliverdin D-ring induced upon the red/far-red light irradiation can regulate the histidine kinase activity. PAS (Per-Arnt-Sim) domain in cytoplasm often functions as the substantial sensor module in TCS and will be discussed later in detail.

## *Response Regulator*

### Receiver Domain

The structures of receiver domains are solved for many RRs such as CheY,[35] Spo0F,[36] NtrC,[37] NarL,[38] CheB,[15] PhoB,[39] FixJ,[40] Spo0A,[41] Etr1,[42] DrrD,[43] DivK,[44] PhoP,[45] CheY2,[46] ArcA,[47] KdpE,[48] TorR[48] and PrrA.[49] The structures exhibit basically similar features, i.e., $\alpha/\beta$ fold, in which central $\beta$-sheets ($\beta$2-$\beta$1-$\beta$3-$\beta$4-$\beta$5) is flanked by two ($\alpha$1 and $\alpha$5) and three ($\alpha$2, $\alpha$3 and $\alpha$4) helices (Fig. 5A).

In some receiver domains, structures of their activated form are determined by using phosphoryl mimics.[50-53] In structural comparison between the active and inactive forms, it was found that conformations of the conserved Thr87 and Tyr106 (numbering as in CheY) are altered, the side chains of which moves from outside to inside upon the activation, as shown in Figures 5A and 5B. This reorientation causes to stabilize the $\alpha$4 helix and $\beta$4-$\alpha$4 loop. The reoriented $\alpha$4-$\beta$5-$\alpha$5 surface makes a new interaction site with the partner protein, e.g., FliM in Figure 5B.[54] In the case of FixJ, its homodimer is formed through interaction at the reoriented site (Fig. 5C).[55] The Asp residue (Asp57 in CheY) located in the C-terminal edge of the $\beta$3 strand is the phosphorylation site of the receiver domain. As are illustrated in Figure 5D, the phosphoryl group can be stabilized by divalent cation, $Mg^{2+}$ or $Mn^{2+}$ and through many hydrogen-bonding interaction.

### Effecter Domain

RRs function as transcription factor have a typical helix-turn-helix motif for binding to DNA as the effecter domain. In the case of NarL, the receiver domain interacts with the effecter domain, masking the DNA-recognition helix ($\alpha$9) in the nonphosphorylated (inactive) form (Fig. 5E).[38,56] This is also the case for other RRs, such as CheB[15] and NtrC1[57]; the active sites of methylesterase and ATPase, respectively, are masked in the nonphosphorylated forms. On the other hand, the

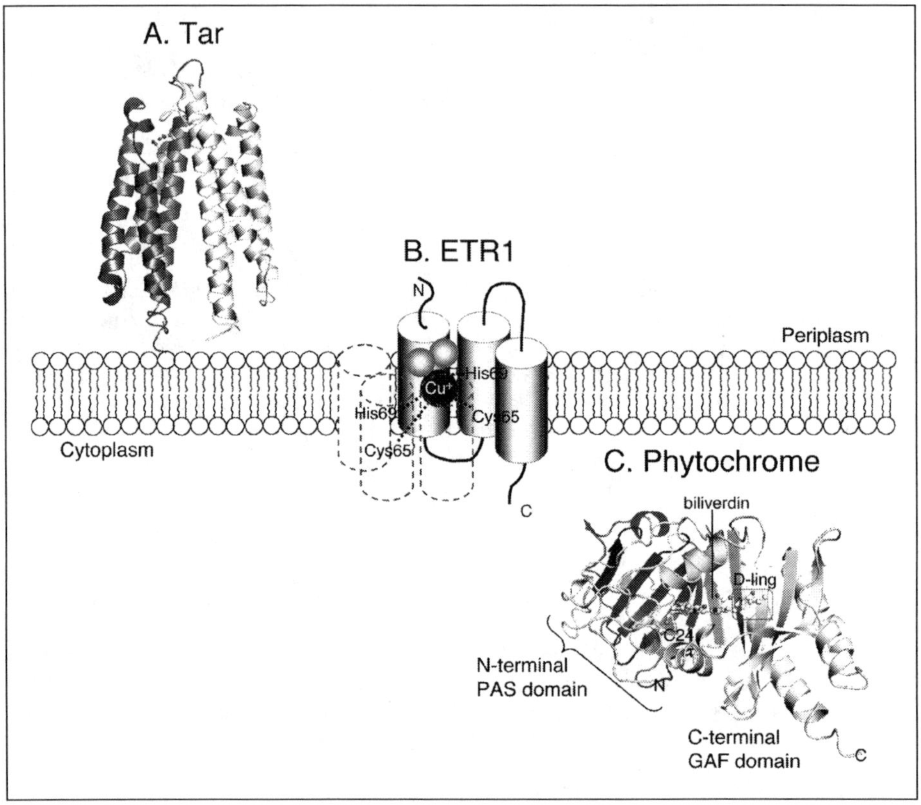


Figure 4. Structures of sensor domains of HK. The lipid bilayer membrane is schematically drawn. A) Periplasmic aspartate receptor, Tar (aspartate bound form, 1vlt.pdb). Each subunit of Tar formed dimer is colored by white and dark-gray, respectively. The ligand, aspartate, is represented as ball-and-stick. B) Schematic representation of ethylene sensor, ETR1. The membrane bound helices are represented as cylinders. Cylinders drawn by gray dashed line indicate the other subunit. The ethylene molecule and $Cu^+$ are represented by light-gray and black spheres, respectively. The black dashed lines indicate the coordination bonds for $Cu^+$. C) Sensor domain of red/far-red light receptor, bacteriophytochrome (1ztu.pdb). The sensor domain consists of N-terminal PAS (residues 38-128, dark-gray) and C-terminal GAF (white) domains. The bilverdin chromophore and the covalent bonding Cys24 are represented as ball-and-stick.

DNA binding form of the isolated effecter domain of NarL was reported (Fig. 5F), in which the domain forms a symmetric dimer and, eventually, the $\alpha 9$ helix can directly interact with the target DNA.[58] Based on these results, it has been now considered that, to expose the DNA recognition site of the effecter domain and to form the RR dimer, RR should dramatically change its conformation upon accepting the phosphoryl group at the Asp site in the receiver domain.

## Signal Transduction in Histidine Kinase

### *Regulation upon Signal Sensing in Quorum Sensor LuxP/LuxQ System*

In the bioluminescent process in *Vibrio harvey*, the bacteria can sense their cell density (quorum sensing) by receiving a small molecule called "autoinducer", which is produced by themselves.[59] LuxP/LuxQ is the receptor/transmitter for the quorum sensing system.[60] LuxP is a soluble protein

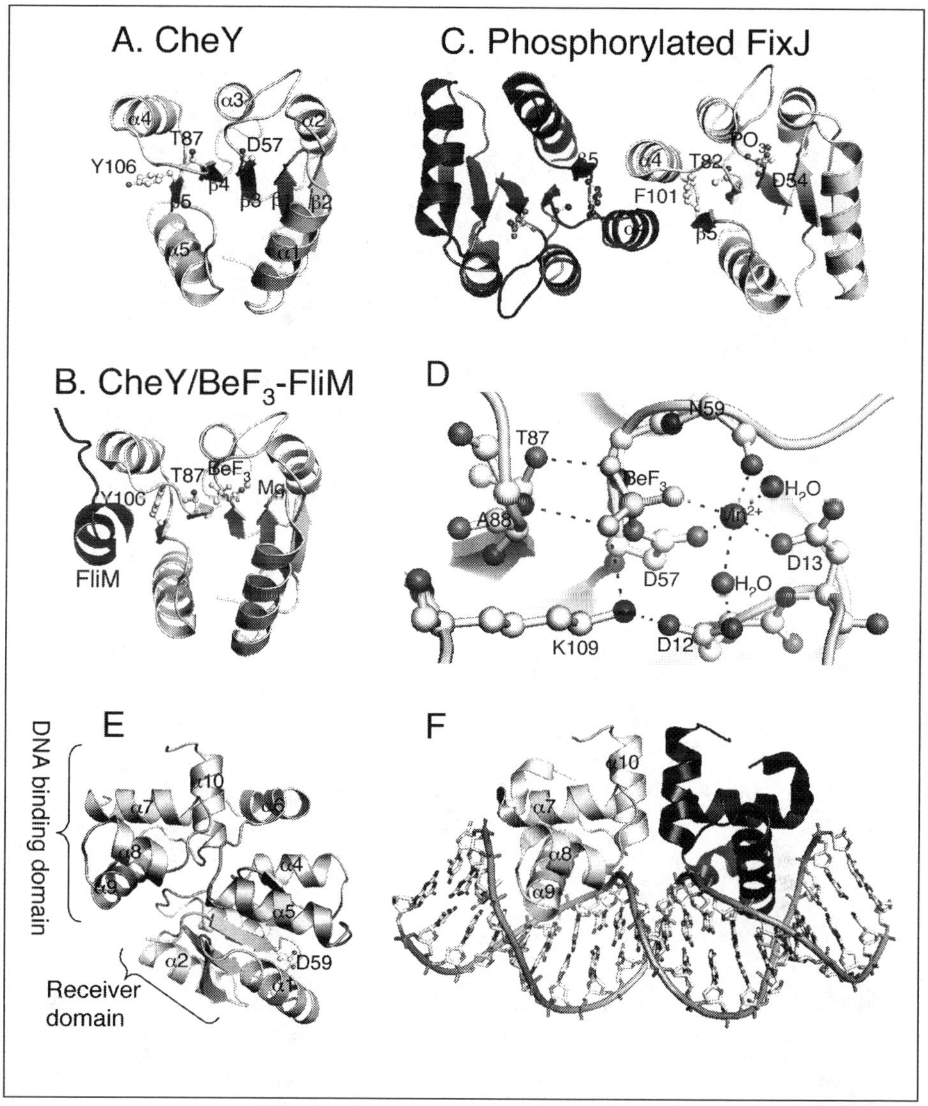


Figure 5. Structures of response regulator. A) CheY (3chy.pdb), B) CheY-$BeF_3$-$Mg^{2+}$ complexed with FliM peptide (1f4v.pdb), C) Receiver domain of FixJ-$PO_3$ (1d5w.pdb). The conserved Asp, Thr and Tyr/Phe residues and ligands are represented as ball-and-stick, respectively. FliM peptide (in Fig. 5B) and dimerized subunit (in Fig. 5C) are colored by dark-gray, respectively. D) Ball-and-stick representation of active site of CheY-$BeF_3$-$Mn^{2+}$ (1fqw.pdb). The dashed lines indicate hydrogen bonds. E) Full-length structure of transcriptional RR, NarL (1rnl.pdb). The α6-α10 helices are designed as DNA binding domain. The α8 and α9 helices are referred as "scaffold" and "recognition" helices, respectively. Asp59, the phosphorylated residue in the receiver domain, is represented as ball-and-stick. F) DNA binding domain of NarL complexed with DNA (1je8.pdb). Each subunit of NarL formed dimer is colored by white and dark-gray, respectively. The white subunit is drawn as same orientation of that as shown in Figure 5E. If the full-length structure (Fig. 5E) is fitted into this model, the receiver domain crashes with DNA.

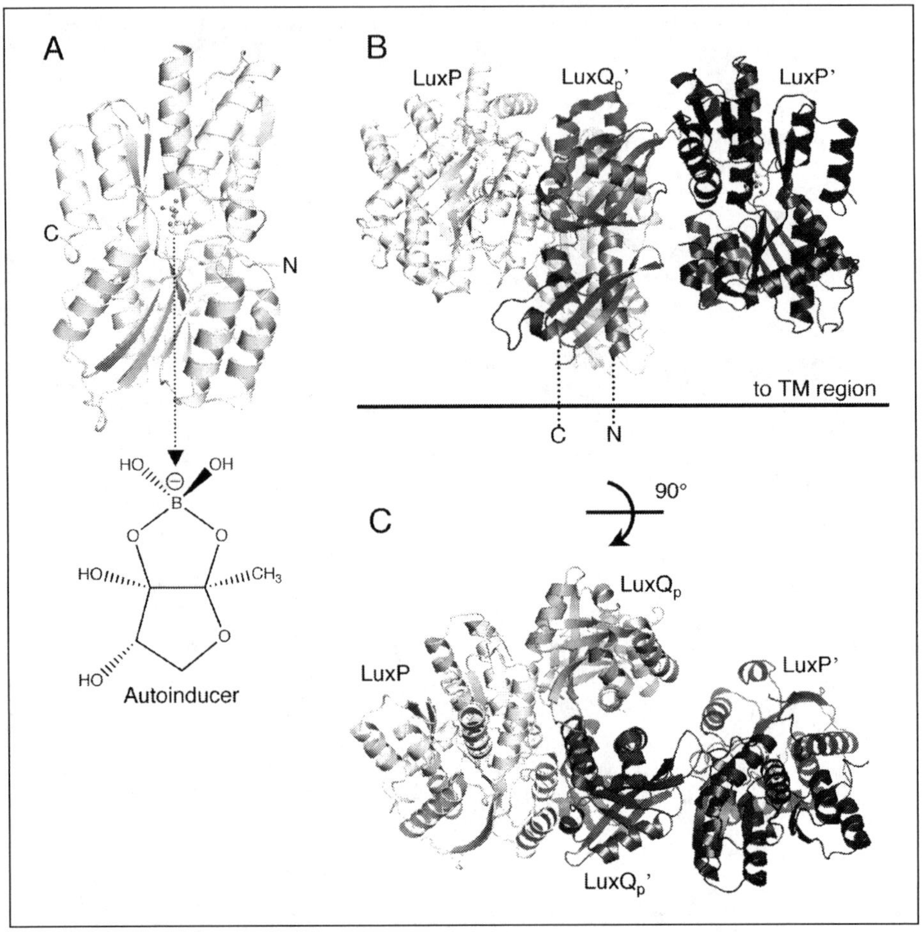


Figure 6. Structures of quorum sensing LuxP/LuxQ system. A) LuxP (1jx6.pdb) and chemical structure of autoinducer. The autoinducer in LuxP is represented as ball-and-stick. B) LuxP-autoinducer complexed with periplasmic domain of LuxQ ($LuxQ_p$) (2hj9.pdb). LuxP and $LuxQ_p$ form a constitutive heterodimer. When autoinducer binds to LuxP, the other heterodimer ($LuxQ_p$'/LuxP') is attracted to LuxP/$LuxQ_p$. LuxP, $LuxQ_p$, $LuxQ_p$' and LuxP' are colored by white, light-gray, dark-gray and black, respectively. The dashed lines of $LuxQ_p$' indicate the trans-membrane (TM) linker extended from N- and C-terminals. C) Rotation of (B) at 90° along horizontal axis. Whereas $LuxQ_p$' has come in contact with LuxP/$LuxQ_p$ complex, the opposite does not occurred. In consequence, two LuxP/$LuxQ_p$ complexes form asymmetry dimer upon binding of autoinducer.

located in the periplasmic region and can bind the autoinducer in the cavity created between two domains (Fig. 6A).[28] LuxQ is a transmembrane HK, the periplasmic region of which is composed of two PAS folds and is associated with LuxP (Fig. 6B).[29] The autoinducer binding to LuxP induces a conformational change to close the crevice.[28] This conformational change of LuxP creates a new interaction site for another subunit of periplasmic region of the LuxQ, forming an asymmetric dimer (Fig. 6C).[30] It has been suggested that this asymmetry of periplasmic region of the LuxQ disrupts the symmetrical dimerization of the cytoplasmic histidine kinase domain and reduces the kinase activity.[30]

## *Autophosphorylation of His in Dimerization Domain by Catalytic Domain*

Most recently, Marina et al reported the crystal structure of HK (TM0853) fragment from *Thermotoga maritima*, consisting of dimerization and catalytic domains (Figs. 7A and 7B).[61] Two monomeric TM0853 molecules contact with each other through some hydrophobic interactions between helices I-II in the dimerization domain to form a dimer, that is related by a crystallographic two-fold symmetry. The structure looks like a butterfly (Fig. 7A). The structural characteristics are apparently similar to that observed in Spo0B, which is a His containing phosphotransferase involved in the spoluration phosphorelay of *Bacillus subtilis*.[62,63] In the crystal structure of thermophilic HK, the ATP binding site in the catalytic domain is far away (>20A) from the

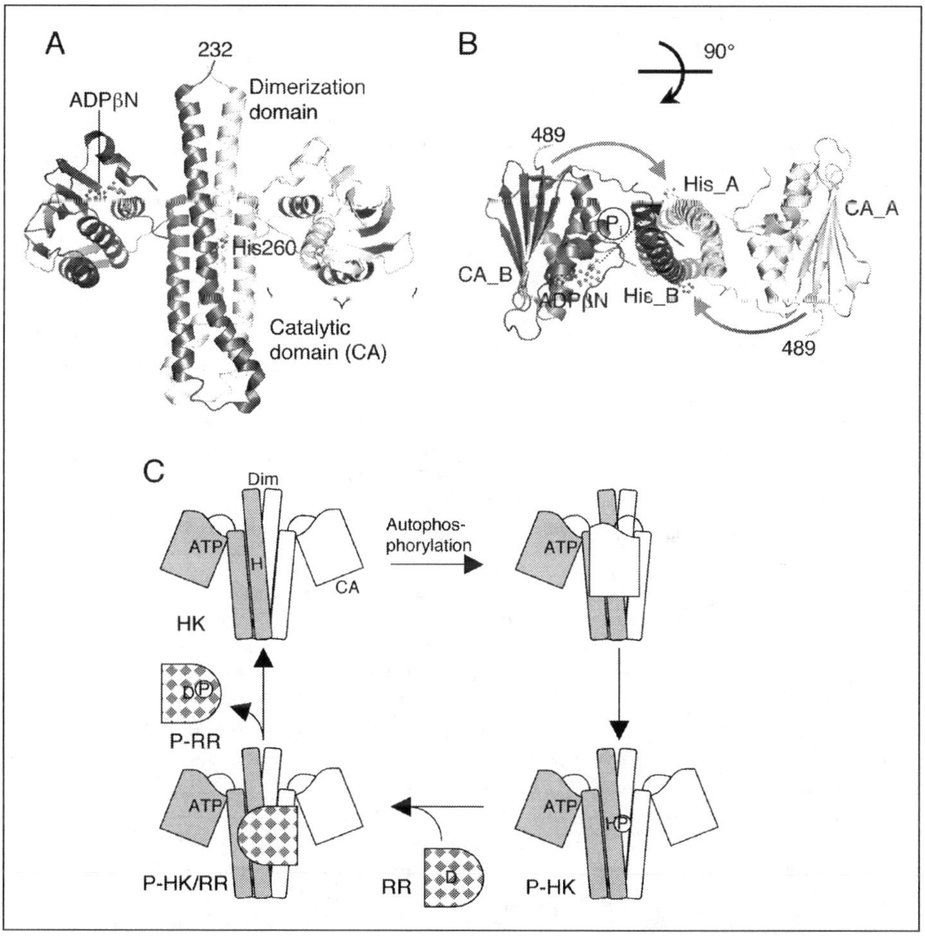


Figure 7. Structures of entire histidine kinase domain. A) histidine kinase domain of HK-TM0853 (2c2a.pdb). Each subunit formed dimer is colored by white and dark-gray, respectively. His260 and ADPβN are represented as ball-and-stick, respectively. B) Rotation of (A) at 90° along horizontal axis. The ADPβN in catalytic domain (CA)_B is far from His_A with the distance of 25 Å. Therefore, the CA domains have to move dynamically as indicated as gray allows for trans-autophosphorylation. C) Schematic diagram of reaction catalyzed by HK. Positions of ATP and phosphorylated His (H) are indicated on an HK dimer (white and gray). Position of phosphorylated Asp (D) is indicated on a RR (white with gray mesh). The phosphoryl group is indicated as a circle.

phosphorylated His residue in the dimerization domain. Therefore, the catalytic domain should rotate around the dimerization domain by 70° for the *trans*-autophosphorylation reaction (Fig. 7B). Based on the crystallographic observation, a possible mechanism for the phosphor-transfer in the TCS was proposed, in which dynamic motion of the catalytic domain should be involved, as is illustrated in Figure 7C.

## Oxygen Sensor FixL/FixJ System

FixL/FixJ system is HK/RR from root nodule bacteria (rhizobia), which lives symbiotically in the root of leguminous plants. The rhizobia have an ability of nitrogen fixation by nitrogenase enzymes, which can convert nitrogen ($N_2$) in air into ammonia ($NH_3$). Since the enzymes are labile to oxygen ($O_2$), the rhizobia should perceive a concentration of oxygen and express the nitrogen fixation enzymes under low $O_2$ tension. The substantial $O_2$ sensor module is the FixL/FixJ system. The sensor domain of FixL contains the heme (Fe-porphyrin complex) as the $O_2$ binding site (Fig. 8A). The $O_2$ binding to the heme iron is in a chemical equilibrium with dissociation constant of approximately 30 μM;[64] under general oxygen concentration, $O_2$ associates to the heme iron, while $O_2$ is dissociated from the iron with lowering the $O_2$ tension. In the $O_2$-bound (oxy) form of the sensor domain, the histidine kinase domain of FixL is inactive, while it is activated in the $O_2$-free (deoxy) form to phosphorylate the cognate RR, FixJ.[65,66]

### *Structure and Oxygen Sensing Mechanism of PAS Sensor Domain of FixL*

To date, the structures of the FixL sensor domain from *Bradyrhizobium japonicum* (*Bj*FixL)[67] and *Sinorhizobium meliloti*[68] are available, showing that three α-helices (αD, αE and αF) are covered with five β-sheets (βB-βA-βI-βG-βH) (Fig. 8A). The structural characteristics are generally called the "PAS fold". The heme is embedded in the cavity constructed between the β-sheet and αF helix. His from the αF helix coordinates to the heme iron as the fifth axial ligand and the opposite sixth site is provided for the external ligand ($O_2$) binding (Fig. 8A). As the case for general heme-containing enzymes/proteins, CO and NO, as well as $O_2$, can bind to the ferrous iron ($Fe^{2+}$) of the heme in FixL. When $O_2$ binds to the heme iron in the sensor domain, the activity of the histidine kinase domain is entirely suppressed. However, upon binding CO and NO, the histidine kinase activity is not inhibited, in contrast to the $O_2$ binding.[69] Therefore, FixL can discriminate $O_2$ from other diatomic molecules.

Gong and coworkers reported the crystal structures of the *Bj*FixL sensor domain in the deoxy, oxy, CO-bound and NO-bound forms and compared these structures with each other to elucidate conformational changes induced by the ligand binding (Figs. 8C to 8F, respectively).[67,70,71] When the sixth position of heme iron is vacant in the deoxy forms, the Arg residue in the G-sheet (Arg220 in *Bj*FixL) is located outside the heme pocket and is hydrogen-bonded with the propionate of the heme (Fig. 8C). However, when $O_2$ binds to heme iron, the Arg residue comes into the heme pocket and is hydrogen-bonded with the iron-bound $O_2$ (Fig. 8D).[71] In addition, the reorientation of the Arg residue induces a conformational change in the FG-loop region between F-helix and G-sheet, as is shown in Figure 8B. On the other hand, such reorientation of the Arg residue was not detected in the CO and NO binding to FixL (Figs. 8E and 8F, respectively).[70,71] On the basis of these crystallographic observations, now it is believed that the movement of Arg upon the $O_2$ association/dissociation would act as a trigger to induce the conformational change in the FG loop region, which should be a first event in the $O_2$ sensing by the FixL sensor domain and could be also responsible for the ligand specificity for the sensing by FixL.

### *Structure of HK/RR Complex: Signaling Pathway in FixL/FixJ System*

Since the full-length FixL structure containing the sensor and histidine kinase domains has been unavailable yet, it is still unclear how the conformational change in the FG loop of the sensor domain could be transferred to histidine kinase domain as an inhibitory signal. To elucidate this issue, the structural analyses were conducted by using a homologous protein of FixL from thermophilic bacterium.[72,73] HK from *Thermotoga maritima*, ThkA (TM1359), exhibits a sequence similarity by 72% (29% identity) to FixL,[73] and is expected to have a PAS sensor domain directly

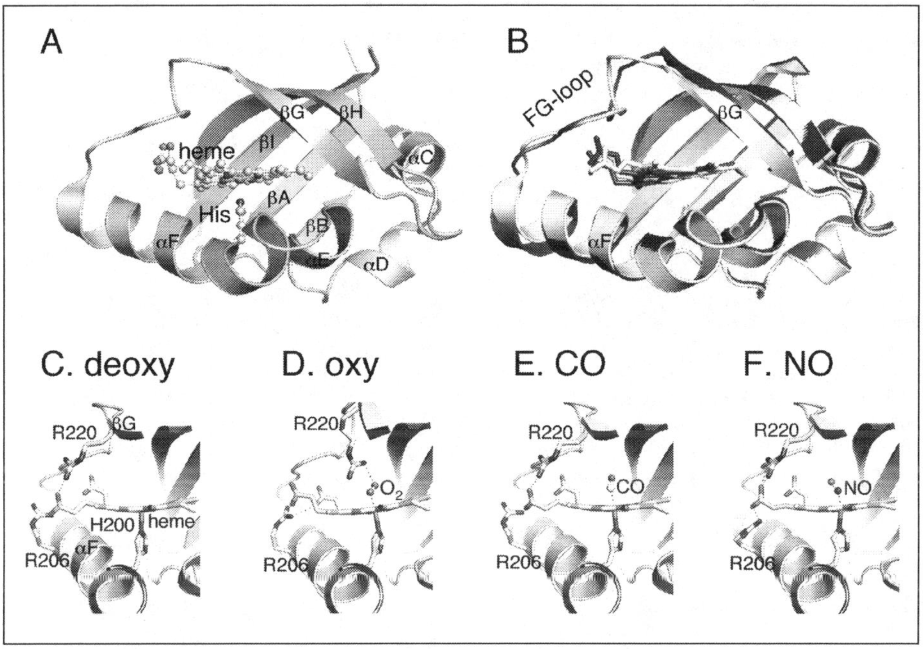


Figure 8. Structures of PAS-sensor domain of oxygen sensor FixL. A) Overall structure of deoxy ($Fe^{2+}$) FixL (1lsw.pdb). The heme and the fifth ligand His residue are represented as ball-and-stick. B) Superimposition of deoxy (white) and oxy ($Fe^{2+}$-$O_2$, dark-gray, 1dp6.pdb) FixL. The FG-loop region shows conformational change upon oxygen binding, whereas overall structures are similar to each other. (C-F) Heme-around structures of deoxy (in C), oxy (in D), CO (in E, $Fe^{2+}$-CO, 1lsv.pdb) and NO (in F, $Fe^{2+}$-NO, 1dp8.pdb), respectively. Arg,220 Arg,206 His200 and heme are represented as stick, respectively. Gas molecules ($O_2$, CO and NO) are represented as ball-and-stick. The dashed lines indicate hydrogen bond or coordination bond.

followed by histidine kinase domain, as the case for FixL. The cognate RR of ThkA is TrrA.[72,74] ThkA retains both autophosphorylation and TrrA-kinase activity.[72,75]

Small angle X-ray scattering (SAXS) technique was applied to elucidate a solution structure of ThkA and its complex with TrrA. From the SAXS measurements, molecular size and shape of ThkA could be estimated and these values are compiled in Table 1. On the basis of these analytical estimation, it was found that ThkA exists in dimer in solution and two sensor domains locate in both ends of the dimeric ThkA (see also caption of Table 1). The titration measurements of ThkA with TrrA revealed the 2:1 binding scheme, 2TrrA + $ThkA_2 \Leftrightarrow ThkA_2$/2TrrA, with the dissociation constant ($K_d$) of $8.2 \times 10^{-11}$ $M^{-2}$ and suggested that two TrrA molecules bind to the central portion of the ThkA dimer (Table 1).

The crystal structure of ThkA/TrrA complex obtained at 4.2 Å resolution revealed the relative orientation of sensor, dimerization and catalytic domains of ThkA and TrrA (Fig. 9). Even in low resolution, the ThkA/TrrA complex structure could provides a possible explanation of the signaling pathway of the FixL/FixJ system (Fig. 9B).

## From Sensor Domain to Catalytic Domain

The most important finding in the structure of the HK/RR complex was that the sensor domain interacts directly with the catalytic domain in the same subunit (Fig. 9), suggesting that the sensor domain regulates the catalytic domain in the same subunit; i.e., the so-called *cis*-acting

***Table 1. Structural parameters of ThkA, TrrA and the complex obtained by SAXS experiments***[72]

| | $I(0)$ [b] (a.u.) | $M_w$ [c] (kDa) | $R_g$ [d] (Å) | $D_{max}$ [e] (Å) | Association State |
|---|---|---|---|---|---|
| ThkA/TrrA | 651 ± 3 | 98 (54.2) | 37.8 ± 0.2 | 105 ± 5 | $(ThkA_2)/2TrrA$ |
| ThkA | 506 ± 4 | 76 (40.9) | 37.3 ± 0.1 | 105 ± 5 | Dimer: $ThkA_2$ |
| ΔThkA[a] | 344 ± 3 | 51 (28.2) | 31.3 ± 0.1 | 88 ± 2 | Dimer: $\Delta ThkA_2$ |
| TrrA | 74 ± 1 | 11 (13.3) | 15.7 ± 0.1 | 42 ± 2 | Monomer: TrrA |

[a]ThkA without sensor domain. [b]Forward scattering intensities. The value of the complex equals to $[I(0)_{ThkA} + 2I(0)_{TrrA}]$ and indicates that two TrrA molecules bind to $ThkA_2$; $2TrrA + ThkA_2 \Leftrightarrow ThkA_2/2TrrA$. [c]Molecular mass estimated by $I(0)$ compared with that obtained from bovine serum albumin. The parentheses indicate the theoretical $M_w$ values calculated from amino acid sequence. Both ThkA and ΔThkA have approximately twice $M_w$ values of the theoretical ones, respectively and therefore exist as dimer in solution. TrrA has the $M_w$ value of 11 kDa similar to the theoretical one and exists as monomer in solution. [d]Radii of gyration. [e]Longest linear distances calculated from pair distribution function. The difference of 17 Å between ThkA and ΔThkA indicates that the two sensor domains locate in both edges of $ThkA_2$. No difference between ThkA/TrrA complex and ThkA indicates that two TrrA molecules bind to the central region of $ThkA_2$.

regulation. When the crystal structure of the sensor domain of ThkA was fitted onto the electron density map, its G-sheet could contact with the catalytic domain. The G-sheet of the ThkA sensor domain corresponds to the FG-loop region of the FixL sensor domain, as was evaluated by the structural comparison (unpublished results). The observation support the assumption that the conformational change of the FG-loop of FixL sensor domain would be a direct switch to regulate the kinase activity of the catalytic domain. The regulation mechanism is apparently different from that with periplasmic sensor domain, as stated in LuxP/LuxQ system. The regulation mechanisms of HK depend on the type of the sensor domain.

### From Catalytic Domain to Dimerization Domain

In the active HK, the catalytic domain can phosphorylate the His residue in the other subunit (*trans*-autophosphorylation, Fig. 9B).[1] The relative orientation of dimerization and catalytic domains of ThkA is similar to that of TM0853 (Figs. 7A, 7B, 9A and 9B) and the distance between the ATP binding site of catalytic domain and the substrate His was estimated at 25 Å (Fig. 9B). Therefore, the dynamic motion of the catalytic domain, rotation around the dimerization domain, would be required in *trans*-autophosphorylation reaction, in good agreement with the proposed mechanism so far (Fig. 7C).

### From Dimerization Domain to RR

TrrA molecule binds to the His-containing helix-I in the dimerization domain (Fig. 9A). The similar situation could be observed in Spo0B system.[62,63] The Nε of His30 in Spo0B and Oγ of Asp in its RR are faced each other with 5.27 Å of the distance in the their complex. In the intermediate of the phosphoryl transfer reaction from His of the HK dimerization domain to Asp of the RR receiver domain, a penta-coordinated phosphorous atom could be formed by binding to both proteins.[63]

Two RRs bind separately to form the HK/RR complex (Fig. 9). Therefore, RRs form its homodimer after accepting a phosphoryl group and dissociation from HK for transcriptional activation. For example, the phosphorylated receiver domain of FixJ contacts with each other through the α4-β5 region (Fig. 5D).[55]

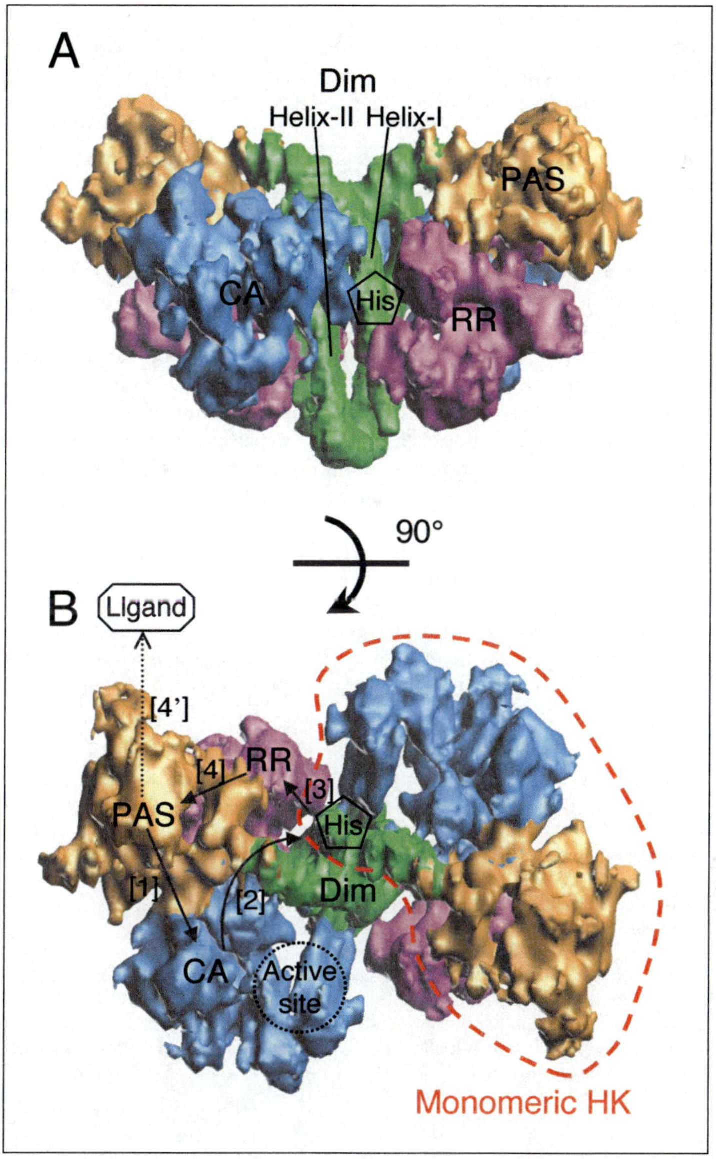


Figure 9. Electron density maps of ThkA/TrrA complex. A) The front view. Sensor (PAS), dimerization (Dim), catalytic (CA) and TrrA (RR) are colored by orange, green, blue and magenta, respectively. The position of phosphorylated His is indicated as pentagon. B) Rotation of (A) at 90° along horizontal axis. The red dashed line and the dotted circle indicate monomeric ThkA and active site, respectively. Allows indicate the direction of signal transduction pathway: [1] *cis*-regulation of PAS to CA, [2] *trans*phosphorylation of CA to Dim, [3] phosphotransfer to RR from Dim, [4] feedback regulation of RR to PAS and [4'] dissociation of ligand enhanced by binding of RR.

**From RR to Sensor Domain**

In the structure of the HK/RR complex of the thermophilic TCS system, it was first finding that RR (TrrA) contacts with not only the HK (ThkA) dimerization domain but also the sensor domain (Fig. 9). This finding implies a possibility that RR could act as the feedback regulator of the HK sensor domain (Fig. 9B) as well as the acceptor of phosphoryl group. Indeed, it was reported that the $O_2$ affinity of the FixL sensor domain is reduced upon the FixJ binding in the presence of ADP and consequently the kinase activity of FixL is promoted.[76]

## Conclusions

We have a progress in understanding the signaling mechanisms of TCS, since a lot of structural information is now available on the proteins and the domains. However, they might have only given a start to describe the mechanism. The discussions will give a route of further structural analysis that should be advanced in next stage. For example, direct comparison between the active and the inactive structures of entire HK molecule will allow us to understand the regulation mechanism of kinase activity. Moreover, it is implied that the dynamic conformation changes are required in the signal transduction process, although the structures that have been obtained are static. If a combination usage of crystallography and solution analysis supplements each other, the researches on dynamics of TCS protein will be progressed. At last, it should be noted that TCS has not been discovered in mammals. The inhibitor for HK or RR may therefore become a new type of antibacterial agents that have little side effect. The detailed mechanism may help to design and synthesis of such efficient and safe drugs.

### *Acknowledgements*

The authors thank to Drs. Hiro Nakamura and Hiroshi Sugimoto in RIKEN SPring-8 Center and Dr. Shuji Akiyama in PRESTO, JST, for proofreading of this manuscript and critical comments. This work was supported by the Special Postdoctoral Researchers Program in RIKEN (to S.Y.).

## References

1. Ninfa EG, Atkinson MR, Kamberov ES et al. Mechanism of autophosphorylation of Escherichia coli nitrogen regulator II (NRII or NtrB): transphosphorylation between subunits. J Bacteriol 1993; 175:7024-7032.
2. Kato M, Mizuno T, Shimizu T et al. Insights into multistep phosphorelay from the crystal structure of the C-terminal HPt domain of ArcB. Cell 1997; 88:717-723.
3. Kato M, Mizuno T, Shimizu T et al. Refined structure of the histidine-containing phosphotransfer (HPt) domain of the anaerobic sensor kinase ArcB from Escherichia coli at 1.57 A resolution. Acta Crystallogr 1999; D55:1842-1849.
4. Mizuno T, Matsubara M. Role of the histidine-containing phosphotransfer domain (HPt) in the multistep phosphorelay through the anaerobic hybrid sensor, ArcB. In: Inouye M, Dutta R, eds. Histidine Kinases in Signal Transduction. San Diego: Academic Press, 2003:165-190.
5. Bilwes AM, Park SY, Quezada CM et al. Structure and function of CheA, the histidine kinase central to bacterial chemotaxis. In: Inouye M, Dutta R, eds. Histidine Kinase in Signal Transduction. San Diego: Academic Press, 2003:47-72.
6. Wadhams GH, Armitage JP. Making sense of it all: bacterial chemotaxis. Nature Rev Mol Cell Biol 2004; 5:1024-1037.
7. Mourey L, Da Re S, Pedelacq JD et al. Crystal structure of the CheA histidine phosphotransfer domain that mediates response regulator phosphorylation in bacterial chemotaxis. J Biol Chem 2001; 276:31074-31082.
8. McEvoy MM, Hausrath AC, Randolph GB et al. Two binding modes reveal flexibility in kinase/response regulator interactions in the bacterial chemotaxis pathway. Proc Natl Acad Sci USA 1998; 95:7333-7338.
9. Welch M, Chinardet N, Mourey L et al. Structure of the CheY-binding domain of histidine kinase CheA in complex with CheY. Nature Struct Biol 1998; 5:25-29.
10. Bilwes AM, Alex LA, Crane BR et al. Structure of CheA, a signal-transducing histidine kinase. Cell 1999; 96:131-141.
11. Bilwes AM, Quezada CM, Croal LR et al. Nucleotide binding by the histidine kinase CheA. Nature Struct Biol 2001; 8:353-360.

12. Park SY, Borbat PP, Gonzalez-Bonet G et al. Reconstruction of the chemotaxis receptor-kinase assembly. Nature Struct Mol Biol 2006; 13:400-407.
13. Stock AM, West AH. Response regulator proteins and their interaction with histidine kinase. In: Inouye M, Dutta R, eds. Histidine Kinase in Signal Transduction. San Diego: Academic Press, 2003:237-271.
14. West AH, Martinez-Hackert E, Stock AM. Crystal structure of the catalytic domain of the chemotaxis receptor methylesterase, CheB. J Mol Biol 1995; 250:276-290.
15. Djordjevic S, Goudreau PN, Xu Q et al. Structural basis for methylesterase CheB regulation by a phosphorylation-activated domain. Proc Natl Acad Sci USA 1998; 95:1381-1386.
16. Tomomori C, Tanaka T, Dutta R et al. Solution structure of the homodimeric core domain of Escherichia coli histidine kinase EnvZ. Nature Struct Biol 1999; 6:729-734.
17. Marina A, Mott C, Auyzenberg A et al. Structural and mutational analysis of the PhoQ histidine kinase catalytic domain. Insight into the reaction mechanism. J Biol Chem 2001; 276:41182-41190.
18. Tanaka T, Saha SK, Tomomori C et al. NMR structure of the histidine kinase domain of the E. coli osmosensor EnvZ. Nature, 1998; 396:88-92.
19. Song Y, Peisach D, Pioszak AA et al. Crystal structure of the C-terminal domain of the two-component system transmitter protein nitrogen regulator II (NRII; NtrB), regulator of nitrogen assimilation in Escherichia coli. Biochemistry 2004; 43:6670-6678.
20. Nowak E, Panjikar S, Morth JP et al. Structural and functional aspects of the sensor histidine kinase PrrB from Mycobacterium tuberculosis. Structure 2006; 14:275-285.
21. Wigley DB, Davies GJ, Dodson EJ et al. Crystal structure of an N-terminal fragment of the DNA gyrase B protein. Nature 1991; 351:624-629.
22. Ban C, Junop M, Yang W. Transformation of MutL by ATP binding and hydrolysis: a switch in DNA mismatch repair. Cell 1999; 97:85-97.
23. Prodromou C, Roe SM, O'Brien R et al. Identification and structural characterization of the ATP/ADP-binding site in the Hsp90 molecular chaperone. Cell 1997; 90:65-75.
24. Jeffery CJ, Koshland DE, Jr. Three-dimensional structural model of the serine receptor ligand-binding domain. Protein Sci 1993; 2:559-566.
25. Yeh JI, Biemann HP, Pandit J et al. The three-dimensional structure of the ligand-binding domain of a wild-type bacterial chemotaxis receptor. Structural comparison to the cross-linked mutant forms and conformational changes upon ligand binding. J Biol Chem 1993; 268:9787-9792.
26. Yeh JI, Biemann HP, Prive GG et al. High-resolution structures of the ligand binding domain of the wild-type bacterial aspartate receptor. J Mol Biol 1996; 262:186-201.
27. Falke JJ, Koshland DE, Jr. Global flexibility in a sensory receptor: a site-directed cross-linking approach. Science 1987; 237:1596-1600.
28. Chen X, Schauder S, Potier N et al. Structural identification of a bacterial quorum-sensing signal containing boron. Nature 2002; 415:545-549.
29. Neiditch MB, Federle MJ, Miller ST et al. Regulation of LuxPQ receptor activity by the quorum-sensing signal autoinducer-2. Mol Cell 2005; 18:507-518.
30. Neiditch MB, Federle MJ, Pompeani AJ et al. Ligand-induced asymmetry in histidine sensor kinase complex regulates quorum sensing. Cell 2006; 126:1095-1108.
31. Schaller GE, Bleecker AB. Ethylene-binding sites generated in yeast expressing the Arabidopsis ETR1 gene. Science 1995; 270:1809-1811.
32. Rodriguez FI, Esch JJ, Hall AE et al. A copper cofactor for the ethylene receptor ETR1 from Arabidopsis. Science, 1999; 283:996-998.
33. Rockwell NC, Su YS, Lagarias JC. Phytochrome structure and signaling mechanisms. Annu Rev Plant Biol 2006; 57:837-858.
34. Wagner JR, Brunzelle JS, Forest KT et al. A light-sensing knot revealed by the structure of the chromophore-binding domain of phytochrome. Nature 2005; 438:325-331.
35. Volz K, Matsumura P. Crystal structure of Escherichia coli CheY refined at 1.7-A resolution. J Biol Chem 1991; 266:15511-15519.
36. Madhusudan, Zapf J, Whiteley JM et al. Crystal structure of a phosphatase-resistant mutant of sporulation response regulator Spo0F from Bacillus subtilis. Structure 1996; 4:679-690.
37. Volkman BF, Nohaile MJ, Amy NK et al. Three-dimensional solution structure of the N-terminal receiver domain of NTRC. Biochemistry 1995; 34:1413-1424.
38. Baikalov I, Schroder I, Kaczor-Grzeskowiak M et al. Structure of the Escherichia coli response regulator NarL. Biochemistry 1996; 35:11053-11061.
39. Sola M, Gomis-Ruth FX, Serrano L et al. Three-dimensional crystal structure of the transcription factor PhoB receiver domain. J Mol Biol 1999; 285:675-687.
40. Birck C, Mourey L, Gouet P et al. Conformational changes induced by phosphorylation of the FixJ receiver domain. Structure 1999; 7:1505-1515.

41. Lewis RJ, Muchova K, Brannigan JA et al. Domain swapping in the sporulation response regulator Spo0A. J Mol Biol 2000; 297:757-770.
42. Muller-Dieckmann HJ, Grantz AA, Kim SH. The structure of the signal receiver domain of the Arabidopsis thaliana ethylene receptor ETR1. Structure 1999; 7:1547-1556.
43. Buckler DR, Zhou Y, Stock AM. Evidence of intradomain and interdomain flexibility in an OmpR/PhoB homolog from Thermotoga maritima. Structure 2002; 10:153-164.
44. Guillet V, Ohta N, Cabantous S et al. Crystallographic and biochemical studies of DivK reveal novel features of an essential response regulator in Caulobacter crescentus. J Biol Chem 2002; 277:42003-42010.
45. Birck C, Chen Y, Hulett FM et al. The crystal structure of the phosphorylation domain in PhoP reveals a functional tandem association mediated by an asymmetric interface. J Bacteriol 2003; 185:254-261.
46. Riepl H, Scharf B, Schmitt R et al. Solution structures of the inactive and BeF3-activated response regulator CheY2. J Mol Biol 2004; 338:287-297.
47. Toro-Roman A, Mack TR, Stock AM. Structural analysis and solution studies of the activated regulatory domain of the response regulator ArcA: a symmetric dimer mediated by the alpha4-beta5-alpha5 face. J Mol Biol 2005; 349:11-26.
48. Toro-Roman A, Wu T, Stock AM. A common dimerization interface in bacterial response regulators KdpE and TorR. Protein Sci 2005; 14:3077-3088.
49. Nowak E, Panjikar S, Konarev P et al. The structural basis of signal transduction for the response regulator PrrA from Mycobacterium tuberculosis. J Biol Chem 2006; 281:9659-9666.
50. Cho HS, Lee SY, Yan D et al. NMR structure of activated CheY. J Mol Biol 2000; 297:543-551.
51. Lee SY, Cho HS, Pelton JG et al. Crystal structure of activated CheY. Comparison with other activated receiver domains. J Biol Chem 2001; 276:16425-16431.
52. Halkides CJ, Zhu X, Phillion DP et al. Synthesis and biochemical characterization of an analogue of CheY-phosphate, a signal transduction protein in bacterial chemotaxis. Biochemistry 1998; 37:13674-13680.
53. Halkides CJ, McEvoy MM, Casper E et al. The 1.9 A resolution crystal structure of phosphono-CheY, an analogue of the active form of the response regulator, CheY. Biochemistry 2000; 39:5280-5286.
54. Lee SY, Cho HS, Pelton JG et al. Crystal structure of an activated response regulator bound to its target. Nature Struct Biol 2001; 8:52-56.
55. Gouet P, Fabry B, Guillet V et al. Structural transitions in the FixJ receiver domain. Structure 1999; 7:1517-1526.
56. Baikalov I, Schroder I, Kaczor-Grzeskowiak M et al. NarL dimerization? Suggestive evidence from a new crystal form. Biochemistry 1998; 37:3665-3676.
57. Lee SY, De La Torre A, Yan D et al. Regulation of the transcriptional activator NtrC1: structural studies of the regulatory and AAA+ ATPase domains. Genes Dev 2003; 17:2552-2563.
58. Maris AE, Sawaya MR, Kaczor-Grzeskowiak M et al. Dimerization allows DNA target site recognition by the NarL response regulator. Nature Struct Biol 2002; 9:771-778.
59. Bassler BL, Wright M, Showalter RE et al. Intercellular signalling in Vibrio harveyi: sequence and function of genes regulating expression of luminescence. Mol Microbiol 1993; 9:773-786.
60. Bassler BL, Wright M, Silverman MR. Multiple signalling systems controlling expression of luminescence in Vibrio harveyi: sequence and function of genes encoding a second sensory pathway. Mol Microbiol 1994; 13:273-286.
61. Marina A, Waldburger CD, Hendrickson WA. Structure of the entire cytoplasmic portion of a sensor histidine-kinase protein. EMBO J 2005; 24:4247-4259.
62. Zapf J, Sen U, Madhusudan et al. A transient interaction between two phosphorelay proteins trapped in a crystal lattice reveals the mechanism of molecular recognition and phosphotransfer in signal transduction. Structure 2000; 8:851-862.
63. Varughese KI, Tsigelny I, Zhao H. The crystal structure of beryllofluoride Spo0F in complex with the phosphotransferase Spo0B represents a phosphotransfer pretransition state. J Bacteriol 2006; 188:4970-4977.
64. Tanaka A, Nakamura H, Shiro Y et al. Roles of the heme distal residues of FixL in $O_2$ sensing: a single convergent structure of the heme moiety is relevant to the downregulation of kinase activity. Biochemistry 2006; 45:2515-2523.
65. Gilles-Gonzalez MA, Ditta GS, Helinski DR. A haemoprotein with kinase activity encoded by the oxygen sensor of Rhizobium meliloti. Nature 1991; 350:170-172.
66. Saito K, Ito E, Hosono K et al. The uncoupling of oxygen sensing, phosphorylation signalling and transcriptional activation in oxygen sensor FixL and FixJ mutants. Mol Microbiol 2003; 48:373-383.
67. Gong W, Hao B, Mansy SS et al. Structure of a biological oxygen sensor: a new mechanism for heme-driven signal transduction. Proc Natl Acad Sci USA, 1998; 95:15177-15182.

68. Miyatake H, Mukai M, Park SY et al. Sensory mechanism of oxygen sensor FixL from Rhizobium meliloti: crystallographic, mutagenesis and resonance Raman spectroscopic studies. J Mol Biol 2000; 301:415-431.
69. Akimoto S, Tanaka A, Nakamura K et al. $O_2$-specific regulation of the ferrous heme-based sensor kinase FixL from Sinorhizobium meliloti and its aberrant inactivation in the ferric form. Biochem Biophys Res Commun 2003; 304:136-142.
70. Hao B, Isaza C, Arndt J et al. Structure-based mechanism of $O_2$ sensing and ligand discrimination by the FixL heme domain of Bradyrhizobium japonicum. Biochemistry 2002; 41:12952-12958.
71. Gong W, Hao B, Chan MK. New mechanistic insights from structural studies of the oxygen-sensing domain of Bradyrhizobium japonicum FixL. Biochemistry, 2000; 39:3955-3962.
72. Yamada S, Akiyama S, Sugimoto H et al. The signaling pathway in histidine kinase and the response regulator complex revealed by X-ray crystallography and solution scattering. J Mol Biol 2006; 362:123-139.
73. Kumita H, Yamada S, Nakamura H et al. Chimeric sensory kinases containing $O_2$ sensor domain of FixL and histidine kinase domain from thermophile. Biochim Biophys Acta 2003; 1646:136-144.
74. Nelson KE, Clayton RA, Gill SR et al. Evidence for lateral gene transfer between Archaea and bacteria from genome sequence of Thermotoga maritima. Nature 1999; 399:323-329.
75. Yamada S, Nakamura H, Kinoshita E et al. Separation of a phosphorylated histidine protein using phosphate affinity polyacrylamide gel electrophoresis. Anal Biochem 2007; 360:160-162.
76. Nakamura H, Kumita H, Imai K et al. ADP reduces the oxygen-binding affinity of a sensory histidine kinase, FixL: the possibility of an enhanced reciprocating kinase reaction. Proc Natl Acad Sci USA, 2004; 101:2742-2746.

# Chapter 4

# The Two-Component Network and the General Stress Sigma Factor RpoS ($\sigma^S$) in *Escherichia coli*

**Regine Hengge***

## Abstract

The general stress sigma factor RpoS ($\sigma^S$) is induced during entry into stationary phase and in response to multiple stress conditions. RpoS is regulated at the levels of transcription, translation, proteolysis and protein activity. A key factor in RpoS control is the two-component response regulator RssB, which acts as a direct recognition and targeting factor for ClpXP-mediated RpoS proteolysis. A major, but not the only phosphodonor for RssB is the complex histidine sensor kinase ArcB. ArcB coordinates RpoS proteolysis with *rpoS* transcription by also phosphorylating the response regulator ArcA, which besides controlling a large regulon, also acts as a transcriptional repressor for *rpoS*. ArcB activity depends on the redox state of the respiratory chain, which links RpoS control to the balance between energy supply and available respiratory electron acceptor. In addition, the BarA/UvrY and Rcs phosphorelay systems can activate *rpoS* transcription and translation, respectively. These systems are involved in the control of motility, biofilm formation and/or virulence, suggesting that further studying a potential role of RpoS in these physiological functions may be rewarding.

## Introduction

The RpoS or $\sigma^S$ protein is the master regulator of the general stress response in *Escherichia coli*.[1] RpoS, which occurs also in other γ-proteobacteria, is a sigma subunit of RNA polymerase (RNAP) present at very low levels in rapidly growing cells not experiencing any stresses. In response to a large variety of stress conditions, however, RpoS is rapidly and often rather dramatically induced (for a review of RpoS regulation, see ref. 2) When present at high levels, RpoS competes with the vegetative (RpoD or $\sigma^{70}$) and other sigma factors for core RNAP and reprograms this enzyme to switch to transcription at RpoS-dependent promoters. At first glance, RpoS-dependent "stress" promoters look very similar to vegetative RpoD-transcribed promoters, yet the combination of specific small deviations from the characteristics of vegetative promoters renders these promoters RpoS-specific (for reviews, see refs. 3,4). Comprehensive microarray-based transcriptome analyses performed under various conditions where RpoS is strongly induced (entry into stationary phase, shifts to high osmolarity and pH 5) have shown that almost 500 genes, i.e., approximately 10% of the genes in the *E. coli* genome are under direct or indirect RpoS control.[5-7]

*Regine Hengge—Institut für Biologie-Mikrobiologie, FB Biologie, Chemie und Pharmazie, Freie. Universität Berlin, Königin-Luise-str.12-16, 14195 Berlin, Germany.
Email: rhenggea@zedat.fu-berlin.de

*Bacterial Signal Transduction: Networks and Drug Targets*, edited by Ryutaro Utsumi.

## Regulation of RpoS

Stress conditions that induce RpoS include carbon, phosphorus, nitrogen or amino acid starvation,[8-10] reductions in growth rate,[9] the classical glucose/lactose diauxic lag phase,[11] shifts to high osmolarity,[9,12] or low pH,[13,14] the classical heat shock procedure (i.e., a shift from 30° to 42°C),[15] but also growth at reduced temperature (e.g., room temperature)[16] (for a summary, see ref.2). Under some conditions, RpoS induction is lasting (e.g., when cells enter stationary phase due to starvation), under others it is transient (e.g., upon pH downshift or during the diauxic lag phase). As a multitude of environmental and cellular signals have to be processed in RpoS control, it is not surprising, that this control occurs at all possible levels, i.e., *rpoS* transcription and translation as well as proteolysis and the activity of RpoS protein[9] (Fig. 1). While *rpoS* transcription can be further stimulated by reduced growth rate (by a mechanism that involves ppGpp,[8,17,18]), there is always a fair amount of *rpoS* mRNA present in the cell. Due to mRNA secondary structure in the translational initiation region, however, this mRNA is not efficiently translated. Yet, translation can be rapidly stimulated by small regulatory RNAs, which together with the RNA-binding protein Hfq trigger alterations in *rpoS* mRNA secondary structure that allow access of ribosomes to the translational initiation site.[19-24] Furthermore, RpoS, which is also synthesized at a small rate even in the absence of stress, is rapidly degraded by a proteolytic recognition factor, RssB and the ATP-dependent ClpXP protease.[25-30] A rapid increase in the cellular level of RpoS can also be achieved by an instantanous inhibition of this degradation. Finally, the activity of RpoS, i.e., its ability to successfully compete with other sigma subunits for core RNAP and thereby to activate its regulon, is controlled by factors such as ppGpp, Rsd and Crl, which interact with RNAP core, RpoD and/or RpoS.[31-34]

Different stress conditions affect different levels of RpoS control, with a tendency of the more severe and potentially lethal stress conditions acting on the most rapid regulatory process in RpoS control, i.e., proteolysis. Stress conditions that most strongly induce RpoS (hyperosmotic shift or pH downshift) stimulate *rpoS* mRNA translation as well as interfere with RpoS proteolysis (Fig. 1). As mentioned above, numerous factors, including small molecules, small regulatory RNAs and

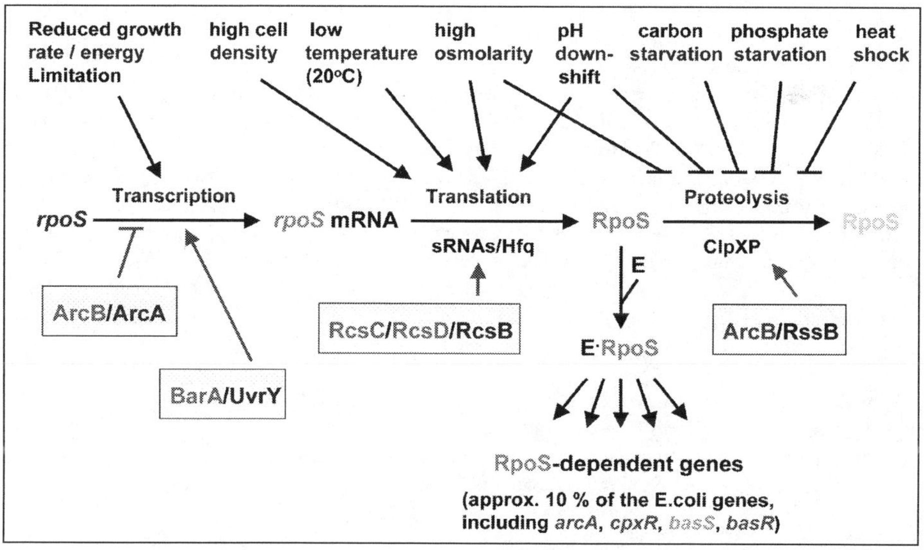


Figure 1. Environmental signal input and two-component systems controlling *rpoS* transcription and translation as well as proteolysis of RpoS protein. Mechanistic details of this summary figure are explained in the text. The figure is an extended version of a previously published figure.[11]

regulatory proteins, have been identified to play a role in this complex regulation of RpoS. However, in almost all of these control mechanisms the actual stimulus or primary signal has remained elusive. Is it a changing extracellular physical parameter itself (as e.g., temperature, external osmolarity or pH) or is it a cellular consequence of this changing parameter, e.g., some cellular damage? Or can an initial response, e.g., the massive influx of potassium ions upon osmotic upshift then trigger secondary and more lasting responses?

As two-component systems are major signal-processing devices in bacteria, it is not surprising that such systems are also involved in the massive information processing in the control of RpoS. The key factor in the proteolytic control of RpoS is the response regulator RssB which acts as a recognition factor that targets RpoS to ClpXP proteases.[25,26,30,35,36] While the role of RssB in RpoS degradation is relatively well understood (see below), it has been recognized only recently, that RssB activation as a proteolytic targeting factor for RpoS (i.e., phosphorylation) is linked to the respiratory status and energy supply of the cell via the ArcB sensor kinase. The ArcB/ArcA phosphorelay also controls *rpoS* transcription and thereby coordinates RpoS expression and degradation.[37] In addition, there is still rather patchy evidence that additional two-component systems such as the RcsC/RcsD/RcsB and BarA/ UvrY(YecB) phosphorelays are involved in RpoS control (Fig. 1). This review attempts to put this currently available information into a more or less consistent perspective that may also guide further research on signal input into RpoS regulation.

## The Response Regulator RssB Acts as a Targeting Factor in RpoS Proteolysis

RpoS is degraded by the general ATP-dependent ClpXP protease.[27,29,30,36] Yet, RpoS alone is not bound by the hexameric $ClpX_6$ ring, which represents the substrate-binding and unfolding chaperone unit of the ClpXP protease, unless phosphorylated RssB is also present. RssB acts as the primary and essential recognition factor for RpoS proteolysis which delivers RpoS to the $ClpX_6$ moiety of the ClpXP protease[25,36] (Fig. 2). Regulation of the activity or availability of RssB therefore confers regulation to RpoS degradation. Phosphorylated RssB, ClpXP and ATP are sufficient to achieve RpoS degradation in vitro.[30]

The function of RssB as a proteolytic recognition or targeting factor is so far unique for a two-component response regulator. It may be that this function evolved only relatively recently, as the phylogenetic distribution of RssB is more narrow than that of RpoS. Also, when RssB is expressed stoichiometrically with RpoS in an *E. coli* strain that lacks the ClpXP protease, RssB acts like an antisigma factor with phosphorylation-controlled activity.[38] Moreover, the RssB output domain may have a common ancestor with the PP2C-type Ser/Thr protein phosphatases (although RssB does not feature all amino acids crucial for phosphatase activity), which in gram-positive bacteria control "partner-switching" modules of gene regulation.[39] All this suggests that the RssB protein was derived from a more primitive regulatory protein that antagonized another protein by direct interaction. The combination with a receiver domain allowed strong regulation of activity and probably its evolution into an antisigma factor that controlled RpoS activity. This antagonistic but still reversible role would have been further accentuated by establishing interactions of RpoS as well as of RssB with ClpXP, resulting in a highly regulated proteolytic targeting system that now irreversibly eliminates RpoS.

The RssB-binding site in RpoS, the "turnover element", is a single alpha helix ($\alpha$2.5/3.0, which is the first of three $\alpha$ helices in domain 3 of RpoS. A specific amino acid, lysine173, is absolutely essential for interaction with RssB.[25] This $\alpha$ helix is exposed in free RpoS, which allows direct and tight binding to phosphorylated RssB in the absence of any other factors.[25] In the complex with RNAP core enzyme, however, RpoS is protected against interaction with RssB.[30] Nevertheless, also in the RNAP holoenzyme, lysine173 is still free to interact with a cytidine, which most RpoS-dependent promoters feature at position-13 and which is a major determinant for RpoS-selectivity of a promoter.[4,7,40] As various attempts to determine the structure of RssB have failed so far (purified RssB tends to aggregate under all conditions tested), the RpoS-binding site in RssB is still unknown. Pull-down assays in vitro as well as complementation experiments in vivo have shown that the

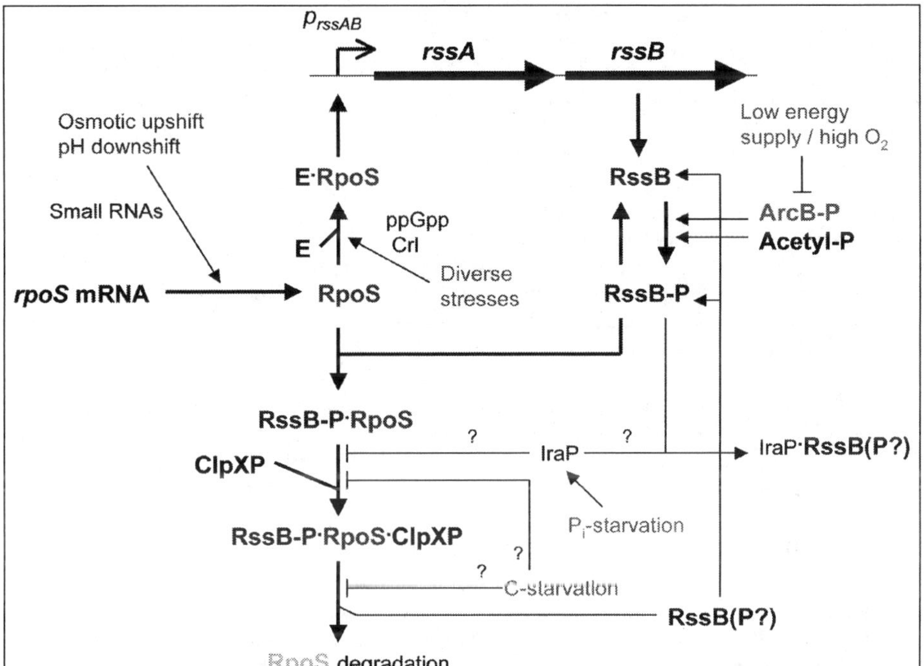


Figure 2. Signal integration into the proteolytic targeting cycle of RpoS. Phosphorylated RssB directly binds to RpoS and delivers it to the ClpXP protease. Alternatively, RpoS can bind to RNAP core enzyme (E), in which it is protected against interaction with RssB-P and then can activate multiple genes, including the *rssAB* operon. The latter results in an adjustment of RssB levels to RpoS levels and represents a negative feedback cycle that allows adaptation after RpoS stabilization due to RssB titration because of a sudden increase in RpoS expression (for details, see text). For the RssB-binding antagonist IraP, it is not yet clear whether it sequesters RssB (not allowing RssB to bind RpoS) or whether it interferes with RssB function also in the RssB-RpoS complex. Also, it has not yet been clarified, whether RssB is dephosphorylated during its release from the RssB-P-RpoS-ClpXP complex. The figure is a modified version of a previously published figure.[45]

isolated N-terminal receiver or C-terminal output domains of RssB are inactive.[26] It is therefore hypothesized that either phosphorylation of the N-terminal receiver domain conformationally activates the C-terminal domain to bind RpoS or that both domains contribute surfaces to the binding site for RpoS, which e.g., might be a cleft formed by both domains which would accommodate the lysine173-containing recognition helix of RpoS.

Binding to RssB appears to trigger a change in RpoS conformation, as this exposes a binding site for the hexameric $ClpX_6$ ring, which is located close to the N-terminus of RpoS. This allows ternary complex formation between RssB, RpoS and $ClpX_6$.[36] In RpoS alone, this $ClpX_6$-binding site seems to be occluded by interaction with the C-terminal part of RpoS, as suggested by the finding that RpoS::LacZ hybrid proteins, which contain the N-terminal region but lack the C-terminal part of RpoS, can be bound (in vivo and in vitro) by $ClpX_6$ in the absence of RssB. Interestingly, in a wildtype background, such $ClpX_6$-binding hybrid proteins are degraded, but only when their RpoS part is long enough to also contain the RssB binding site (i.e., $\alpha$2.5/3.0 located approximately in the middle of RpoS). A point mutation in this site (K173E) of such a long hybrid protein again abolishes its proteolysis.[36] This indicates that RssB plays a dual role, first by inducing a change in RpoS conformation that allows $ClpX_6$ to bind (this step is bypassed in the hybrid proteins lacking the C-terminal part of RpoS) and then by triggering another step that occurs after initial $ClpX_6$ binding, which may affect the structure of the ternary complex and/or a subsequent step in targeting RpoS for degradation by ClpXP protease. For the conformational change in RpoS that exposes the $ClpX_6$ binding site, a direct interaction between RssB and $ClpX_6$ would in principle not be necessary. Nevertheless, such an interaction has been assumed on the basis of sequence similarity at the C-termini of RssB and SspB, a known adaptor for the ClpXP protease, which binds to the $ClpX_6$ ring even in the absence of substrate and provides it with high selectivity for specific substrates (e.g., SsrA-tagged proteins).[41-43] While RssB in the absence of its substrate RpoS does not bind tightly to $ClpX_6$[26] and therefore should be regarded as a targeting factor rather than a protease adaptor protein, an interaction between RssB and $ClpX_6$ may play a role in the above mentioned second function of RssB in initiating RpoS proteolysis by ClpXP. RssB itself is not degraded,[26] which is unlike the fate of various proteolytic adaptor proteins in *Bacillus subtilis*, which get degraded as soon as there is no more substrate for them to deliver to the protease and thereby "make room" for other adaptors.[44] RssB, by contrast, leaves the proteolytic complex upon delivery of its substrate and most likely initiates the next round of targeting.[26] Whether RssB is dephosphorylated during this process, has not yet been shown directly. In any case, RssB seems to act catalytically (Fig. 2), which is consistent with its cellular levels being 10 to 20fold lower than that of its substrate RpoS.[26,38]

## The Role of RssB in the Regulation of RpoS Proteolysis by Environmental Signals

While the proteolytic targeting cycle catalyzed by RssB operates with a minimum of components (phosphorylated RssB, ClpXP and ATP) and therefore seems simple, it features an amazing range of possibilities for regulation, which explains why and how so many environmental signals can be integrated in the control of RpoS degradation (Fig. 2; see also ref. 45):

1. The activity of RssB can be controlled by phosphorylation or dephosphorylation, which raises the question of the phosphodonors for the response regulator RssB and their regulation of activity; the in vivo and in vitro data so far available indicate that the ArcB histidine sensor kinase [37] as well as the small phosphodonor acetyl phosphate[46] phosphorylate RssB and that energy supply controls RssB phosphorylation levels (for details, see below).
2. As RssB is the limiting component for the overall cellular rate of RpoS degradation, a sudden strong increase in RpoS synthesis (as e.g., upon osmotic upshift or pH downshift) can titrate the proteolytic recognition machinery, which results in RpoS stabilization and accumulation; therefore, RpoS proteolysis is also sensitive to the expression levels of RssB (under conditions where RssB is phosphorylated).[35]
3. The availability or function of RssB can also be controlled by a specific antagonist, e.g., by a small protein, IraP, which interacts with RssB and interferes with RpoS proteolysis specifically under conditions of phosphate starvation.[47]
4. There is in vivo evidence that under conditions of carbon starvation, at least the RssB-RpoS complex is still formed, but remains unproductive for proteolysis, suggesting that some not yet identified factor may interfere with successful targeting of RpoS to ClpXP protease under these conditions.[38]
5. As RpoS is protected against RssB when itself bound to RNAP, any factors that stimulate RpoS activity by affecting sigma factor competition for RNAP core, can also result in altered degradation of RpoS; this has recently been demonstrated for Crl, a protein that binds to RpoS and stimulates its activity.[34,48] Crl expression is under the control of quorum sensing[49] and is also induced upon nitrogen starvation (V. Carabetta and T. Silhavy, pers. comm.).

From this list, it is obvious that phosphorylation and dephosphorylation of the response regulator RssB is just one of several possibilities to control the rate of RpoS degradation. This explains, why mutant strains expressing variants of RssB, in which the normally phosphorylated aspartic acid residue at position 56 has been replaced by specific amino acids (proline or alanine), which allow basal activity of RssB, still exhibit regulation of RpoS proteolysis.[26,50] As this chapter is about the role of two-component systems in RpoS control, only the first three mechanisms mentioned above that affect RssB activity, expression and/or availability will be discussed in detail here.

## The ArcB/ArcA/RssB "Three-Component" System Coordinates RpoS Expression and Proteolysis with Energy Metabolism

RssB is a typical "orphan" response regulator, as no gene encoding a histidine sensor kinase is located anywhere close to the *rssB* gene. *rssB* is in an operon with *rssA*[35,51] (Fig. 2), but *rssA* encodes for a special phospholipase (M. Marquardt, P. Auraß, A. Flieger and R. Hengge, unpublished data). Extensive mutant searches for a cognate sensor kinase for RssB, using screens for reduced RpoS proteolysis and/or increased RpoS levels, remained unsuccessful, indicating functional redundancy in the phosphorylation of RssB. The small phosphodonor acetyl phosphate was found to highly efficiently phosphorylate RssB and an acetyl phosphate-free *ackA pta* mutant indeed exhibits reduced RpoS turnover.[46] Consistently, cellular acetyl phosphate levels are highly variable depending on the kind and concentration of carbon/energy souce supplied.[52]

While these findings indicated that acetyl phosphate plays a physiological role in RssB phosphorylation, RpoS proteolysis was not completely abolished in the acetyl phosphate-free mutants suggesting the presence of additional phosphodonors. Finally, targeted knock-out of specific candidate sensor kinase genes further clarified the phosphorylation of RssB. Specifically, mutations in the ArcB-ArcA phosphorelay system resulted in increased RpoS levels in growing cells[37,53,54] and it was observed that the ArcB sensor kinase can phosphorylate RssB in vitro.[37,55]

Detailed genetic and biochemical analysis revealed that the role of the Arc phosphorelay system is not only physiologically relevant but is also much more complex than initially assumed.[37] The phosphorylated response regulator ArcA acts as a direct repressor of *rpoS* transcription by binding to two sites flanking the major *rpoS* promoter (with the upstream site overlapping an activating cAMP-CRP binding site). The sensor kinase ArcB not only phosphorylates ArcA but also RssB and via the latter stimulates RpoS proteolysis (Fig. 3A). Thus ArcB/ArcA/RssB represent a branched "three-component" system that coordinates RpoS synthesis and proteolysis and thereby keeps RpoS levels low in rapidly growing cells.[37] This occurs as long as ArcB autophosphorylates, which raises the question of the nature of the signal input sensed by ArcB. It is well established that, together with the FNR protein, the ArcB/ArcA phosphorelay is a global control system during aerobic-anaerobic transitions.[56] However, the effects of ArcB and ArcA on RpoS are observed in aerated cultures, indicating that this system is not only active in the absence of oxygen. The signal directly sensed by ArcB is the redox state of the quinones, with oxidized quinones inducing intermolecular disulfide bond formation within the ArcB dimer which eliminates its autophosphorylation activity.[57,58] The redox state of the quinones is a parameter that integrates information about energy supply (i.e., electrons entering the respiratory chain) and the concentration of the electron acceptor (i.e., oxygen in aerated cultures). Thus, in only moderately aerobic standard laboratory cultures, the transition from full supplementation to carbon- and energy starvation should result in quinone oxidation and therefore inactivated ArcB. This in turn reduces phosphorylation of ArcA and RssB and therefore derepresses *rpoS* transcription and reduces RpoS degradation[37] (Fig. 3B; under these conditions, *rpoS* expression is also stimulated by cAMP-CRP which responds to carbon limitation and the BarA-UvrY phosphorelay system; see below). Thus, the Arc system seems to control RssB activity and RpoS degradation in response to energy supply signals.[37] In addition, anaerobic cultures exhibit reduced RpoS levels in stationary phase (R. Hengge, unpublished results), consistent with the Arc system being hyperactive in the absence of oxygen.

The ArcB/ArcA/RssB phosphotransfer network exhibits additional characteristics involved in fine regulation. The kinetics of phosphorylation of the two response regulators are different: phosphotransfer to ArcA is approximately 10fold faster than to RssB. With this "kinetic advantage", ArcA competes very efficiently with RssB for phosphorylation by ArcB, but not vice versa.[37] As a consequence, even relatively small alterations in ArcA levels may have a strong impact on the phosphorylation state of RssB and therefore RpoS degradation. As *arcA* expression is further

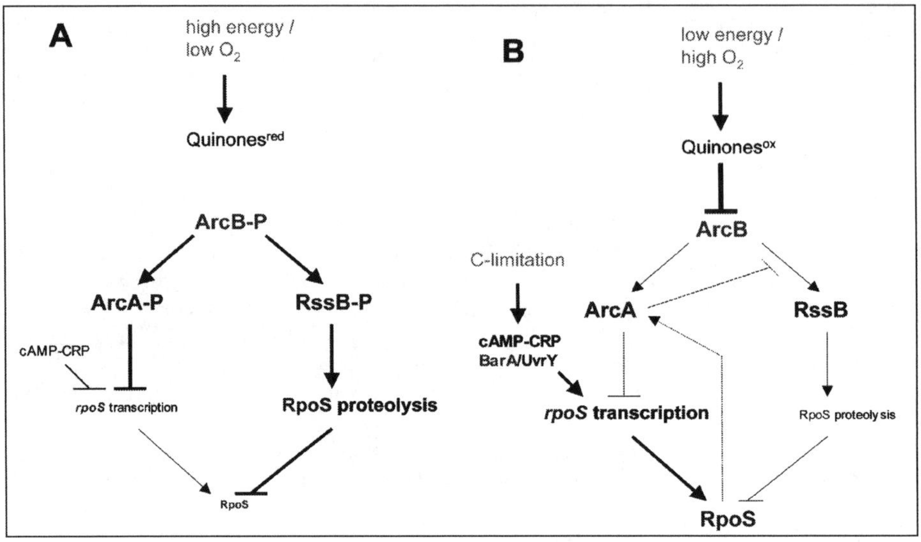


Figure 3. Coordination of *rpoS* transcription and RpoS proteolysis by the branched ArcB/ArcA/RssB "three-component" system in response to the redox state of the respiratory chain. A: a high energy/oxygen balance maintains the quinones in the reduced state that does not interfere with ArcB activity; this results in high ArcA and RssB phosphorylation and therefore low *rpoS* transcription and rapid RpoS proteolysis and consequently low RpoS levels. B: a low energy/oxygen balance results in oxidized quinones, which interfere with ArcB autophosphorylation, thereby shutting down ArcA and RssB phosphorylation; this leads to high *rpoS* transcription and reduced RpoS proteolysis and consequently to high RpoS levels. RpoS further downregulates its own proteolysis by activating *arcA* expression, as ArcA highly efficiently competes with RssB for residual phosphorylation by ArcB (for further details see text). The figure is a modified version of a previously published figure.[37]

induced in a RpoS-dependent manner during entry into stationary phase, a positive feedback loop is established, in which RpoS-induced ArcA results in reduced RssB phosphorylation and therefore further stabilization and accumulation of RpoS. Consistently, an *arcA* mutant does not stabilize RpoS to the same extent as an otherwise wildtype strain.[37] Moreover, dephosphorylation of ArcA via an ArcB-mediated reverse phosphorelay has been shown.[59] Phosphorylated RssB, however, is very stable. Neither does reverse phosphoflow take place,[37] nor does RssB contain intrinsic autophosphatase activity.[26] Dephosphorylation and therefore also signal decay may be linked to the proteolytic targeting process catalyzed by RssB.

Finally, the question arises, whether other sensor kinases or small phosphodonor molecules besides ArcB and acetyl phosphate also contribute to RssB phosphorylation. This is clearly the case, as an *arcB ackA pta* mutant still exhibits some RpoS degradation (F. Mika and R. Hengge, unpublished results). Moreover, how strongly the *arcB* mutation affects RpoS levels and degradation, depends on medium composition and the specific state in the growth cycle, indicating that other sensor kinases are also involved and the relative contributions of all phosphodonors is subject to regulation.[37] As a consequence of such a conditional use of several sensor kinases, major signal input into RssB-mediated regulation of RpoS proteolysis may be different under different conditions. Unfortunately, these additional sensor kinases have remained unidentified. At least under standard growth conditions in liquid media, mutations in some other phosphorelay sensor kinase genes (*rcsC, barA, evgS*) do not affect RpoS proteolysis (F. Mika and R. Hengge, unpublished results). In vitro, the UhpB sensor kinase, which responds to external glucose-6-phosphate (i.e., an excellent carbon and energy source), as well as the chemotaxis sensor kinase CheA have been found to phosphorylate RssB.[55] Based on its physiological context,[60] an involvement of UhpB in RssB/RpoS control seems conceivable but has not been tested. A role for CheA, however, appeared rather unlikely in view of its highly dynamic phosphorylation and dephosphorylation operating on a millisecond-to-second scale,[61] and a *cheA* mutation does not affect RpoS degradation in vivo (E. Klauck and R. Hengge, unpublished results).

## Role of the RpoS/RssB Feedback Cycle

In growing *E. coli* cells, RpoS is rapidly degraded by the RssB/ClpXP machinery. Nevertheless, cellular levels of RssB are low (10-20 fold lower than that of RpoS) and limiting for the overall cellular rate of RpoS degradation.[35] This is not only a prerequisite of RpoS degradation being sensitive to even small changes in the number of phosphorylated RssB molecules in the cell, but this situation also allows titration under conditions that result in rapid and strong increases in *rpoS* mRNA translation, e.g., osmotic upshift and pH downshift. Thus, RpoS stabilization can be a secondary consequence of a sudden increase in RpoS synthesis. However, the expression of the *rssAB* operon is driven by a RpoS-dependent promoter (Fig. 2), i.e., the cellular levels of RpoS and RssB are readjusted which resets the threshold for titration. As readjustment via this feedback cycle occurs with slower kinetics than titration, this mechanism provides the system with adaptation.[35] Upon pH downshift (from pH 7 to 5), this adaptation is complete, as the RpoS "peak" is transient and the system returns to prestimulus RpoS levels and half-lifes even when the cells remain and grow at pH5 (M. Pruteanu, J. Heuveling and R. Hengge, unpublished data). In hyperosmotically shifted cells, adaptation is only partial,[12] suggesting that the cellular RssB pool might not only be titrated but, perhaps in the somewhat longer run, also be increasingly dephosphorylated.

System properties conferred by the RpoS/RssB feedback loop have also been studied by a systems biology approach that combined quantitative measurements in vivo with mathematical modeling. These studies demonstrated that the negative RpoS/RssB feedback loop, by which RpoS induces the expression of its own "destruction" factor RssB, speeds up the reaction kinetics of the system and extends the range of adjustment of RpoS levels and half-lifes to different rates of RpoS synthesis[62] (and M. Pruteanu. A. Possling and R. Hengge, unpublished data).

## Role of the IraP Protein as an Antagonist of RssB

Another way of regulating RssB availability is the appearance of an "antiRssB" factor. Such a protein, IraP (YaiB), has recently been identified. IraP binds RssB in vitro, overproduction of IraP in vivo interferes with RpoS proteolysis and an *iraP* mutant is defective in RpoS stabilization specifically under conditions of phosphate starvation, i.e., conditions where IraP is induced physiologically. Based on these data, it has been suggested that IraP sequesters RssB and thereby interferes with RssB-RpoS interaction.[47] However, the alternative scenario, that IraP may bind to RssB also when the latter is in a complex with RpoS and may then interfere with targeting of RpoS to ClpXP, has not been excluded experimentally. It is tempting to speculate that additional proteins with an IraP-like function might be induced under some other stress conditions too, but evidence is lacking so far.[47]

## Effects of RssB on Other Two-Component Systems

Several observations indicate that the RssB response regulator can affect phosphotransfer in other two-component systems without being itself a recipient or donor of phosphate in the respective reactions. With purified components in vitro, the presence of RssB induced dephosphorylation of several sensor kinases (BaeS, DcuS, HydH, NarQ, NtrB and RstA) and therefore interfered with phosphorylation of the corresponding response regulators (BaeR, DcuR, HydG, NarP, NtrB and RstB).[55] In such cases, RssB may act as a competitive inhibitor, which interacts with the sensor kinase although it is not phosphorylated. If so, this may have interesting regulatory implications, as unphosphorylated RssB is present under various stress conditions where RpoS is stabilized and accumulates (and then further activates *rssB* expression). RssB-mediated interference may therefore reduce the activity of the above mentioned two-component systems under such conditions.

On the other hand, when RssB is present at *substoichiometric* concentrations with ArcA, ArcB and ATP in an in vitro transphosphorylation assay, it even somewhat stimulates phosphorylation of ArcA (with RssB being hardly phosphorylated), as if small amounts of RssB had a positive allosteric effect.[37] The mechanistic basis of this effect is completely unknown, but it raises the question of whether a sensor kinase dimer can simultanously interact with two different response regulator receiver domains.

Although these effects remain to be characterized in detail, they indicate that interference between two-component systems can be more complex than simple "cross-talk", i.e., "noncognate" phosphotransfer and may include competitive-inhibitory and even allosteric effects especially when an apparent "orphan" component like RssB may have relatively relaxed specificity for interaction with other two-component proteins combined to slow phosphorylation kinetics or efficiency.

## Role of the BarA/UvrY Phosphorelay System in *RpoS* Transcription

RpoS expression during entry into stationary phase was found to be reduced (though not abolished) in a mutant deficient in the complex sensor kinase gene *barA*,[63] even before it was clear that the response regulator UvrY(YecB; SirA in *Salmonella*) is phosphorylated by BarA.[64] A *uvrY* mutation has a similar reducing effect on *rpoS* transcription and this effect seems to be direct, as observed in gel retardation experiments with purified UvrY protein and *rpoS* promoter DNA fragments although a clear footprint pattern could not be obtained (F. Mika and R. Hengge, unpublished results). Although the BarA/UvrY system is clearly not essential for induction of RpoS during entry into stationary phase, it appears to have a positive modulatory effect.

The data available so far do not yet allow a clear picture of the physiological and molecular role of the BarA/UvrY system in *E. coli*. Microarray studies indicated that this system plays a global regulatory role.[53] It affects carbon metabolism (i.e., switching between glycolytic and gluconeogenic carbon sources), downregulates motility, stimulates biofilm formation and contributes to virulence of uropathogenic *E. coli* strains.[65-67] Its effect on carbon metabolism, motility and in part on biofilm formation are due to BarA/UvrY activating the expression of the small CsrB RNA,[65,68] which sequesters CsrA, a global regulator and RNA-binding protein that post-transcriptionally controls these processes.[69] The positive role of BarA/UvrY in virulence gene regulation and part of its role in biofilm formation are independent of the Csr system.[66] The connection to the BarA/UvrY system suggests that a putative role of RpoS in virulence, motility and biofilm formation may warrant further study, especially since the expression of several diguanylate cyclases (GGDEF proteins) was recently found to be under RpoS control.[70] These enyzmes synthesize the signaling molecule cyclic-di-GMP known to be involved in the inverse control of biofilm formation and motility.[71,72]

## Role of the Rcs Phosphorelay System in RpoS Translation

Several reports have indicated that also the RcsC/RcsD/RcsB phosphorelay plays multiple positive but indirect roles in RpoS expression (this system features three components, as the HPT domain is present on a separate polypeptide, RcsD (YojN).[73]) This is especially interesting, as the Rcs phosphorelay has a global spectrum of activities reminiscent of the BarA/UvrY system: it downregulates motility (by repressing the master regulator operon *flhDC*[74]) and activates the *cps* genes required for the synthesis of colanic acid, i.e., a matrix component in biofilms.[75] As flagella and motility are important during the first stages of biofilm formation, whereas matrix components are involved in biofilm maturation, the Rcs system is believed to play a role in the proper timing of the different steps in biofilm formation. Further complexity of this system comes from the ability of the RcsB response regulator to either act alone or in concert with the RcsA protein. The RcsC-activating stimuli include hyperosmolarity and desiccation, as well as growth on solid surfaces and/or cell surface perturbations (for a recent review, see ref. 73).

Although the actual molecular signal, that activates the RcsC sensor and therefore the target genes of the Rcs pathway, is still a mystery, it is known that the small RprA RNA, which can strongly activate *rpoS* translation, is under positive control of the Rcs system.[21,76] In addition, the Rcs system can also stimulate *rpoS* translation by repressing the expression of LrhA which in turn interferes with the expression of RprA and probably other small RNAs that activate *rpoS* translation.[77] On the other hand, LrhA is known to act as repressor in flagellar gene expression.[78] A negative role of the Rcs system in LrhA expression[77] would therefore attenuate repression of flagellar genes by LrhA and replace it by its by own direct repression of the flagellar master regulator operon *flhDC*[74] as mentioned above. The physiological conditions where these interdependencies (which represent an interesting variant of incoherent feedforward loop[79]) become relevant, are still unknown. Preliminary data suggest that the entire Rcs phosphorelay becomes essential for hyperosmotic induction of *rpoS* translation in certain mutant backgrounds, suggesting that (i) hyperosmotic shift is indeed among the conditions sensed by the Rcs system and that (ii) there is redundant signal input in *rpoS* translational control under osmotic upshift conditions (M. Pruteanu and R. Hengge, unpublished data).

## RpoS-Regulated Two-Component Systems

While it is clear that two-component systems and in particular the response regulator RssB, have a strong impact in the control of the RpoS general stress sigma factor, only few two-component systems are under RpoS control. RpoS-mediated stationary phase induction of the *arcA* gene and its role in a positive feedback loop that further stabilizes RpoS has been mentioned above. The only other two-component genes which exhibited reduced expression in an *rpoS* mutant in a genome-wide microarray analysis,[7] were *cpxR* encoding a response regulator involved in an extracytoplasmic stress response and the *basRS* (*pmrAB*) operon which affects LPS composition and therefore the barrier properties of the outer membrane in response to iron and zinc.[80,81] Yet ratios of RpoS dependence of these genes were only small (between 2.1 and 2.5 fold) and did not exceed the statistically reliable cut-off (of 2.0) under all RpoS-inducing stress conditions tested. This suggests that the expression of none of the *E. coli* two-component systems is under strong RpoS control and therefore confined to the general stress and post-exponential conditions under which RpoS becomes decisive for overall cellular transcription. Rather, two-component systems seem to modulate vegetative gene expression and thereby orchestrate appropriate responses to a multitude of specific stresses in growing cells. Thus, in the overall cellular regulatory network, several two-component systems and their sensory input signals are positioned upstream of RpoS, but two-component systems do not seem to be relevant parts of the regulatory cascades downstream of RpoS (which are abundant as about 8% of the RpoS-dependent genes encode regulatory and/or signal transduction proteins).[7]

## Conclusions and Perspectives

As detailed in this chapter, the response regulator RssB, which serves as a direct recognition and proteolytic targeting factor for RpoS, is absolutely central to RpoS regulation. By phosphorylating and thereby activating both RssB and the ArcA response regulator, the complex histidine sensor kinase ArcB coordinates RpoS proteolysis with *rpoS* transcription in response to the redox state of the respiratory chain, which in turn depends on the balance between energy supply (electron source) and terminal electron acceptor (electron consumption). Available evidence, however, indicates that acetyl phosphate as well as other not yet identified sensor kinases susceptible to other stimuli also contribute to RssB phosphorylation and therefore RpoS degradation. Together with elaborate control mechanisms that affect RssB expression and availability, these systems allow the integration of multiple environmental signals in the control of RpoS proteolysis.

The other two-component systems found to contribute to RpoS regulation, i.e., the BarA/UvrY and the Rcs phosphorelay systems, seem to play auxiliary roles in *rpoS* transcription and translation under the conditions tested so far. As both systems have been implicated in downregulation of motility and the formation of biofilms, i.e., the transition into a sedentary bacterial lifestyle found at surfaces with cells mostly in post-exponential phase, a potential role of these signaling systems as well as of RpoS itself under these conditions certainly deserves further study.

It is noteworthy that all three two-component systems involved in RpoS control are complex phosphorelay systems that catalyze serial phosphotransfer reactions between transmitter, receiver and HPT domains. Such systems are especially well suited to integrate numerous signals and to form phosphotransfer networks. Moreover, the response regulator most crucial for RpoS control, i.e., RssB, seems to interfere with other two-component systems in novel ways that may have global regulatory effects and should be explored in the future.

### *Acknowledgements*

I would like to thank Joachim Reidl and Kürsad Turgay for helpful comments on the manuscript. Funding of research performed in the author's laboratory and mentioned in this review has been provided by the Deutsche Forschungsgemeinschaft and the Fonds der Chemischen Industrie.

## References

1. Hengge-Aronis R. The general stress response in Escherichia coli. In: Storz G, Hengge-Aronis R, eds. Bacterial Stress Responses. Washington, D.C.: ASM Press, 2000:161-178.
2. Hengge-Aronis R. Signal transduction and regulatory mechanisms involved in control of the $\sigma^S$ subunit of RNA polymerase in Escherichia coli. Microbiol Molec Biol Rev 2002; 66:373-395.
3. Hengge-Aronis R. Stationary phase gene regulation: what makes an Escherichia coli promoter $\sigma^S$-dependent? Curr Op Microbiol 2002; 5:591-595.
4. Typas A, Hengge R. The molecular basis of selectivity of stress-regulated promoters for σ-containing RNA polymerase. Mol Microbiol 2007; 63:1296-1306.
5. Lacour S, Landini P. $\sigma^S$-dependent gene expression at the onset of stationary phase in Escherichia coli: function of $\sigma^S$-dependent genes and identification of their promoter sequences. J Bacteriol 2004; 186:7186-7195.
6. Patten CL, Kirchhhof MG, Schertzberg MR et al. Microarray analysis of RpoS-mediated gene expression in Escherichia coli K-12. Mol Genet Genomics 2004; 272:580-591.
7. Weber H, Polen T, Heuveling J et al. Genome-wide analysis of the general stress response network in Escherichia coli: $\sigma^S$-dependent genes, promoters and sigma factor selectivity. J Bacteriol 2005; 187:1591-1603.
8. Gentry DR, Hernandez VJ, Nguyen LH et al. Synthesis of the stationary-phase sigma factor $\sigma^S$ is positively regulated by ppGpp. J Bacteriol 1993; 175:7982-7989.
9. Lange R, Hengge-Aronis R. The cellular concentration of the $\sigma^S$ subunit of RNA-polymerase in Escherichia coli is controlled at the levels of transcription, translation and protein stability. Genes Dev 1994; 8:1600-1612.
10. Lange R, Hengge-Aronis R. Identification of a central regulator of stationary-phase gene expression in Escherichia coli. Mol Microbiol 1991; 5:49-59.
11. Fischer D, Teich A, Neubauer P et al. The general stress sigma factor $\sigma^S$ of Escherichia coli is induced during diauxic shift from glucose to lactose. J Bacteriol 1998; 180:6203-6206.
12. Muffler A, Traulsen DD, Lange R et al. Posttranscriptional osmotic regulation of the $\sigma^S$ subunit of RNA polymerase in Escherichia coli. J Bacteriol 1996; 178:1607-1613.
13. Lee IS, Lin J, Hall HK et al. The stationary-phase sigma factor $\sigma^S$ (RpoS) is required for a sustained acid tolerance response in virulent Salmonella typhimurium. Mol Microbiol 1995; 17:155-167.
14. Bearson SMD, Benjamin Jr. WH, Swords WE et al. Acid shock induction of RpoS is mediated by the mouse virulence gene mviA of Salmonella typhimurium. J Bacteriol 1996; 178:2572-2579.
15. Muffler A, Barth M, Marschall C et al. Heat shock regulation of $\sigma^S$ turnover: a role for DnaK and relationship between stress responses mediated by $\sigma^S$ and $\sigma^{32}$ in Escherichia coli. J Bacteriol 1997; 179:445-452.
16. Sledjeski DD, Gupta A, Gottesman S. The small RNA, DsrA, is essential for the low temperature expression of RpoS during exponential growth in E. coli. EMBO J 1996; 15:3993-4000.
17. Lange R, Fischer D, Hengge-Aronis R. Identification of transcriptional start sites and the role of ppGpp in the expression of rpoS, the structural gene for the $\sigma^S$ subunit of RNA-polymerase in Escherichia coli. J Bacteriol 1995; 177:4676-4680.
18. Teich A, Meyer S, Lin HY et al. Growth rate related concentration changes of the starvation response regulators $\sigma^S$ and ppGpp in glucose-limited fed-batch and continuous cultures of Escherichia coli. Biotechnol Progr 1999; 15:123-129.
19. Brown L, Elliott T. Mutations that increase expression of the rpoS gene and decrease its dependence on hfq function in Salmonella typhimurium. J Bacteriol 1997; 179:656-662.
20. Majdalani N, Cunning C, Sledjeski D et al. DsrA RNA regulates translation of RpoS message by an antiantisense mechanims, independent of its action as an antisilencer of transcription. Proc Natl Acad Sci USA 1998; 95:12462-12467.

21. Majdalani N, Hernandez D, Gottesman S. Regulation and mode of action of the second small RNA activator of RpoS translation, RprA. Mol Microbiol 2002; 46:813-826.
22. Muffler A, Fischer D, Hengge-Aronis R. The RNA-binding protein HF-I, known as a host factor for phage Qβ RNA replication, is essential for the translational regulation of rpoS in Escherichia coli. Genes Dev 1996; 10:1143-1151.
23. Repoila F, Majdalani N, Gottesman S. Small noncoding RNAs, co-ordinators of adaptation processes in Escherichia coli: the RpoS paradigm. Mol Microbiol 2003; 48:855-861.
24. Worhunsky DJ, Godek K, Litsch S et al. Interactions of the noncoding RNA DsrA and RpoS mRNA with the 30 S ribosomal subunit. J Biol Chem 2003; 278:15815-15824.
25. Becker G, Klauck E, Hengge-Aronis R. Regulation of RpoS proteolysis in Escherichia coli: The response regulator RssB is a recognition factor that interacts with the turnover element in RpoS. Proc Natl Acad Sci USA 1999; 96:6439-6444.
26. Klauck E, Lingnau M, Hengge-Aronis R. Role of the response regulator RssB in $\sigma^S$ recognition and initiation of $\sigma^S$ proteolysis in Escherichia coli. Mol Microbiol 2001; 40:1381-1390.
27. Muffler A, Fischer D, Altuvia S et al. The response regulator RssB controls stability of the $\sigma^S$ subunit of RNA polymerase in Escherichia coli. EMBO J 1996; 15:1333-1339.
28. Pratt LA, Silhavy TJ. The response regulator, SprE, controls the stability of RpoS. Proc Natl Acad Sci USA 1996; 93:2488-2492.
29. Schweder T, Lee K-H, Lomovskaya O et al. Regulation of Escherichia coli starvation sigma factor ($\sigma^S$) by ClpXP protease. J Bacteriol 1996; 178:470-476.
30. Zhou Y, Gottesman S, Hoskins JR et al. The RssB response regulator directly targets $\sigma^S$ for degradation by ClpXP. Genes Dev 2001; 15:627-637.
31. Jishage M, Ishihama A. Transcriptional organization and in vivo role of the Escherichia coli rsd gene, encoding the regulator of RNA polymerase sigma D. J Bacteriol 1999; 181:3768-3776.
32. Jishage M, Kvint K, Shingler V et al. Regulation of sigma factor competition by the alarmone ppGpp. Genes Dev 2002; 16:1260-1270.
33. Pratt LA, Silhavy TJ. Crl stimulates RpoS activity during stationary phase. Mol Microbiol 1998; 29:1225-1236.
34. Typas A, Barembruch C, Hengge R. Stationary phase reorganisation of the E. coli transcription machinery by Crl protein, a fine-tuner of $\sigma^S$ activity and levels. EMBO J 2007; 26:1569-1578.
35. Pruteanu M, Hengge-Aronis R. The cellular level of the recognition factor RssB is rate-limiting for $\sigma^S$ proteolysis: Implications for RssB regulation and signal transduction in $\sigma^S$ turnover in Escherichia coli. Mol Microbiol 2002; 45:1701-1714.
36. Stüdemann A, Noirclerc-Savoye M, Klauck E et al. Sequential recognition of two distinct sites in $\sigma^S$ by the proteolytic targeting factor RssB and ClpX. EMBO J 2003; 22:4111-4120.
37. Mika F, Hengge R. A two-component phosphotransfer network involving ArcB, ArcA and RssB coordinates synthesis and proteolysis of $\sigma^S$ in E. coli. Genes Dev 2005; 19:2770-2781.
38. Becker G, Klauck E, Hengge-Aronis R. The response regulator RssB, a recognition factor for $\sigma^S$ proteolysis in Escherichia coli, can act like an anti$\sigma^S$ factor. Mol Microbiol 2000; 35:657-666.
39. Galperin MY. Structural classification of bacterial response regulators: diversity of output domains and domain combinations. J Bacteriol 2006; 188:4169-4182.
40. Becker G, Hengge-Aronis R. What makes an Escherichia coli promoter $\sigma^S$-dependent? Role of the -13/-14 nucleotide promoter positions and region 2.5 of $\sigma^S$. Mol Microbiol 2001; 39:1153-1165.
41. Dougan DA, Weber-Ban E, Bukau B. Targeted delivery of an ssrA-tagged substrate by the adaptor protein SspB to its cognate AAA+ protein ClpX. Mol Cell 2003; 12:373-380.
42. Flynn JM, Levchenko I, Seidel M et al. Overlapping recognition determinants within the ssrA degradation tag allow modulation of proteolysis. Proc Natl Acad Sci 2001; 98:10584-10589.
43. Wah DA, Levchenko I, Baker TA et al. Characterization of a specificity factor for an AAA+ ATPase. Assembly of SspB dimers with ssrA-tagged proteins and the ClpX hexamer. Chem Biol 2002; 9:1237-1245.
44. Kirstein J, Schlothauer T, Dougan DAd et al. Adaptor protein controlled oligomerization activates the AAA+ protein ClpC. EMBO J 2006; 25:1481-1491.
45. Jenal U, Hengge-Aronis R. Regulation by proteolysis in bacterial cells. Curr Opin Microbiol 2003; 6:163-172.
46. Bouché S, Klauck E, Fischer D et al. Regulation of RssB-dependent proteolysis in Escherichia coli: a role for acetyl phosphate in a response regulator-controlled process. Mol Microbiol 1998; 27:787-795.
47. Bougdour A, Wickner S, Gottesman S. Modulating RssB activity: IraP, a novel regulator of $\sigma^S$ stability in Escherichia coli. Genes Dev 2006; 20:884-897.
48. Bougdour A, Lelong C, Geiselmann J. Crl, a low temperature-induced protein in Escherichia coli that binds directly to the stationary phase sigma subunit of RNA polymerase. J Biol Chem 2004; 279:19540-19550.
49. Gonzalez-Barrios AF, Zuo R, Hashimoto Y et al. Autoinducer 2 controls biofilm formation in Escherichia coli through a novel motility quorum-sensing regulator (MqsR, B3022). J Bacteriol 2006; 188:305-316.

50. Peterson CN, Ruiz N, Silhavy TJ. RpoS proteolysis is regulated by a mechanism that does not require the SprE (RssB) response regulator phosphorylation site. J Bacteriol 2004; 186:7403-7410.
51. Ruiz N, Peterson CN, Silhavy TJ. RpoS-dependent transcriptional control of sprE: regulatory feedback loop. J Bacteriol 2001; 183:5974-5981.
52. McCleary WR, Stock JB. Acetyl phosphate and the activation of two-component response regulators. J Biol Chem 1994; 269:31567-31572.
53. Oshima T, Aiba H, Masuda Y et al. Transcriptome analysis of all two-component regulatory system mutants of Escherichia coli. Mol Microbiol 2002; 46:281-291.
54. Sugiura M, Aiba H, Mizuno T. Identification and classification of two-component systems that affect rpoS expression in Escherichia coli. Biosci Biotechnol Biochem 2003; 67:1612-1615.
55. Yamamoto K, Hirao K, Oshima T et al. Functional characterization in vitro of all two-component signal transduction systems from Escherichia coli. J Biol Chem 2005; 280:1448-1456.
56. Malpica R, Sandoval GR, Rodriguez C et al. Signaling by the arc two-component system provides a link between the redox state of the quinone pool and gene expression. Antioxid Redox Signal 2006; 8:781-795.
57. Georgellis D, Kwon O, Lin ECC. Quinones as the redox signal for the Arc Two-component system of bacteria. Science 2001; 292:2314-2315.
58. Malpica R, Franco B, Rodriguez C et al. Identification of quinone-sensitive redox switch in the ArcB sensor kinase. Proc Natl Acad Sci USA 2004; 101:13318-13323.
59. Georgellis D, Kwon O, De Wulf P et al. Signal decay through a reverse phosphorelay in the Arc two-component signal transduction system. J Biol Chem 1998; 273:32864-32869.
60. Wright JS, Olekhnovich IN, Touchie G et al. The histidine kinase domain of UhpB inhibits UhpA action at the Escherichia coli uhpT promoter. J Bacteriol 2000; 182:6279-6286.
61. Sourjik V. Receptor clustering and signal processing in E. coli chemotaxis. Trends Microbiol 2005; 12:569-576.
62. Backfisch T, Pruteanu M, Hengge R et al. Mathematical modeling of RpoS regulation in E. coli. Found Systems Biol Engineer 2005:239-242.
63. Mukhopadhyay S, Audia JP, Roy RN et al. Transcriptional induction of the conserved alternative sigma factor RpoS in Escherichia coli is dependent on BarA, a probable two-component regulator. Mol Microbiol 2000; 37:371-381.
64. Pernestig A-K, Melefors Ö, Georgellis D. Identification of UvrY as the cognate response regulator for the BarA sensor kinase in Escherichia coli. J Biol Chem 2001; 276:225-231.
65. Pernestig A-K, Georgellis D, Romeo T et al. The Escherichia coli BarA-UvrY two-component system is needed for efficient switching between glycolytic and gluconeogenic carbon sources. J Bacteriol 2003; 185:843-853.
66. Teplitski M, Goodier RI, Ahmer BM. Pathways lieading from BarA/SirA to motility and virulence gene expression in Salmonella. J Bacteriol 2003;185:7257-7265.
67. Tomenius H, Pernestig AK, Jonas K et al. The Escherichia coli BarA-UvrY two-component system is a virulence determinant in the urinary tract. BMC Microbiol 2006; 6:27-.
68. Suzuki K, Wang X, Weilbacher T et al. Regulatory circuitry of the CsrA/CsrB and BarA/UvrY systems of Escherichia coli. J Bacteriol 2002; 184:5130-5140.
69. Romeo T. Global regulation by the small RNA-binding protein CsrA and the noncoding RNA molecule CsrB. Mol Microbiol 1998; 29:1321-1330.
70. Weber H, Pesavento C, Possling A et al. Cyclic-di-GMP-mediated signaling within the $\sigma^S$ network of Escherichia coli. Mol Microbiol 2006; 62:1014-1034.
71. Jenal U. Cyclic di-guanosine-monophosphate comes of age: a novel secondary messenger involved in modulating cell surface structures in bacteria? Curr Opin Microbiol 2004; 7:185-191.
72. Simm R, Morr M, Kader A et al. GGDEF and EAL domains inversely regulate cyclic di-GMP levels and transition from sessility to motility. Mol Microbiol 2004; 53:1123-1134.
73. Majdalani N, Gottesman S. The Rcs phosphorelay: a complex signal transduction system. Annu Rev Microbiol 2005; 599:379-405.
74. Francez-Charlot A, Laugel B, Van Gemert A et al. RcsCDB His-Asp phosphorelay system negatively regulates the flhDC operon in Escherichia coli. Mol Microbiol 2003; 49:823-832.
75. Gottesman S, Trisler P, Torres-Cabassa AS. Regulation of capsular polysaccharide synthesis in Escherichia coli K12: characterization of three regulatory genes. J Bacteriol 1985; 162:1111-1119.
76. Majdalani N, Chen S, Murrow J et al. Regulation of RpoS by a novel small RNA: the characterization of RprA. Mol Microbiol 2001; 39:1382-1394.
77. Peterson CN, Carabetta VJ, Chowdhury T et al. LrhA regulates rpoS translation in response to the Rcs phosphorelay system in Escherichia coli. J Bacteriol 2006; 188:3175-3181.
78. Lehnen D, Blumer C, Polen T et al. LrhA as a new transcriptional key regulator of flagella, motility and chemotaxis genes in Escherichia coli. Mol Microbiol 2002; 45:521-532.

79. Mangan S, Alon U. Structure and function of the feed-forward loop network motif. Proc Natl Acad Sci USA 2003; 100:11980-11985.
80. Froelich JM, Tran K, Wall D. A pmrA constitutive mutant sensitizes Escherichia coli to deoxycholic acid. J Bacteriol 2006; 188:1180-1183.
81. Hagiwara D, Yamashino T, Mizuno T. A genome-wide view of the Escherichia coli BasS-BasR two-component system implicated in iron-responses. Biosci Biotechnol Biochem 2004; 68:1758-1768.

# Chapter 5

# Small RNAs Controlled by Two-Component Systems

**Claudio Valverde* and Dieter Haas**

## Abstract

Two-component systems (TCSs) allow bacteria to monitor diverse environmental cues and to adjust gene expression accordingly at the transcriptional level. It has been recently recognized that prokaryotes also regulate many genes and operons at a posttranscriptional level with the participation of small, noncoding RNAs which serve to control translation initiation and stability of target mRNAs, either directly by establishing antisense interactions or indirectly by antagonizing RNA-binding proteins. Interestingly, the expression of a subset of these small RNAs is regulated by TCSs and in this way, the small RNAs expand the scope of genetic control exerted by TCSs. Here we review the regulatory mechanisms and biological relevance of a number of small RNAs under TCS control in Gram-negative and -positive bacteria. These regulatory systems govern, for instance, porin-dependent permeability of the outer membrane, quorum-sensing control of pathogenicity, or biocontrol activity. Most likely, this emerging and rapidly expanding field of molecular microbiology will provide more and more examples in the near future.

## Introduction

Two-component systems (TCSs) provide a major regulatory framework by which many prokaryotes and a few eukaryotes adjust gene expression in response to changing environmental conditions and, in some instances, to increasing cell population densities.[1] Typical TCSs consist of a sensor kinase and a response regulator. When a sensory histidine kinase interacts with external or internal stimuli, it undergoes conformational changes, resulting in its autophosphorylation. Subsequent phosphotransfer to a conserved aspartate residue in the cognate response regulator leads to activation of that protein. About two thirds of all sequenced prokaryotic response regulators have DNA-binding domains and are transcriptional regulators. Only few (<1%) response regulators have an RNA-binding domain and presumably are involved in transcription antitermination.[2] Thus, the major task of TCSs is to regulate transcription of target genes. However, while transcriptional control of gene expression is clearly of primary importance in prokaryotes, these organisms also need regulatory mechanisms to control translation initiation and mRNA stability. To achieve this, bacteria as well as archaea can rely on the expression of small, noncoding RNAs. These molecules typically consist of 60 to 300 nt. Most of the ~100 known small RNAs of bacteria have been discovered in *Escherichia coli* and many of them seem to have posttranscriptional regulatory functions.[3-6]

Two major types of small RNAs having posttranscriptional control activity have been characterized in bacteria. The first and probably most frequent type acts by base-pairing with 5′ leader

*Corresponding Author: Claudio Valverde—Programa Interacciones Biológicas, Departamento de Ciencia y Tecnología, Universidad Nacional de Quilmes, Roque Sáenz Peña 352, Bernal B1876BXD, Argentina. Email: cvalver@unq.edu.ar

*Bacterial Signal Transduction: Networks and Drug Targets*, edited by Ryutaro Utsumi.

sequences of target mRNAs. The genes for the small regulatory RNAs and the target genes usually do not overlap and in this case, base pairing is often imperfect and can involve several stretches of 6 to 9 paired nucleotides. The resulting RNA duplex can form around the ribosome binding site and then will block translation initiation. Alternatively, when a small RNA binds to an mRNA segment upstream of the ribosome binding site, access of the ribosomes can be enhanced, resulting in activation of translation. In Gram-negative bacteria, most if not all small RNAs of the base-pairing type bind Hfq (host factor for Qß replicase), a chaperone protein that facilitates the interaction of small RNAs with complementary stretches of mRNAs. The hexameric Hfq protein[7,8] preferentially binds to AU-rich RNA sequences.[4,9,10] In several cases examined, the formation of a double-stranded RNA region can stimulate mRNA degradation by RNase E or RNase III.[11-13] This process is formally similar to RNA silencing pathways in eukaryotes. Although prokaryotes do not have the equivalent of the eukaryotic RNA-induced silencing complex (RISC),[14] the scaffold provided by Hfq during the pairing interaction of small RNAs with their target mRNA can be viewed as a mechanistic RISC precursor.[3] Not all small RNAs of bacteria interact with 5′ leader sequences. There is also an example of a small RNA that stabilizes a target mRNA by binding to the 3′ end.[15]

The second type of small RNAs with posttranscriptional control activity has a high affinity for RNA-binding proteins that usually cause translational repression. The prototype of such an RNA-binding protein is CsrA (carbon storage regulator) in enteric bacteria or its homolog RsmA (regulator of secondary metabolism) in *Erwinia* and *Pseudomonas* spp. The dimeric CsrA/RsmA protein[16-18] has a strong affinity for ANGGA motifs (where N is any nucleotide) in 5′ leader sequences of mRNAs; when one of several of these motifs coincides with the region of the ribosome binding site and the translation start codon, translational repression results.[19-21] Small RNAs with multiple ANGGA or GGA motifs in single-stranded parts mimic or even overemphasize the peculiarity of mRNA leader sequences that bind CsrA/RsmA.[22-25] Consequently, small RNAs of this type avidly bind CsrA/RsmA proteins and thereby relieve translational repression caused by to these regulators.[26,27] During translational repression by CsrA/RsmA, some mRNAs become destabilized.[28] CsrA/RsmA proteins can also have a positive effect on the translational expression of some target mRNAs, probably by stabilizing the mRNA.[29] Whether small RNAs can reverse this effect, remains to be determined.

The picture that has emerged during the last five years is that small RNAs of both types have a major role in modulating translation efficiency and mRNA stability in prokaryotes. None of these small RNAs is essential. Yet their expression can profoundly alter the cell's physiology. It is therefore important to consider how the expression of these small RNA is regulated at the transcriptional level. It turns out that many of them are transcribed under the control of TCSs whereas other small RNAs provide fine-tuning of expression of genes that are primarily regulated at the transcriptional level, e.g., via TCSs. In the following we will illustrate this by some well-documented examples.

## Antisense Control of Translation Initiation and mRNA Stability

### *Porin Expression in* E. coli

The outer membrane shields Gram-negative bacteria from toxic metabolites and proteins. However, in order to be taken up by the cells, nutrients must first be able to traverse the outer membrane through general or specific water-filled channels. Structural outer membrane proteins (Omps) known as porins form these channels. *E. coli* has several porins, three of which we will consider in some detail: OmpA, a monomeric protein with a 1-nm diameter forming a gated, general solute channel, OmpF, a trimeric protein with a 1.2-nm diameter, which allows passage of large (~350 Da) hydrophilic and hydrophobic solutes and OmpC, a trimeric protein with a 1.1-nm diameter, which is fairly impermeable to large, hydrophobic or anionic compounds.[30] An *E. coli ompA* mutant is more sensitive than the parental strain to cholate (bile salt), sodium dodecyl sulfate and serum, suggesting that OmpA may be important for the stability of the outer membrane.[31] An *ompC ompF* double mutant is sensitive to osmotic stress above pH 8, but is unaffected by high osmotic pressure at neutral pH.[32]

The OmpA protein is translated from an unusually stable mRNA having a half-life of ~15 min in growing cells. In slowly growing or stationary phase cells the half-life is reduced to ~4 min. The regulated mRNA stability depends on Hfq and on an untranslated 5′ leader sequence to which a small RNA termed MicA (= SraD; 78 nt) binds by a base-pairing mechanism. MicA becomes strongly expressed in rich media when cells enter the stationary phase of growth. High levels of MicA block ribosome binding to *ompA* mRNA, which facilitates cleavage by RNase E and subsequent decay of *ompA* mRNA. Hfq facilitates MicA binding to *ompA* mRNA.[13,33] Thus, MicA downregulates the expression of *ompA* in early stationary phase (Fig. 1).

The EnvZ/OmpR TCS inversely regulates the expression of the *ompC* and *ompF* genes. Conditions of high salt activate autophosphorylation of the EnvZ osmosensor, which in turn results in phosphorylation of the OmpR response regulator. At low osmolarity, EnvZ functions as a phosphatase, removing the phosphoryl group from OmpR. Thus, the intracellular level of phosphorylated OmpR (OmpR-P) positively correlates with the osmolarity of the growth medium. High levels of OmpR-P activate the transcription of *ompC* but repress that of *ompF* (Fig. 1). Because OmpC pores are relatively impermeable to bile salts, this regulation is believed to confer tolerance of *E. coli* to bile salts at high osmolarity, such as in an animal gut. At intermediate osmolarity, the abundance of OmpR is reduced and then OmpR-P represses *ompC* but activates *ompF*. This will favor nutrient uptake through the OmpF pore (which is larger than OmpC) in poor media, e.g., outside the gut. Under low osmotic pressure, OmpR is not phosphorylated and this prevents transcription of both *ompC* and *ompF*.[34] Two small RNAs, MicC (109 nt) and MicF (93 nt), modulate the expression of OmpC and OmpF at the translational level (Fig. 1).[35-37] MicC, which is synthesized preferentially in minimal media and at low temperature, base-pairs with the *ompC*

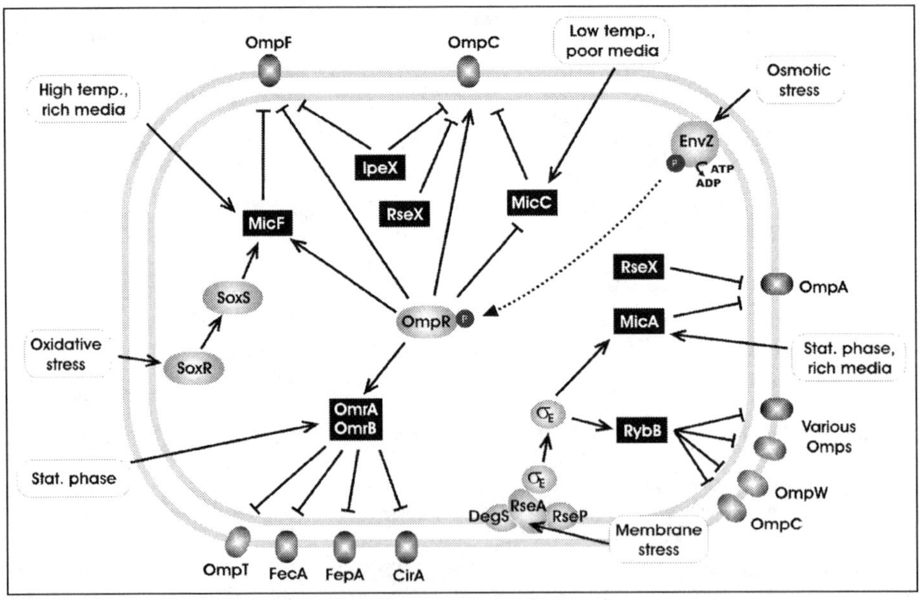


Figure 1. Control of porin expression in *Escherichia coli* by multiple antisense small RNAs. Posttranscriptional modulation of porin levels is finely tuned by a set of sRNAs (boxed in black) as a result of different stimuli channeled through the osmosensory EnvZ/OmpR TCS, the SoxR/SoxS system for monitoring oxidative stress and the extracytoplasmic stress responsive pathway that determines the intracellular levels of the sigma factor $\sigma_E$. Other factors like growth phase, nutrient availability and temperature also influence the expression of small RNA genes. Positive control of gene expression is denoted with arrows whereas negative effects are denoted with blunt ended lines. Dotted lines indicate phosphotransfer.

leader mRNA in the presence of Hfq and thereby blocks ribosome binding. In an *ompR* mutant, MicC levels are higher than in the wild type, suggesting that OmpR directly or indirectly represses *micC* expression.[36]

In a similar way, the small RNA MicF regulates *ompF* expression posttranscriptionally. In parallel with *ompC*, *micF* is induced by high osmolarity under EnvZ/OmpR control. MicF RNA is complementary to the ribosome-binding site and the AUG start codon on *ompF* mRNA, resulting in inhibition of translation. Like MicA and MicC, MicF binds Hfq.[38] Various environmental factors, in addition to osmotic stress, can induce *micF* expression: oxidative stress, increasing temperature, rich nutrients, bile salts, weak acids and certain antibiotics.[37] Oxidative stress activates the transcription factor SoxR, which switches on the transcription of the transcription factor SoxS. Positive control by SoxS affects *micF*, among a range of other target genes.[39] Furthermore, expression of both *ompC* and *ompF* appears to be negatively controlled by the IpeX small RNA.[40] In this complex network of regulatory interactions (Fig. 1), the dual input of the EnvZ/OmpR TCS stands out: under conditions of osmotic stress, this TCS favors the expression of *ompC* over that of *ompF* at both the transcriptional and translational level.

Overexpression of Omps is deleterious to the cell and triggers the extracytoplasmic stress (ECS) response pathway, which is essential in *E. coli*. Overproduction of OmpC or OmpF stimulates the DegS protease located in the inner membrane. DegS truncates RseA, an inner membrane protein which, in the absence of ECS, sequesters the ECS sigma factor $\sigma^E$ (= RpoE). Truncated RseA is cleaved by a second inner membrane protease, RseP and this cleavage delivers $\sigma^E$ to the cytoplasm (Fig. 1; see chapter 6 for further details). When both OmpA and OmpC are removed from the cell by deletion of the corresponding structural genes, RseP and probably DegS as well are no longer essential.[41] In wild type *E. coli*, a small RNA termed RseX (91 nt), which is complementary to the ribosome binding sites of the *ompA* and *ompC* mRNAs, negatively regulates translation of *ompA* and *ompC* in an Hfq-dependent way. In RseP-depleted cells, artificial RseX overexpression rescues cell viability by quenching the production of OmpA and OmpC.[41] $\sigma^E$ activates transcription of *micA* and *rybB*, a gene for yet another small RNA (80 nt), which targets mRNAs specifying OmpC, OmpW and other Omps. Thus, stress-induced expression of MicA and RybB results in down-regulation of OmpA, OmpC, OmpW and minor Omps (Fig. 1).[42]

Posttranscriptional regulation of Omps is further illustrated by the discovery of OmrA and OmrB, two small RNAs that bind Hfq and limit the translation of *ompT* mRNA in *E. coli*.[43] OmpT is a surface protease; whether it also acts as a porin is not clear. In exponential phase, the expression of OmrA and OmrB completely depends on the EnvZ/OmpR TCS (Fig. 1). In stationary phase, EnvZ and OmpR are also important, but RpoS and other unidentified elements contribute to the transcription of the two small RNAs.[43] Further outer membrane proteins regulated by OmrAB include FecA, FepA and CirA (Fig. 1), which form gated channels for iron-siderophore complexes.[43] The picture that emerges from the studies on Omp regulation is that the composition of Omps varies in response to extracellular factors but the total concentration of Omps in the outer membrane is carefully maintained at the posttranscriptional level; small RNAs have a key role in this balance.[44-45]

## *Quorum Sensing and Pathogenicity in* Staphylococcus aureus

The Gram-positive bacterium *Staphylococcus aureus* can be a harmless commensal when it colonizes mucosal membranes, e.g., in the human nose, or a severe pathogen once it has penetrated the barrier of the skin or the mucosa. *S. aureus* has an impressive number of so-called accessory genes coding for surface proteins (such as the anti-immune surface protein A, fibronectin-binding proteins and fibrinogen-binding proteins) and for extracellular proteins (such as hemolysins, enterotoxins, exfoliatins and lytic enzymes). The *agr* (accessory gene regulator) system determines the relative levels of surface proteins, which are colonization factors, and of the extracellular proteins, which are virulence factors. When cell population densities increase, the *agr* system upregulates most virulence factors and downregulates many colonization factors. As a consequence, the *agr* system is required for acute infection by *S. aureus* (e.g., abscess formation, endocarditis). The *agr* system produces a cell population density-related ("quorum sensing") signal and transduces this signal via a TCS (Fig. 2). The AgrD product is a propeptide, which is processed, modified and exported by the AgrB membrane protein. The mature quorum sensing signals of various strains are peptides consisting of 7 to 9 amino acids and contain a thiolactone structure formed between an internal cysteine and a C-terminal phenylalanine, methionine or leucine. The mature cognate signal binds to the N-terminal domain of the AgrC sensor kinase and triggers autophosphorylation. Phosphorylated AgrC transfers its phosphate residue to the response regulator AgrA. Phosphorylated AgrA (AgrA-P) then activates transcription of the *agrBDCA* operon and of the divergently transcribed RNA III locus (Fig. 2). RNA III (514 nt) is the major regulator of pathogenicity. Should we call it a "small" regulatory RNA? As we will explain below, RNA III provides antisense regulation of translation initiation of multiple mRNA targets and to this end combines several different base-pairing segments in one relatively large and stable RNA molecule. At the same time, RNA III codes for a δ-hemolysin.[46-48]

RNA III consists of 14 stem-loop structures (Fig. 2). The 5′ end region of RNA III is partially complementary to the 5′ leader of *hla* (α-hemolysin) mRNA. This interaction positively regulates *hla* translation by preventing a secondary structure of *hla* mRNA in which the ribosome binding site would be occluded.[49] Other examples of how small RNAs can stimulate translation initiation

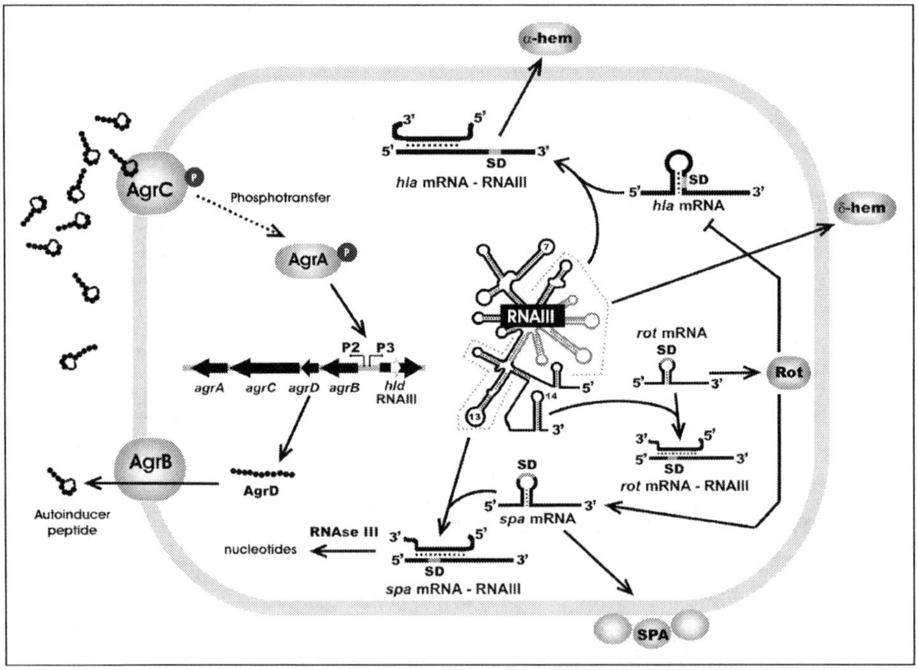


Figure 2. Production of colonization and virulence factors in *Staphylococcus aureus* is balanced by RNA III, an unusual small RNA encoded in the agr locus. The agr gene products constitute a quorum-sensing system that responds to a cyclopeptide autoinducer leading to phosphorylation of the transcriptional activator AgrA, which boosts transcription of the agr genes from promoter P2 and of RNA III from P3. By establishing antisense interactions with different regions of the molecule, the multifunctional RNA III modulates translation of mRNAs for surface protein A (a colonization factor), for alpha-hemolysin (a virulence factor) and for Rot (a transcriptional regulator of *spa* and *hla*). In addition, a central portion of RNA III (grey segments) is translated into the delta-hemolysin peptide. SD, Shine-Dalgarno sequence (ribosome binding-site); other symbols are the same as in Figure 1.

are described in the context of RpoS regulation in *E. coli* (see chapter 4). The 3′ end region of RNA III is partially complementary to the 5′ leader of *spa* (surface protein A) mRNA. Annealing of RNA III to the *spa* leader begins with a "kissing" contact between the loop of hairpin 13 (Fig. 2) and a loop containing the ribosome binding site of *spa* mRNA. The subsequent formation of an extensive RNA duplex buries the *spa* initiation codon and inhibits *spa* translation. Moreover, this RNA duplex is degraded by the double strand-specific RNase III, rendering the inhibition process irreversible.[50] Furthermore, the loops of hairpins 7 and 14 engage in a kissing interaction with two unpaired regions in *rot* (repressor of toxin) mRNA. The RNA III—*rot* mRNA duplex formed strongly represses translation of *rot* mRNA and includes loops 9 and 13 of RNA III.[51] This posttranscriptional regulation of the Rot protein, a transcription factor, has pleiotropic consequences. As Rot represses *hla* and induces *spa*,[52] downregulation of *rot* by RNA III enhances expression of *hla* and diminishes that of *spa*.[46,51] Overall this means that RNA III has a strong positive effect on *hla* and a distinct negative effect on *spa*. It is possible that Hfq facilitates the interactions of RNA III with target mRNAs in *S. aureus*. However, this role of Hfq appears to be less important than that in *E. coli*.[50,51]

## Sequestration of RNA-Binding Proteins

### *Biocontrol Activity of* Pseudomonas fluorescens CHA0 *and Pathogenicity of* Pseudomonas syringae *Pathovars*

*P. fluorescens* CHA0 is a Gram-negative soil bacterium. It effectively colonizes the roots of different crop plants and protects these from fungal pathogens. The biocontrol properties of strain CHA0 are complex; they depend on a blend of antifungal secondary metabolites (such as 2,4-diacetylphloroglucinol [Phl], pyoluteorin, pyrrolnitrin and hydrogen cyanide [HCN]), extracellular lytic enzymes and poorly characterized elicitors of induced systemic resistance (ISR), a mechanism that renders the host plant less susceptible to pathogens.[53,54] Most of the biocontrol activity of strain CHA0 and similar fluorescent pseudomonads is lost in null mutants of a TCS termed GacS/GacA (for global activation of antibiotics and cyanide) because most secondary metabolites and extracellular enzymes are produced at very low levels in these mutants, by comparison with the wild type.[53-60] GacA, whose role as a global regulator was discovered in strain CHA0,[55] has important functions in many Gram-negative bacteria (Table 1).

There are bacterial and plant-derived signals that activate the function of the GacS/GacA system in *pseudomonads*. The CHA0 signals are soluble in organic solvents, heat-stable and unrelated to well-known quorum sensing signal molecules, such as N-acyl-homoserine lactones of Gram-negative bacteria, but their chemical characterization has been hampered by their very low concentrations in culture supernatants and biofilm extracts.[58,60] Signal production is autoregulated by the GacS/GacA system and requires a sufficient supply of thiamine.[61] Cross-talk between different *Pseudomonas* and *Vibrio* species is possible via similar signals.[62] Concerning the plant signals, there is indirect evidence that phenolic glucosides (arbutin, salicin) may act as inducers of the GacS/GacA signal transduction pathway in the plant pathogen *Pseudomonas syringae* pv. *syringae*.[63] Furthermore, root exudates of sugar beet have been observed to stimulate the synthesis of the cyclic lipopeptide amphisin in *Pseudomonas* sp. DSS73, but not in a *gacS* mutant of this strain.[64]

The GacS/GacA TCS is strictly required for the expression of three small RNAs termed RsmX (119 nt), RsmY (118 nt) and RsmZ (127 nt) in strain CHA0 (Fig. 3). A conserved upstream sequence element (consensus TGTAAGN$_6$CTTACA) is found in the *rsmX*, *rsmY* and *rsmZ* promoters as well as in promoters of other GacA-controlled small RNA genes in Gram-negative bacteria and might be a GacA binding site.[25,65-67] The three small RNAs have a high affinity for two small RNA-binding proteins, RsmA and RsmE. Both proteins act as translational repressors of target mRNAs, e.g., the mRNAs carrying *hcnA* (for a subunit of HCN synthase), *aprA* (for an extracellular metalloprotease) or *phlA* (for a subunit of the Phl biosynthetic enzyme complex).[68,69] GacA-driven expression of the three small RNAs sequesters the regulatory proteins RsmA and RsmE; this leads to the translation of the target mRNAs and hence to the synthesis of biocontrol

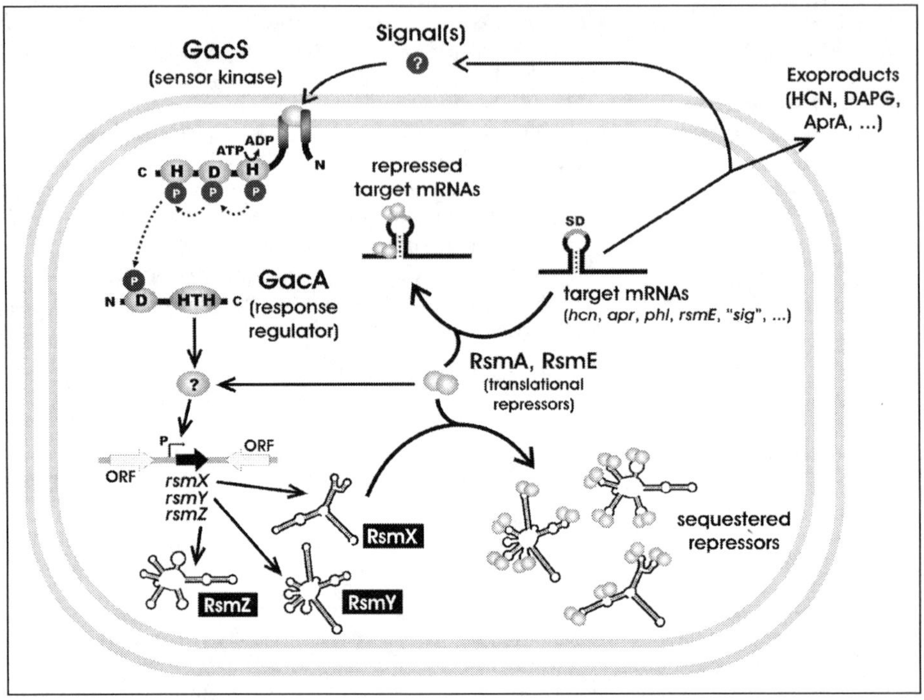


Figure 3. The GacS/GacA TCS and a triad of effector small RNAs determine expression of extracellular factors and biocontrol activity in *Pseudomonas fluorescens* strain CHA0. The GacS/GacA TCS, which responds to uncharacterized autoinducing signals, promotes translation of various mRNAs involved in exoproduct formation and control of plant root pathogens, by relieving a translational blockage caused by the RNA-binding RsmA and RsmE proteins. This is achieved by increasing, either directly or indirectly, the intracellular levels of three small RNAs (RsmX, RsmY and RsmZ) that sequester RsmA and RsmE. The symbols are the same as in Figure 1.

factors (Fig. 3). Simultaneous deletion of *rsmX*, *rsmY* and *rsmZ* is necessary to obtain the same phenotype as that of a *gacS* or *gacA* mutant. Inactivation of only one or two small RNA genes has a minimal effect.[66] Thus, these small RNAs can functionally replace one another. Nevertheless, they might have distinct physiological functions in that their expression patterns during growth in batch culture are different. In the same vein, simultaneous deletion of *rsmA* and *rsmE* is necessary to overcome (suppress) the defects of *gacS* and *gacA* mutants. The expression of *rsmA* increases slightly with increasing cell population densities whereas *rsmE* is expressed under positive GacA control.[69] The GacS/GacA TCS not only regulates the formation of exoproducts, it also positively controls the expression of the stress sigma factor $\sigma^S$ (RpoS) and, as a consequence, the resistance of strain CHA0 to oxidative stress in stationary phase.[70]

In the plant-pathogens *Pseudomonas syringae* pv. *syringae* and *P. syringae* pv. *tomato* DC3000 the GacS/GacA TCS is essential for the production of toxins (syringomycin and coronatine, respectively) and for pathogenicity on bean and tomato, respectively.[63,71-74] The GacS sensor, which was originally called LemA (for lesion manifestation), was discovered in *P. syringae* pv. *syringae*.[71] In both pv. *tomato* and pv. *syringae* the GacA-controlled transcription factor SalA regulates toxin production.[74,75] In pv. *tomato*, but not in pv. *syringae*, GacS/GacA also has a positive effect on HrpL, an alternative sigma factor controlling the type III secretion system (TTSS). Although the chain of command in the GacS/GacA cascade of *P. syringae* pathovars has not been identified conclusively,

***Table 1. Diversity of functions controlled by homologous GacS/GacA TCSs in Gram-negative bacteria***

| Strain | GacS/GacA Orthologs | Controlled Phenotypes | GacS/A Effect | GacS/GacA-Dependent sRNAs Involved | References |
|---|---|---|---|---|---|
| *Acinetobacter baumannii* | GacS/GacA | Growth on citrate | + | RsmX, RsmY, RsmZ (*) | 152 |
| *Azotobacter vinelandii* | GacS/GacA | Alginate and polyhydroxybutyrate production; growth yield; encystment | + | | 58; 101 |
| *Erwinia carotovora* subsp. *carotovora* AH2572 | ExpS/ExpA | Extracellular protease and cellulase production; pathogenicity on potato tubers | + | | 58 |
| *Erwinia carotovora* subsp. *carotovora* Ecc71 | ExpS/ExpA | Harpin; extracellular pectate lyase, polygalacturonase, protease and cellulase production | + | RsmB | 137 |
| *Erwinia carotovora* subsp. *carotovora* SCC3193 | ExpS/ExpA | Pectate lyase, polygalacturonase and cellulase production | + | | 58 |
| *Escherichia coli* K12, DS17 and D517 | BarA/UvrY | Central carbon metabolism, biofilm formation, motility and adaptive responses in K12; attachment and iron regulation sensor in strain D517; virulence in monkey kidney in strain DS17 | + | CsrB, CsrC | 58; 110; 116; 121; 134 |
| *Legionella pneumophila* | LetS/LetA | Motility, cytotoxicity, sodium resistance, macrophage infection and persistence | + | RsmY, RsmZ (*) | 153; 154 |
| *Pseudomonas aureofaciens* 30-84 | LemA/GacA | AHL, phenazine, HCN and exoprotease production | + | | 58 |
| *Pseudomonas aeruginosa* PAO | GacS/GacA | C4-HSL, HCN, pyocyanin, elastase, lipase, chitinase and chitin binding protein production; *dsbA* expression; virulence on leukopenic mice | + | RsmY, RsmZ | 58; 83 |
| | | Swarming, azurin release, early biofilm development | - | RsmY, RsmZ | 83 |
| *Pseudomonas aeruginosa* UCBPP-PA14 | GacS/GacA | Biofilm formation and virulence in plants and animals | + | RsmY, RsmZ (**) | 58 |

***continued on next page***

**Table 1. Continued**

| Strain | GacS/GacA Orthologs | Controlled Phenotypes | GacS/A Effect | GacS/GacA-Dependent sRNAs Involved | References |
|---|---|---|---|---|---|
| *Pseudomonas chlororaphis* O6 | GacS/GacA | 2R,3R-butanediol-mediated SAR; *rpoS* mRNA levels, catalase and peroxidase activity; resistance to oxidative stress | + | | 155; 156 |
| *Pseudomonas chlororaphis* PCL1391 | GacS/GacA | Phenazine-1-carboxamide, AHL, chitinase and exoprotease | + | | 58; 157; 158 |
| *Pseudomonas chlororaphis* PCL1445 | GacS/GacA | Biosurfactant production (putisolvins I and II) | + | | 159 |
| *Pseudomonas chlororaphis* SPR044 | GacS/GacA | C4-AHL and extracellular protease production; growth inhibition of *Bacillus subtilis*; RpoS levels | + | | 160 |
| *Pseudomonas entomophila* L48 | GacS/GacA | Exoprotease and hemolysin production; infection, persistence and pathogenicity on insect gut | + | RsmY, RsmZ (**) | 161 |
| *Pseudomonas fluorescens* BL915 | LemA/GacA | PRN, chitinase, HCN, 2-hexyl-5-propyl-resorcinol and exoprotease production; inhibition and biocontrol of *Rhizoctonia solani* | + | | 58 |
| *Pseudomonas fluorescens* CHA0 | GacS/GacA | Production of Gac autoinducer, HCN, DAPG, PLT, PRN, IAA, exoprotease AprA, phospholipase C, TSO; biocontrol of various plant diseases; ecological fitness; survival in stationary phase and to oxidative stress; RpoS levels; toxicity on protists | + | RsmX, RsmY, RsmZ | 25; 58; 60; 61; 65; 66; 69; 70; 162 |
| | | Pyochelin, salicylic acid, unidentified fluorescent compound | - | | 58 |
| *Pseudomonas fluorescens* F113 | GacS/GacA | DAPG, HCN, exoprotease | + | PrrB | 58 |
| | | Swimming and swarming motility; flagellum length, *fliC* expression; pyoverdin production | - | | 163; 164 |

**continued on next page**

***Table 1. Continued***

| Strain | GacS/GacA Orthologs | Controlled Phenotypes | GacS/A Effect | GacS/GacA-Dependent sRNAs Involved | References |
|---|---|---|---|---|---|
| *Pseudomonas fluorescens* Pf-5 | ApdA/GacA | PRN, PLT, DAPG, HCN, exoprotease and TSO production; stress resistance and inhibition of *Rhizoctonia solani* | + | RsmX, RsmY, RsmZ (**) | 58; 66 |
| *Pseudomonas marginalis* CY091 | LemA/GacA | Pectate lyase, polygalacturonase and pyoverdine production; soft rot pathogenicity on potato slices and pepper fruits | + | | 58 |
| *Pseudomonas* sp. DSS73 | GacS/GacA | Amphisin, HCN, extracellular protease and chitinase production, growth on glycyl-glutamate | + | | 64 |
| *Pseudomonas* sp. M18 | Gacs/GacA | PLT production | + | | 165 |
| | | Phenazine-1-carboxylic acid production | - | | |
| *Pseudomonas* sp. PCL1171 | GacS/GacA | Biocontrol of wheat take-all decline, motility; biosurfactant, extracellular chitinase, protease and lipase production; *rpoS* expression; PLT production | + | | 59; 166 |
| | | IAA production | - | | 167 |
| *Pseudomonas syringae* pv. *coronafaciens* Pc27R | LemA/GacA | Exoprotease and tabtoxin production; chlorosis formation on oat leaves | + | | 58 |
| *Pseudomonas syringae* pv. *syringae* B301D | GacS/GacA | Lipodepsipeptidic phytotoxin production; expression of outer membrane protein gene *pseA* and *dnaK* | + | | 63; 168; 169 |
| *Pseudomonas syringae* pv. *syringae* B728a | LemA/GacA | Exoprotease, syringomycin, syringolin, AHL and alginate production; swarming; ecological fitness; brown spot disease on bean | + | RsmX, RsmY, RsmZ (**) | 58; 170 |
| *Pseudomonas syringae* BR2R | LemA/GacA | Exoprotease and tabtoxin production | + | | 58 |
| *Pseudomonas tolaasii* NCPPB1116 | PheN | Tolaasin and exoprotease production; brown blotch formation on *Agaricus* mushrooms | + | | 58 |
| | | Motility and production of a fluorescent compound | - | | 58 |

***continued on next page***

*Table 1. Continued*

| Strain | GacS/GacA Orthologs | Controlled Phenotypes | GacS/A Effect | GacS/GacA-Dependent sRNAs Involved | References |
|---|---|---|---|---|---|
| *Pseudomonas tolaasii* PT814 | RtpA/? | Tolaasin and exoprotease production; brown blotch formation on *Pleurotus* fruiting bodies | + | | 58 |
| *Pseudomonas viridiflava* PJ-08-6A and SF312A | RepA/RepB | Pectate lyase, exoprotease, alginate and fluorescent siderophore production; soft rot pathogenicity | + | | 58 |
| *Salmonella enterica* serovar Typhimurium | BarA/SirA | HilA-dependent regulation of pathogenicity island SPI1, swarming motility, TTSS apparatus and invasion endocytosis of epithelial cells | + | CsrB, CsrC | 58; 117; 119; 120 |
| *Serratia plymuthica* IC1270 | GrrS/GrrA | Extracellular chitinase and protease, PRN production, biocontrol activity | + | | 171 |
| *Serratia* sp. 39006 | PigW/PigQ | Prodigiosin production | + | | 172 |
| *Vibrio cholerae* | VarS/VarA | Cholera toxin and toxin-coregulated pilus production; infant mice colonization | + | CsrB, CsrC, CsrD | 58; 76 |
| *Vibrio fischeri* ES114 | GacS/GacA | Growth yield and substrate utilization, bioluminescence, siderophore production, squid colonization | + | CsrB1, CsrB2 (*) | 173 |
| | | Number of flagella per cell | - | | |

Abbreviations: HCN, hydrogen cyanide; PRN, pyrrolnitrin; PLT, pyoluteorin; DAPG, 2,4-diacetylphloroglucinol; TTSS, type III secretion system; AHL, N-acyl-homoserine lactone; IAA, indolacetic acid; TSO, tryptophan side-chain oxidase; SAR, systemic acquired resistance. (*) Predicted sRNAs[67] (**) Identified by BlastN

there is evidence from bioinformatics studies that strain DC3000 has three GacA-controlled small RNAs (*rsmX, rsmY, rsmZ*) and four RsmA-like RNA-binding proteins.[18,67] The existence of two of these small RNAs has been demonstrated experimentally.[74]

The GacS/GacA TCS occurs in many different Gram-negative bacteria (Table 1). The genes and pathways whose expression the GacS/GacA system regulates vary greatly and further examples will be discussed below. As a common feature of GacS/GacA signal transduction cascades it is observed that artificial overexpression of any sequestering small RNA will suppress the effects caused by *gacS* or *gacA* mutations in various microorganisms.[22,25,65,66,76] Maybe the first observation of this kind was made in *P. syringae* pv. *phaseolicola*, a producer of phaseolotoxin and pathogen of bean: Certain phaseolotoxin-negative mutants (unmapped but presumably mutated in *gacS* or *gacA*) were functionally complemented by a 0.4-kb locus termed TRR (for thermoregulatory region) when this locus was inserted into a multi-copy plasmid.[77] Although at the time of publication (1993) the TRR product had not been identified, in retrospect it is evident that TRR codes for an *rsmY*-like small RNA.[65] Overexpression of TRR not only restores phaseolotoxin production in the mutant background but also overrides temperature control of phaseolotoxin production. The wild type synthesizes the toxin at 20°C, but not at 28°C, whereas the suppressed mutants produce the toxin at 20°C and 28°C.[77]

## *Quorum Sensing and Pathogenicity in* Pseudomonas aeruginosa

Quorum sensing regulation in *P. aeruginosa* is highly complex and has been reviewed elsewhere.[78-80] Here we will only discuss the impact of the GacS/GacA TCS on quorum sensing, virulence and biofilm formation. Small sequestering RNAs are expressed under strict GacA control in *P. aeruginosa*, much like in *P. fluorescens*. However, in strains PAO and PA14 of *P. aeruginosa*, only two such RNAs occur, RsmY and RsmZ, and their simultaneous absence in mutant strains is needed to reproduce a GacA-negative phenotype. RsmY and RsmZ antagonize the action of one RsmA protein, by avidly binding to it (Fig. 4).[81-83] RsmZ does not bind Hfq whereas RsmY does although the significance of the latter observation is not clear.[84]

From a proteomic analysis it appears that the RsmA/RsmY/RsmZ triad mediates most important functions of the GacS/GacA TCS.[83] In strain PA14 and, to a lesser extent, in strain PAO, mutations in either *gacS* or *gacA* attenuate virulence of *P. aeruginosa* for different host organisms including burnt mice, nematodes, insects and plants.[85-87] This loss of virulence in *gacS/gacA* mutants is due to a markedly diminished expression of virulence factors that are regulated positively by quorum sensing such as HCN, phenazines and lytic exoenzymes.[83,87-90] The underlying mechanisms are complex. On the one hand, the GacS/GacA system positively influences the expression of the quorum sensing machinery, in particular that of the *rhlI* gene, which encodes the enzyme for the biosynthesis of the quorum sensing signal N-butanoyl-homoserine lactone.[83,88,91] Then this signal, together with its hierarchically superior companion signal, N-(3-oxododecanoyl)-homoserine lactone, positively controls the transcription of a wide range of extracellular virulence factors, via the transcriptional regulators RhlR and LasR, respectively.[78,79] On the other hand, the GacS/GacA system also regulates the expression of certain virulence factors directly at the level of translation (Fig. 4), in the same way as shown for *P. fluorescens*. In the case of the expression of the *hcnABC* genes (for HCN synthase), the input of the GacS/GacA system appears to be stronger in the direct translational control pathway than in the quorum sensing pathway.[83,92]

Microscopical examination of *P. aeruginosa* biofilms has revealed a role of GacA in biofilm maturation.[93] The activity of the GacS/GacA system is modulated by two membrane-bound sensor kinases, RetS (for regulation of exopolysaccharides and type III secretion) and LadS (for lost adhesion).[94,95] The cognate response regulators for RetS and LadS have not been identified conclusively. In the following, we will assume that GacA is the partner for RetS, GacS and LadS (Fig. 4) as current experimental data are consistent with this model. However, additional response regulators depending on RetS and LadS may exist. RetS acts as an antagonist of GacS. When RetS is inactivated by mutation, both *rsmY* and *rsmZ* are expressed at very high levels (B. Humair and D. Haas, unpublished results)[94,96] and this leads to effective sequestration of RsmA protein. The

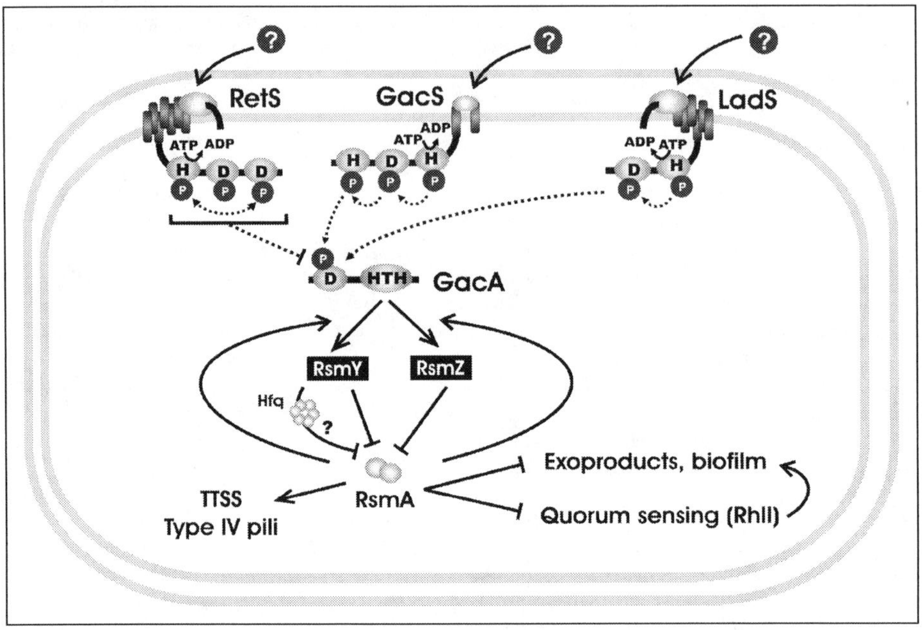


Figure 4. Multiple sensory systems converge to control expression of protein-sequestering small RNAs which regulate quorum sensing and pathogenicity in *Pseudomonas aeruginosa*. Recent evidence suggests that at least three TCS sensors, GacS, LadS and RetS, channel their inputs through the response regulator GacA, which is required for transcription of RsmY and RsmZ small RNAs. The levels of these regulatory RNAs determine the availability of the RNA-binding protein RsmA, which posttranscriptionally regulates the synthesis of RhlI-mediated N-butanoyl-homoserine lactone, exoproduct formation, biofilm development, type 3 secretion system (TTSS) and type IV pilus formation. The symbols are the same as in Figure 1.

consequences are a small colony phenotype due to strong cell-cell aggregation, overproduction of extracellular products (especially polysaccharides), enhanced biofilm formation, lack of the TTSS and reduced twitching motility due to downregulation of type IV pili.[94,97,98] Not surprisingly, *rsmA* and *retS* mutants have similar phenotypes.[81,91,99] By contrast, when LadS is mutationally inactivated, *rsmZ* expression is very low, adhesion and biofilm formation are reduced, whereas TTSS expression is enhanced (Fig. 4).[95] It will be interesting to investigate phosphotransfer in the GacS—RetS—LadS—GacA system. GacS has a kinase domain (with a conserved histidine residue), a response regulator domain (with a conserved aspartate residue) and a histidine phosphotransfer (Hpt) domain (with a conserved histidine residue). Studies on the *E. coli* GacS homolog BarA and on the *P. fluorescens* GacS protein have shown that all three domains are essential for biological activity.[60,102,103] RetS has an unusual arrangement that consists of a periplasmic domain and a transmembrane docking domain followed by a central kinase domain (with a conserved histidine residue) and two distinct receiver domains 1 and 2 (each with a conserved aspartate residue). The kinase domain and receiver 2 are essential for activity; receiver 1 might have a modulating function.[96] The periplasmic domains of GacS and RetS appear to be relatively unimportant for activity.[58,60,96] LadS, which also seems to have a cytoplasmic membrane location, has a histidine kinase and one response regulator domain whose functions await exploration.[95] RetS and LadS occur not only in *P. aeruginosa*, but also in other pseudomonads, according to genome sequence data. RetS homologs are present in *P. putida*, *P. syringae*, *P. fluorescens* and *P. entomophila* and display an overall identity of ≥55%, resulting from 75 to 80% identity along the 550 amino acid residues downstream of the transmembrane domain and a highly variable segment in the 400 *N*-terminal residues. Moreover, in *Azotobacter vinelandii* strain AvOP, a RetS protein of 933 residues having 71% identity in the C-terminal part is predicted from the genome sequence. In *A. vinelandii* the GacS/GacA system controls alginate production.[100,101] LadS homologs are predicted only in pseudomonads, with an average of 55 to 65% identity spread along the polypeptides. However, the *P. putida*, *P. entomophila* and *P. syringae* LadS homologs lack a putative response regulator domain at the C-terminus as they are about 120 to 130 residues shorter than the *P. aeruginosa* and *P. fluorescens* LadS proteins.

According to the model presented in Figure 4, there are three sensors (RetS, GacS, LadS) in the Gac/Rsm signal transduction pathway of *P. aeruginosa*. They transduce the input of several chemically unidentified signals by influencing the level of phosphorylation of GacA; possibly additional response regulators are involved in this process. The integrated input determines the level of transcription of the *rsmY* and *rsmZ* genes. After this divergence, the signal transduction pathway converges again at the level of RsmA. The further posttranscriptional effects of RsmA can be positive or negative; many mechanistic details remain to be elucidated, particularly in the branch that positively regulates TTSS and pilus biogenesis.

## *Virulence and Carbon Metabolism in Enteric Bacteria*

Enteric bacteria have homologs of the GacS/GacA TCS and functional analogs of the sequestering small RNAs RsmX/RsmY/RsmZ. In strain K-12 of *E. coli*, the GacS homolog BarA (for bacterial adaptive responses) was originally recognized as a multi-copy suppressor of an *envZ* deletion mutation. However, there is no evidence for a cross-talk between BarA and OmpR under natural conditions.[104] The GacA homolog UvrY owes its name to the location of the *uvrY* gene, which lies immediately upstream of the UV repair gene *uvrC* and forms a *uvrYC* operon. However, UvrY has no apparent DNA repair function and is not part of the SOS regulon.[105,106] A transcriptome analysis of all TCS mutants conducted in *E. coli* K-12 has revealed that an *uvrY* mutation profoundly influences transcript levels of peripheral carbon utilization pathways.[107] Moreover, *barA* and *uvrY* mutants are more sensitive to hydrogen peroxide than is the wild type.[102,108] In a fimbriated uropathogenic strain of *E. coli*, the BarA/UvrY system contributes to the fitness of the microorganism in the bladder.[109,110] In the enteropathogen *Salmonella enterica* serovar Typhimurium, the cognate response regulator of BarA is named SirA (for *Salmonella* invasion regulator), as *sirA* mutants are defective for invasion of epithelial cells and attenuated in a gastroenteritis model.[111,112] SirA also has a repressive effect on flagellar gene expression.[113]

In *E. coli* and *S. enterica*, UvrY and SirA, respectively, drive the expression of two small RNAs termed CsrB and CsrC (Fig. 5). The CsrB (~350 nt) small RNA is similar in both microorganisms and so is the CsrC (~250 nt) RNA. However, CsrB and CsrC have no apparent sequence homology with the RsmX, RsmY and RsmZ RNAs of pseudomonads. CsrB and CsrC have about 16 and 9 predicted stem-loop structures, respectively and repeated AGGA or ANGGA motifs are found mostly in single-stranded RNA segments.[22,114-117] CsrB and CsrC RNAs strongly bind to the CsrA protein,[22,116] but not to Hfq.[118] In *S. enterica*, a double mutant deleted for *csrB* and *csrC* behaves like *barA* and *sirA* mutants, in terms of loss of invasion capacity.[117] SirA has been reported to bind to the *csrB* promoter of *S. enterica* in vitro[119,120] and UvrY has been found to activate the expression of a *csrB-lacZ* fusion in a coupled transcription-translation system of *E. coli*.[121] These experiments provide evidence for a direct interaction of the SirA/UvrY response regulator with a *csrB* promoter element. The signals activating the BarA sensor have not been identified. However, it has been proposed that repression of invasion genes by bile salts requires the BarA/SirA TCS.[122] Interestingly, a *barA* mutant of *S. enterica* is non-invasive but the absence of BarA can be functionally compensated by feeding acetate, suggesting that acetyl-phosphate derived from acetate might directly phosphorylate SirA without BarA.[115]

The CsrA protein has multiple functions and targets. In *S. enterica*, CsrA differentially regulates the expression of flagella, TTSS and invasiveness (Fig. 5).[114,123] Furthermore, CsrA positively controls the expression of CsrB and CsrC.[117] The mechanisms by which CsrA exerts these effects in the BarA/SirA cascade are not entirely clear, as the target mRNAs to which CsrA binds have not

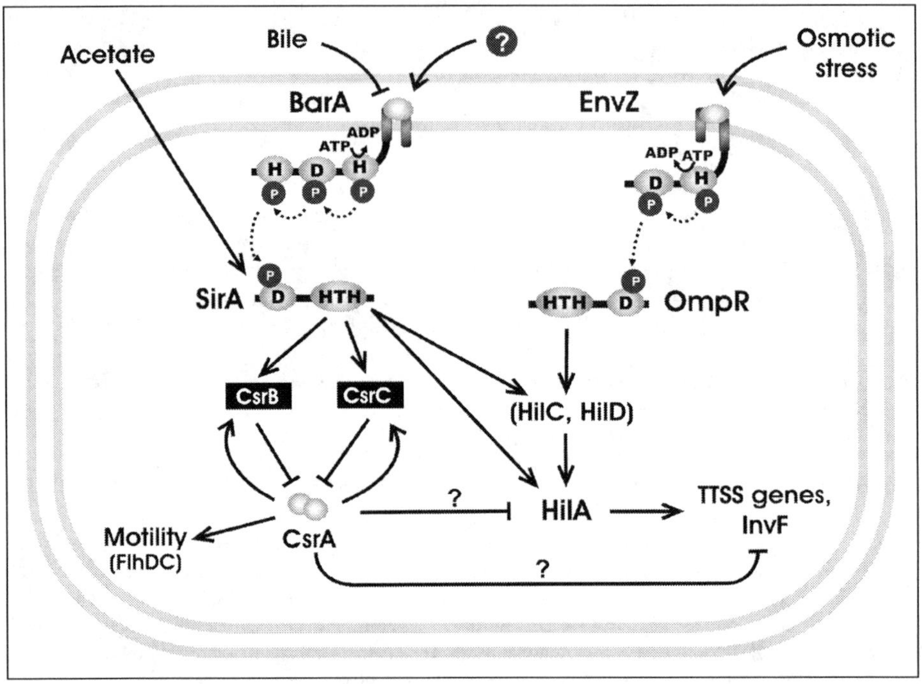


Figure 5. TCSs and small RNAs control virulence in *Salmonella enterica* serovar Typhimurium. The BarA/SirA TCS and the small RNAs CsrB and CsrC of enteric bacteria are functional homologs of the Gac and Rsm elements described in pseudomonads. HilA, HilC and HilD are transcription factors. For an explanation of symbols see Figure 1.

been identified. At any rate, the regulatory proteins HilC, HilD and HilA are intermediates in the signal transduction from BarA/SirA to TTSS genes and InvF, the major regulator of invasion genes (Fig. 5).[124,125] Among a number of regulatory systems (not shown), the EnvZ/OmpR TCS when activated by high salt conditions contributes to the activation of the HilCDA pathway (Fig. 5).

In *E. coli*, several target mRNAs of CsrA have been characterized and this allows rationalization of some of the pleiotropic effects of the BarA/UvrY cascade. The glycogen biosynthetic enzymes ADP-glucose pyrophosphorylase (GlgC) and glycogen synthase (GlgA) and the catabolic enzyme glycogen phosphorylase (GlgP) are encoded by the *glgCAP* operon. CsrA binds to the 5′ leader of *glgC* mRNA; this prevents ribosome binding and translation and facilitates *glgCAP* mRNA decay. These results explain why UvrY has a positive effect on glycogen metabolism.[28,126,127] The *pgaABCD* operon codes for a polysaccharide consisting of ß-1,6-N-acetyl-D-glucosamine units. This adhesin mediates cell adhesion to polystyrene and biofilm formation. CsrA cooperatively binds to at least six sites in the 5′ leader of *pgaA* mRNA; the 5th and 6th sites overlap the Shine-Dalgarno sequence and the start codon, respectively. Thus, CsrA competes with the 30S ribosomal subunit for binding and this results in enhanced *pgaA* transcript decay in vivo. As a result, a *uvrY* mutant has a diminished ability to form biofilms.[128,129] CsrA also blocks translation of the *cstA* gene, which encodes a carbon starvation-inducible peptide transporter. CsrA binding is observed at three or four sites in the 5′ leader of *cstA*. Again one of these sites lies in the ribosome binding site.[130] As mentioned before, a CsrA consensus binding site, ANGGA, emerges from these studies. In *Pseudomonas* spp., RsmA/RsmE binding appears to have the same specificity (K. Lapouge and D. Haas, unpublished data). Negative effects of CsrA on gluconeogenesis in *E. coli* may involve similar mechanisms.[26]

CsrA can act as a positive regulator in *E. coli*. For instance, CsrA binding to *flhDC* mRNA (the master regulator of flagellar motility) enhances the half-life and the translation levels by about 3-fold.[29] CsrA activates biofilm dispersal[131] and the expression of several enzymes in glycolysis[26] and in the Entner-Doudoroff pathway.[132] CsrA also stimulates the transcription of the *csrB* and *csrC* genes.[116,121,133] Some of these effects are probably indirect and involve transcriptional regulators. Multiple activities of CsrA are reflected in competition experiments conducted, for instance, in glucose medium, where a *uvrY* mutant wins over the wild type during growth, but loses in late stationary phase: in the *uvrY* mutant glycolysis is expected to be favored over gluconeogenesis as long as glucose is available.[134]

In the plant pathogen *Erwinia carotovora* subsp. *carotovora*, which causes soft rot in a wide variety of plants, the GacS/GacA TCS (alternatively named ExpS/ExpA) positively controls the expression of major virulence determinants (Fig. 6).[135-137] Only one, relatively large, GacA-dependent, regulatory RNA termed RsmB (479 nt) is found in this organism.[67,138,139] RsmB has 24 GGA motifs, endowing it with a high binding capacity for the CsrA-like regulatory protein RsmA.[140] RsmB is processed to an 'RsmB species of 259 nt which has RsmA-binding activity as well, but whose biological significance is not clear.[139] There are three transcriptional regulators that repress the *rsmB* gene: RsmC, KdgR and HexA. Of these, RsmC also activates *rsmA* expression (Fig. 6).[141-143] N-Acyl-homoserine lactones modulate *rsmA* expression: they bind to the transcription regulators ExpR1 and ExpR2 and thereby prevent transcriptional activation of the *rsmA* promoter by these regulators.[144] RsmA is a repressor of pathogenicity genes (Fig. 6), i.e., *pel* for pectate lyase, *peh* for polygalacturonase, *cel* for cellulase, *prt* for protease, *nip* for a necrosis factor and *hrpL* for the alternative sigma factor controlling TTSS.[145-148] The TTSS of *E. carotovora* is involved in the manifestation

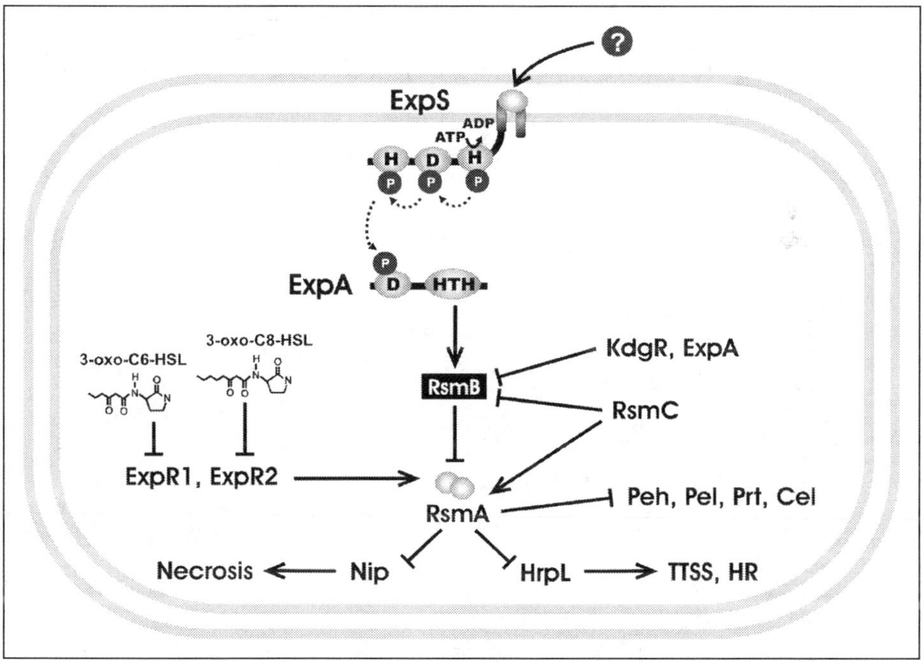


Figure 6. In the plant pathogen *Erwinia carotovora* subsp. *carotovora* the expression of major virulence determinants is positively regulated by the GacS/GacA-homolog TCS designated ExpR/ExpA. A single small RNA (RsmB) and a single RNA binding protein (RsmA) serve to control the expression of several pathogenicity factors at a posttranscriptional level. For an explanation of symbols see Figure 1.

of a hypersensitive reaction in tobacco. In agreement with this model, mutations in *gacS*, *gacA* or *rsmB* cause loss of virulence, whereas mutation in *rsmA* brings about hypervirulence.[136,138,146] Direct interaction between RsmA and target mRNAs remains to be demonstrated.

## Combination of Antisense and Sequestration Mechanisms

Both types of posttranscriptional control that we have discussed above (i.e., antisense base-pairing and mRNA mimicry with protein sequestration) can be combined in a single organism. The human pathogen *Vibrio cholerae* provides an example. In this Gram-negative bacterium the central transcriptional regulator HapR activates the production of haemagglutinating protease and represses the synthesis of virulence factors (including cholera toxin and the toxin-coregulated pilus) and biofilm formation (Fig. 7). The input from three signal transduction pathways and at least seven small RNAs determine *hapR* expression. These pathways act in parallel. The first system responds to the quorum sensing signal CAI-1 (for cholera autoinducer 1) whose chemical structure is unknown. CAI-1 activates the membrane-bound sensor kinase CqsS. The second system senses AI-2 (for autoinducer 2), a furanosyl borate diester, via the periplasmic binding protein LuxP and the membrane-bound sensor kinase LuxQ. At low cell densities, when the concentrations of both signal molecules are low, CqsS and LuxQ have kinase activity and transfer phosphate to the LuxU protein. LuxU-phosphate passes the phosphate to the response regulator LuxO. LuxO-phosphate and the nucleoid protein Fis assist the alternative σ factor RpoN to transcribe four small RNA genes called *qrr-1*, *qrr-2*, *qrr-3* and *qrr-4* (for quorum regulatory RNA). In the presence of Hfq, the

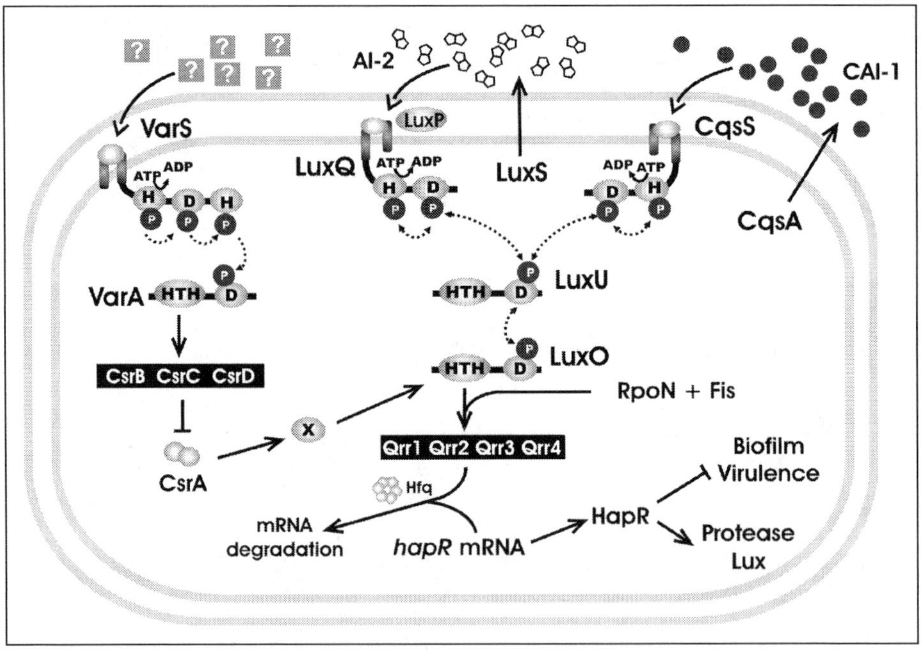


Figure 7. TCS and small RNAs in quorum sensing of *Vibrio cholerae*. Four base-pairing sRNAs (Qrr1-4) promote degradation of *hapR* mRNA, encoding the master regulator of virulence HapR. The levels of Qrr RNAs depend on the amount of transcriptionally active (i.e., phosphorylated) LuxO protein, which is determined by the input from two convergent quorum-sensing systems as well as by the concentration of free CsrA protein. CsrA is sequestered by 3 small RNAs (CsrB, CsrC and CsrD) under the positive control of the VarS/VarA TCS. Thus, LuxO~P prevails at low cell densities whereas inactive LuxO is more abundant in dense populations. For an explanation of symbols see Figure 1.

four small RNAs bind to the ribosome binding site of *hapR* mRNA by a base-pairing mechanism and this interaction promotes *hapR* mRNA degradation.[149,150] At high cell densities, when CAI-1 and AI-2 are present, CqsS and LuxQ have phosphatase activity. Reverse phosphotransfer from LuxO-phosphate to the sensor proteins results in transcriptionally inactive LuxO. As the Qrr's cease to be produced, *hapR* mRNA can be translated and the HapR protein directs virulence factor expression and biofilm maturation. Simultaneous deletion of all four *qrr* genes (or of the *hfq* gene) is necessary to obtain constitutive HapR levels.[149]

Mutations in the VarS/VarA TCS (homologs of GacS/GacA) cause greatly diminished production of cholera toxin and the toxin-coregulated pilus.[151] The VarS/VarA signal transduction pathway consists of three VarA-dependent small RNAs (CsrB, CsrC and CsrD), the RNA-binding protein CsrA and an unknown component (Fig. 7). The three RNAs are homologs of *E. coli* CsrB.[67,76] At low cell densities, the VarS/VarA system is assumed not to be phosphorylated and the small RNAs are not expressed. In this mode, the CsrA protein contributes to the activation of LuxO, by an unknown mechanism.[76] At high cell densities, the VarS/VarA system activates the transcription of the three small RNAs, which then sequester CsrA, resulting in low LuxO activity (Fig. 7). Thus, the VarS/VarA regulatory pathway is a branch of the quorum sensing regulon in *V. cholerae*.[76]

## Concluding Remarks

The biological functions of many prokaryotic small RNAs have been detected in bioinformatic or genetic screens (Table 2 gives some further examples); however, many functions still need to be discovered.[3,5,6] Among those prokaryotic small RNAs that have been studied in experimental detail, most have a role in posttranscriptional regulation and the antisense and sequestering small RNAs discussed here clearly represent the predominant types. As illustrated by *omp* gene regulation, their function can be to modulate gene expression in response to changing environmental conditions and to prevent overshooting of protein production, which would be deleterious to the cell. Another important function of small RNAs is to validate gene expression at the translational level,

**Table 2.** ***Additional reports of TCSs that control expression of small RNAs***

| Strain | TCS | Small RNAs Under TCS Control | Functions of Small RNAs | References |
|---|---|---|---|---|
| *Escherichia coli* K12 | Rcs phosphorelay (RcsC/RcsB via RcsD) | RprA (105 nt) | Promotion of *rpoS* translation | 4; 174; 175 |
| *Clostridium perfringens* | VirR/VirS | VR-RNA (386 nt) | Upregulation of chromosomally encoded alpha- (*colA*) and kappa- (*plc*) toxins and downregulation of exonuclease *cadA*. Upregulation of plasmid-encoded beta-2-toxin (*cpb2*) and downregulation of collagen adhesin (*cna*) | 176; 177; 178 |
| *Bacillus subtilis* | Spo0F/Spo0B/Spo0A | SurA (280 nt) | Not determined yet. Expression under sporulation control. | 179 |
| *Streptococcus pneumoniae* | FasBC/A | FasX (~290 nt) | Upregulation of virulence factors | 180 |

following major decisions taken by the cells at the transcriptional level. When this validation step involves degradation of mRNAs, it is essentially irreversible and thus can provide a commitment to a developmental process, e.g., biofilm formation or expression of virulence factors. As cellular regulators, small RNAs have the advantage that they are energetically less expensive to make and simpler to degrade than are proteins.

Redundancy of small RNA genes is often observed. This implies that loss of one particular small RNA gene can often be compensated by the remaining functionally equivalent small RNAs. In such a situation, mutants lacking one small RNA do not have an obvious phenotype and are difficult to obtain in a conventional screen. Redundancy, however, may only be apparent under laboratory conditions. Differential transcription of functionally related small RNAs may in fact allow fine-tuning of target gene expression. In this way, multiple signals can be integrated at the level of translation and mRNA stability.

## *Acknowledgements*

We would like to express our sincere gratitude to Dr. Cornelia Reimmann (Université de Lausanne, Switzerland) and Dr. Stephan Heeb (University of Nottingham, UK) for their time and dedication to make a careful review of this chapter and for their suggestions. Research in the authors' laboratories was supported by the Swiss National Foundation for Scientific Research (Switzerland) and by CONICET, ANPCyT, Banco RIO and Universidad Nacional de Quilmes (Argentina).

## References

1. Cashin P, Goldsack L, Hall D et al. Contrasting signal transduction mechanisms in bacterial and eukaryotic gene transcription. FEMS Microbiol Lett 2006; 261:155-164.
2. Galperin MY. Structural classification of bacterial response regulators: diversity of output domains and domain combinations. J Bacteriol 2006; 188:4169-4182.
3. Gottesman S. Micros for microbes: noncoding regulatory RNAs in bacteria. Trends Genet 2005; 21:399-404.
4. Majdalani N, Vanderpool CK, Gottesman S. Bacterial small RNA regulators. Crit Rev Biochem Mol Biol 2005; 40:93-113.
5. Hüttenhofer A, Vogel J. Experimental approaches to identify noncoding RNAs. Nucleic Acids Res 2006; 34:635-646.
6. Livny J, Brencic A, Lory S et al. Identification of 17 Pseudomonas aeruginosa sRNAs and prediction of sRNA-encoding genes in 10 diverse pathogens using the bioinformatic tool sRNAPredict. Nucleic Acids Res 2006; 34:3484-3493.
7. Schumacher MA, Pearson RF, Møller T et al. Structures of the pleiotropic translational regulator Hfq and an Hfq-RNA complex: a bacterial Sm-like protein. EMBO J 2002; 21:3546-3556.
8. Nikulin A, Stolboushkina E, Perederina A et al. Structure of Pseudomonas aeruginosa Hfq protein. Acta Crystallogr D Biol Crystallogr 2005; 61:141-146.
9. Valentin-Hansen P, Eriksen M, Udesen C. The bacterial Sm-like protein Hfq: a key player in RNA transactions. Mol Microbiol 2004; 51:1525-1533.
10. Kawamoto H, Koide Y, Morita T et al. Base-pairing requirement for RNA silencing by a bacterial small RNA and acceleration of duplex formation by Hfq. Mol Microbiol 2006; 61:1013-1022.
11. Vogel J, Argaman L, Wagner EG et al. The small RNA IstR inhibits synthesis of an SOS-induced toxic peptide. Curr Biol 2004; 14:2271-2276.
12. Morita T, Maki K, Aiba H. RNase E-based ribonucleoprotein complexes: mechanical basis of mRNA destabilization mediated by bacterial noncoding RNAs. Genes Dev 2005; 19:2176-2186.
13. Udekwu KI, Darfeuille F, Vogel J et al. Hfq-dependent regulation of OmpA synthesis is mediated by an antisense RNA. Genes Dev 2005; 19:2355-2366.
14. Zamore PD. Ancient pathways programmed by small RNAs. Science 2002; 296:1265-1269.
15. Opdyke JA, Kang JG, Storz G. GadY, a small-RNA regulator of acid response genes in Escherichia coli. J Bacteriol 2004; 186:6698-6705.
16. Gutierrez P, Li Y, Osborne MJ et al. Solution structure of the carbon storage regulator protein CsrA from Escherichia coli. J Bacteriol 2005; 187:3496-3501.
17. Rife C, Schwarzenbacher R, McMullan D et al. Crystal structure of the global regulatory protein CsrA from Pseudomonas putida at 2.05 Å resolution reveals a new fold. Proteins 2005; 61:449-453.
18. Heeb, S, Kuehne SA, Bycroft M et al. Functional analysis of the posttranscriptional regulator RsmA reveals a novel RNA-binding site. J Mol Biol 2006; 355:1026-1036.

19. Baker CS, Morozov I, Suzuki K et al. CsrA regulates glycogen biosynthesis by preventing translation of glgC in Escherichia coli. Mol Microbiol 2002; 44:1599-1610.
20. Wang X, Dubey AK, Suzuki K et al. CsrA posttranscriptionally represses pgaABCD, responsible for synthesis of a biofilm polysaccharide adhesin of Escherichia coli. Mol Microbiol 2005; 56:1648-1663.
21. Dubey AK, Baker CS, Romeo T et al. RNA sequence and secondary structure participate in high-affinity CsrA-RNA interaction. RNA 2005; 11:1579-1587.
22. Liu MY, Gui G, Wei B et al. The RNA molecule CsrB binds to the global regulatory protein CsrA and antagonizes its activity in Escherichia coli. J Biol Chem 1997; 272:17502-17510.
23. Liu Y, Cui Y, Mukherjee A et al. Characterization of a novel RNA regulator of Erwinia carotovora ssp. carotovora that controls production of extracellular enzymes and secondary metabolites. Mol Microbiol 1998; 29:219-234.
24. Aarons S, Abbas A, Adams C et al. A regulatory RNA (PrrB RNA) modulates expression of secondary metabolite genes in Pseudomonas fluorescens F113. J Bacteriol 2000; 182:3913-3919.
25. Heeb S, Blumer C, Haas D. Regulatory RNA as mediator in GacA/RsmA-dependent global control of exoproduct formation in Pseudomonas fluorescens CHA0. J Bacteriol 2002; 184:1046-1056.
26. Romeo T. Global regulation by the small RNA-binding protein CsrA and the noncoding RNA molecule CsrB. Mol Microbiol 1998; 29:1321-1330.
27. Valverde C, Lindell M, Wagner EG et al. A repeated GGA motif is critical for the activity and stability of the riboregulator RsmY of Pseudomonas fluorescens. J Biol Chem 2004; 279:25066-25074.
28. Romeo, T. Posttranscriptional regulation of bacterial carbohydrate metabolism: evidence that the gene product CsrA is a global mRNA decay factor. Res Microbiol 1996; 147:505-512.
29. Wei BL, Brun-Zinkernagel AM, Simecka JW et al. Positive regulation of motility and flhDC expression by the RNA-binding protein CsrA of Escherichia coli. Mol Microbiol 2001; 40:245-256.
30. Nikaido H. Porins and specific channels of bacterial outer membranes. Mol Microbiol 1992; 6:435-442.
31. Wang Y. The function of OmpA in Escherichia coli. Biochem Biophys Res Commun 2002; 292:396-401.
32. Kaeriyama M, Machida K, Kitakaze A et al. OmpC and OmpF are required for growth under hyper-osmotic stress above pH 8 in Escherichia coli. Lett Appl Microbiol 2006; 42:195-201.
33. Rasmussen AA, Eriksen M, Gilany K et al. Regulation of ompA mRNA stability: the role of a small regulatory RNA in growth phase-dependent control. Mol Microbiol 2005; 58:1421-1429.
34. Yoshida T, Qin L, Egger LA et al. Transcription regulation of ompF and ompC by a single transcription factor, OmpR. J Biol Chem 2006; 281:17114-17123.
35. Aiba H, Matsuyama S, Mizuno T et al. Function of micF as an antisense RNA in osmoregulatory expression of the ompF gene in Escherichia coli. J Bacteriol 1987; 169:3007-3012.
36. Chen S, Zhang A, Blyn LB et al. MicC, a second small-RNA regulator of Omp protein expression in Escherichia coli. J Bacteriol 2004; 186:6689-6697.
37. Delihas N, Forst S. MicF: an antisense RNA gene involved in response of Escherichia coli to global stress factors. J Mol Biol 2001; 313:1-12.
38. Zhang A, Wassarman KM, Rosenow C et al. Global analysis of small RNA and mRNA targets of Hfq. Mol Microbiol 2003; 50:1111-1124.
39. Demple B. Redox signaling and gene control in the Escherichia coli soxRS oxidative stress regulon—a review. Gene 1996; 179:53-57.
40. Castillo-Keller M, Vuong P, Misra R. Novel mechanism of Escherichia coli porin regulation. J Bacteriol 2006; 188:576-586.
41. Douchin V, Bohn C, Bouloc P. Down-regulation of porins by a small RNA bypasses the essentiality of the regulated intramembrane proteolysis protease RseP in Escherichia coli. J Biol Chem 2006; 281:12253-12259.
42. Johansen J, Rasmussen AA, Overgaard M et al. Conserved small noncoding RNAs that belong to the sigmaE regulon: role in down-regulation of outer membrane proteins. J Mol Biol 2006; 364:1-8.
43. Guillier M, Gottesman S. Remodelling of the Escherichia coli outer membrane by two small regulatory RNAs. Mol Microbiol 2006; 59:231-247.
44. Guillier M, Gottesman S, Storz G. Modulating the outer membrane with small RNAs. Genes Dev 2006; 20:2338-2348.
45. Vogel J, Papenfort K. Small noncoding RNAs and the bacterial outer membrane. Curr Opin Microbiol 2006; 9:1-7.
46. Novick RP. Autoinduction and signal transduction in the regulation of staphylococcal virulence. Mol Microbiol 2003; 48:1429-1449.
47. Otto M. Quorum-sensing control in Staphylococci—a target for antimicrobial drug therapy? FEMS Microbiol Lett 2004; 241:135-141.

48. Romby P, Vandenesch F, Wagner EG. The role of RNAs in the regulation of virulence-gene expression. Curr Opin Microbiol 2006; 9:229-236.
49. Morfeldt E, Taylor D, von Gabain A et al. Activation of alpha-toxin translation in Staphylococcus aureus by the trans-encoded antisense RNA, RNAIII. EMBO J 1995; 14:4569-4577.
50. Huntzinger E, Boisset S, Saveanu C et al. Staphylococcus aureus RNAIII and the endoribonuclease III coordinately regulate spa gene expression. EMBO J 2005; 24:824-835.
51. Geisinger E, Adhikari RP, Jin R et al. Inhibition of rot translation by RNAIII, a key feature of agr function. Mol Microbiol 2006; 61:1038-1048.
52. Said-Salim B, Dunman PM, McAleese FM et al. Global regulation of Staphylococcus aureus genes by Rot. J Bacteriol 2003; 185:610-619.
53. Haas D, Keel C. Regulation of antibiotic production in root-colonizing Pseudomonas spp. and relevance for biological control of plant disease. Annu Rev Phytopathol 2003; 41:117-153.
54. Haas D, Défago G. Biological control of soil-borne pathogens by fluorescent pseudomonads. Nat Rev Microbiol 2005; 3:307-319.
55. Laville J, Voisard C, Keel C et al. Global control in Pseudomonas fluorescens mediating antibiotic synthesis and suppression of black root rot of tobacco. Proc Natl Acad Sci USA 1992; 89:1562-1566.
56. Gaffney TD, Lam ST, Ligon J et al. Global regulation of expression of antifungal factors by a Pseudomonas fluorescens biological control strain. Mol Plant Microbe Interact 1994; 7:455-463.
57. Corbell N, Loper JE. A global regulator of secondary metabolite production in Pseudomonas fluorescens Pf-5. J Bacteriol 1995; 177:6230-6236.
58. Heeb S, Haas D. Regulatory roles of the GacS/GacA two-component system in plant-associated and other gram-negative bacteria. Mol Plant Microbe Interact 2001; 14:1351-1363.
59. van den Broek D, Chin-A-Woeng TF, Eijkemans K et al. Biocontrol traits of Pseudomonas spp. are regulated by phase variation. Mol Plant Microbe Interact 2003; 16:1003-12.
60. Zuber S, Carruthers F, Keel C et al. GacS sensor domains pertinent to the regulation of exoproduct formation and to the biocontrol potential of Pseudomonas fluorescens CHA0. Mol Plant Microbe Interact 2003; 16:634-644.
61. Dubuis C, Rolli J, Lutz M et al. Thiamine-auxotrophic mutants of Pseudomonas fluorescens CHA0 are defective in cell-cell signaling and biocontrol factor expression. Appl Environ Microbiol 2006; 72:2606-2613.
62. Dubuis C, Haas D. Cross-species GacA-controlled induction of antibiosis in pseudomonads. Appl Environ Microbiol 2007; 73:650-654.
63. Wang N, Lu SE, Wang J et al. The expression of genes encoding lipodepsipeptide phytotoxins by Pseudomonas syringae pv. syringae is coordinated in response to plant signal molecules. Mol Plant Microbe Interact 2006; 19:257-269.
64. Koch B, Nielsen TH, Sorensen D et al. Lipopeptide production in Pseudomonas sp. strain DSS73 is regulated by components of sugar beet seed exudate via the Gac two-component regulatory system. Appl Environ Microbiol 2002; 68:4509-4516.
65. Valverde C, Heeb S, Keel C et al. RsmY, a small regulatory RNA, is required in concert with RsmZ for GacA-dependent expression of biocontrol traits in Pseudomonas fluorescens CHA0. Mol Microbiol 2003; 50:1361-1379.
66. Kay E, Dubuis C, Haas D. Three small RNAs jointly ensure secondary metabolism and biocontrol in Pseudomonas fluorescens CHA0. Proc Natl Acad Sci USA 2005; 102:17136-17141.
67. Kulkarni PR, Cui X, Williams JW et al. Prediction of CsrA-regulating small RNAs in bacteria and their experimental verification in Vibrio fischeri. Nucleic Acids Res 2006; 34:3361-3369.
68. Blumer C, Heeb S, Pessi G et al. Global GacA-steered control of cyanide and exoprotease production in Pseudomonas fluorescens involves specific ribosome binding sites. Proc Natl Acad Sci USA 1999; 96:14073-14078.
69. Reimmann C, Valverde C, Kay E et al. Posttranscriptional repression of GacS/GacA-controlled genes by the RNA-binding protein RsmE acting together with RsmA in the biocontrol strain Pseudomonas fluorescens CHA0. J Bacteriol 2005; 187:276-285.
70. Heeb S, Valverde C, Gigot-Bonnefoy C et al. Role of the stress sigma factor RpoS in GacA/RsmA-controlled secondary metabolism and resistance to oxidative stress in Pseudomonas fluorescens CHA0. FEMS Microbiol Lett 2005; 243:251-258.
71. Hrabak EM, Willis DK. The lemA gene required for pathogenicity of Pseudomonas syringae pv. syringae on bean is a member of a family of two-component regulators. J Bacteriol 1992; 174:3011-3020.
72. Hrabak EM, Willis DK. Involvement of the lemA gene in production of syringomycin and protease by Pseudomonas syringae pv. syringae. Mol Plant Microbe Interact 1993; 6:368-375.
73. Rich JJ, Kinscherf TG, Kitten T et al. Genetic evidence that the gacA gene encodes the cognate response regulator for the lemA sensor in Pseudomonas syringae. J Bacteriol 1994; 176:7468-7475

74. Chatterjee A, Cui Y, Yang H et al. GacA, the response regulator of a two-component system, acts as a master regulator in Pseudomonas syringae pv. tomato DC3000 by controlling regulatory RNA, transcriptional activators and alternate sigma factors. Mol Plant Microbe Interact 2003; 16:1106-1117.
75. Kitten T, Kinscherf TG, McEvoy JL et al. A newly identified regulator is required for virulence and toxin production in Pseudomonas syringae. Mol Microbiol 1998; 28:917-929.
76. Lenz DH, Miller MB, Zhu J et al. CsrA and three redundant small RNAs regulate quorum sensing in Vibrio cholerae. Mol Microbiol 2005; 58:1186-1202.
77. Rowley KB, Clements DE, Mandel M et al. Multiple copies of a DNA sequence from Pseudomonas syringae pathovar phaseolicola abolish thermoregulation of phaseolotoxin production. Mol Microbiol 1993; 8:625-635.
78. Lazdunski AM, Ventre I, Sturgis JN. Regulatory circuits and communication in Gram-negative bacteria. Nat Rev Microbiol 2004; 2:581-592.
79. Juhas M, Eberl L, Tümmler B. Quorum sensing: the power of cooperation in the world of Pseudomonas. Environ Microbiol 2005; 7:459-471.
80. Diggle SP, Cornelis P, Williams P et al. 4-quinolone signalling in Pseudomonas aeruginosa: old molecules, new perspectives. Int J Med Microbiol 2006; 296:83-91.
81. Heurlier K, Williams F, Heeb S et al. Positive control of swarming, rhamnolipid synthesis and lipase production by the posttranscriptional RsmA/RsmZ system in Pseudomonas aeruginosa PAO1. J Bacteriol 2004; 186:2936-2945.
82. Burrowes E, Abbas A, O'Neill A et al. Characterisation of the regulatory RNA RsmB from Pseudomonas aeruginosa PAO1. Res Microbiol 2005; 156:7-16.
83. Kay E, Humair B, Dénervaud V et al. Two GacA-dependent small RNAs modulate the quorum-sensing response in Pseudomonas aeruginosa. J Bacteriol 2006; 188:6026-633.
84. Sonnleitner E, Schuster M, Sorger-Domenigg T et al. Hfq-dependent alterations of the transcriptome profile and effects on quorum sensing in Pseudomonas aeruginosa. Mol Microbiol 2006; 59:1542-1558.
85. Mahajan-Miklos S, Rahme LG, Ausubel FM. Elucidating the molecular mechanisms of bacterial virulence using nonmammalian hosts. Mol Microbiol 2000; 37:981-988.
86. Jander G, Rahme LG, Ausubel FM. Positive correlation between virulence of Pseudomonas aeruginosa mutants in mice and insects. J Bacteriol 2000; 182:3843-3845.
87. Gallagher LA, Manoil C. Pseudomonas aeruginosa PAO1 kills Caenorhabditis elegans by cyanide poisoning. J Bacteriol 2001; 183:6207-6014.
88. Reimmann C, Beyeler M, Latifi A et al. The global activator GacA of Pseudomonas aeruginosa PAO positively controls the production of the autoinducer N-butyryl-homoserine lactone and the formation of the virulence factors pyocyanin, cyanide and lipase. Mol Microbiol 1997; 24:309-319.
89. Mahajan-Miklos S, Tan MW, Rahme LG et al. Molecular mechanisms of bacterial virulence elucidated using a Pseudomonas aeruginosa-Caenorhabditis elegans pathogenesis model. Cell 1999; 96:47-56.
90. Reimmann C, Ginet N, Michel L et al. Genetically programmed autoinducer destruction reduces virulence gene expression and swarming mobility in Pseudomonas aeruginosa PAO1. Microbiology 2002; 148: 923-932.
91. Pessi G, Williams F, Hindle Z et al. The global posttranscriptional regulator RsmA modulates production of virulence determinants and N-acylhomoserine lactones in Pseudomonas aeruginosa. J Bacteriol 2001; 183:6676-6683.
92. Pessi G, Haas D. Dual control of hydrogen cyanide biosynthesis by the global activator GacA in Pseudomonas aeruginosa PAO1. FEMS Microbiol Lett 2001; 200:73-78.
93. Parkins MD, Ceri H, Storey DG. Pseudomonas aeruginosa GacA, a factor in multihost virulence, is also essential for biofilm formation. Mol Microbiol 2001; 40:1215-1226.
94. Goodman AL, Kulasekara B, Rietsch A et al. A signaling network reciprocally regulates genes associated with acute infection and chronic persistence in Pseudomonas aeruginosa. Dev Cell 2004; 7:745-754.
95. Ventre I, Goodman AL, Vallet-Gely I et al. Multiple sensors control reciprocal expression of Pseudomonas aeruginosa regulatory RNA and virulence genes. Proc Natl Acad Sci USA 2006; 103:171-176.
96. Laskowski MA, Kazmierczak BI. Mutational analysis of RetS, an unusual sensor kinase-response regulator hybrid required for Pseudomonas aeruginosa virulence. Infect Immun 2006; 74:4462-4473.
97. Zolfaghar I, Angus AA, Kang PJ et al. Mutation of retS, encoding a putative hybrid two-component regulatory protein in Pseudomonas aeruginosa, attenuates multiple virulence mechanisms. Microbes Infect 2005; 7:1305-1316.
98. Zolfaghar I, Evans DJ, Ronaghi R et al. Type III secretion-dependent modulation of innate immunity as one of multiple factors regulated by Pseudomonas aeruginosa RetS. Infect Immun 2006; 74:3880-3889.
99. Castañeda M, Guzman J, Moreno S et al. The GacS sensor kinase regulates alginate and poly-beta-hydroxybutyrate production in Azotobacter vinelandii. J Bacteriol 2000; 182:2624-2628.

100. Castañeda M, Sánchez J, Moreno S et al. The global regulators GacA and sigma(S) form part of a cascade that controls alginate production in Azotobacter vinelandii. J Bacteriol 2001; 183:6787-6793.
101. Mulcahy H, O'Callaghan J, O'Grady EP et al. The posttranscriptional regulator RsmA plays a role in the interaction between Pseudomonas aeruginosa and human airway epithelial cells by positively regulating the type III secretion system. Infect Immun 2006; 74:3012-3015.
102. Pernestig AK, Melefors O, Georgellis D. Identification of UvrY as the cognate response regulator for the BarA sensor kinase in Escherichia coli. J Biol Chem 2001; 276:225-231.
103. Tomenius H, Pernestig AK, Mendez-Catala CF et al. Genetic and functional characterization of the Escherichia coli BarA-UvrY two-component system: point mutations in the HAMP linker of the BarA sensor give a dominant-negative phenotype. J Bacteriol 2005; 187:7317-7324.
104. Nagasawa S, Tokishita S, Aiba H et al. A novel sensor-regulator protein that belongs to the homologous family of signal-transduction proteins involved in adaptive responses in Escherichia coli. Mol Microbiol 1992; 6:799-807.
105. Moolenaar GF, van Sluis CA, Backendorf C et al. Regulation of the Escherichia coli excision repair gene uvrC—overlap between the uvrC structural gene and the region coding for a 24 kD protein. Nucleic Acids Res 1987; 15:4273-4289.
106. Ronen M, Rosenberg R, Shraiman BI et al. Assigning numbers to the arrows: parameterizing a gene regulation network by using accurate expression kinetics. Proc Natl Acad Sci USA 2002; 99:10555-10560.
107. Oshima T, Aiba H, Masuda Y et al. Transcriptome analysis of all two-component regulatory system mutants of Escherichia coli K-12. Mol Microbiol 2002; 46:281-291.
108. Mukhopadhyay S, Audia JP, Roy RN et al. Transcriptional induction of the conserved alternative sigma factor RpoS in Escherichia coli is dependent on BarA, a probable two-component regulator. Mol Microbiol 2000; 37:371-381.
109. Zhang JP, Normark S. Induction of gene expression in Escherichia coli after pilus-mediated adherence. Science 1996; 273:1234-1236
110. Tomenius H, Pernestig AK, Jonas K et al. The Escherichia coli BarA-UvrY two-component system is a virulence determinant in the urinary tract. BMC Microbiol 2006; 6:27.
111. Johnston C, Pegues DA, Hueck CJ et al. Transcriptional activation of Salmonella typhimurium invasion genes by a member of the phosphorylated response-regulator superfamily. Mol Microbiol 1996; 22:715-727.
112. Ahmer BM, van Reeuwijk J, Watson PR et al. Salmonella SirA is a global regulator of genes mediating enteropathogenesis. Mol Microbiol 1999; 31:971-982.
113. Goodier RI, Ahmer BM. SirA orthologs affect both motility and virulence. J Bacteriol 2001; 183:2249-2258.
114. Altier C, Suyemoto M, Lawhon SD. Regulation of Salmonella enterica serovar typhimurium invasion genes by csrA. Infect Immun 2000; 68:6790-6797.
115. Lawhon SD, Maurer R, Suyemoto M et al. Intestinal short-chain fatty acids alter Salmonella typhimurium invasion gene expression and virulence through BarA/SirA. Mol Microbiol 2002; 46:1451-1464.
116. Weilbacher T, Suzuki K, Dubey AK et al. A novel sRNA component of the carbon storage regulatory system of Escherichia coli. Mol Microbiol 2003; 48:657-670.
117. Fortune DR, Suyemoto M, Altier C. Identification of CsrC and characterization of its role in epithelial cell invasion in Salmonella enterica serovar typhimurium. Infect Immun 2006; 74:331-339.
118. Wassarman KM, Repoila F, Rosenow C et al. Identification of novel small RNAs using comparative genomics and microarrays. Genes Dev 2001; 15:1637-1651.
119. Teplitski M, Goodier RI, Ahmer BM. Pathways leading from BarA/SirA to motility and virulence gene expression in Salmonella. J Bacteriol 2003; 185:7257-7265.
120. Teplitski M, Goodier RI, Ahmer BM. Catabolite repression of the SirA regulatory cascade in Salmonella enterica. Int J Med Microbiol 2006; 296:449-466.
121. Suzuki K, Wang X, Weilbacher T et al. Regulatory circuitry of the CsrA/CsrB and BarA/UvrY systems of Escherichia coli. J Bacteriol 2002; 184:5130-5140.
122. Prouty AM, Gunn JS. Salmonella enterica serovar typhimurium invasion is repressed in the presence of bile. Infect Immun 2000; 68:6763-6769.
123. Lawhon SD, Frye JG, Suyemoto M et al. Global regulation by CsrA in Salmonella typhimurium. Mol Microbiol 2003; 48:1633-1645.
124. Ellermeier CD, Ellermeier JR, Slauch JM. HilD, HilC and RtsA constitute a feed forward loop that controls expression of the SPI1 type three secretion system regulator hilA in Salmonella enterica serovar typhimurium. Mol Microbiol 2005; 57:691-705.
125. Altier C. Genetic and environmental control of Salmonella invasion. J Microbiol 2005; 43:85-92.
126. Liu MY, Romeo T. The global regulator CsrA of Escherichia coli is a specific mRNA-binding protein. J Bacteriol 1997; 179:4639-4642.

127. Baker CS, Morozov I, Suzuki K et al. CsrA regulates glycogen biosynthesis by preventing translation of glgC in Escherichia coli. Mol Microbiol 2002; 44:1599-1610.
128. Wang X, Preston JF 3rd, Romeo T. The pgaABCD locus of Escherichia coli promotes the synthesis of a polysaccharide adhesin required for biofilm formation. J Bacteriol 2004; 186:2724-2734.
129. Wang X, Dubey AK, Suzuki K et al. CsrA posttranscriptionally represses pgaABCD, responsible for synthesis of a biofilm polysaccharide adhesin of Escherichia coli. Mol Microbiol 2005; 56:1648-1663.
130. Dubey, AK, Baker CS, Suzuki K et al. CsrA regulates translation of the Escherichia coli carbon starvation gene, cstA, by blocking ribosome access to the cstA transcript. J Bacteriol 2003; 185:4450-4460.
131. Jackson DW, Suzuki K, Oakford L et al. Biofilm formation and dispersal under the influence of the global regulator CsrA of Escherichia coli. J Bacteriol 2002; 184:3406-3410.
132. Murray EL, Conway T. Multiple regulators control expression of the Entner-Doudoroff aldolase (Eda) of Escherichia coli. J Bacteriol 2005; 187:991-1000.
133. Gudapaty S, Suzuki K, Wang X et al. Regulatory interactions of Csr components: the RNA binding protein CsrA activates csrB transcription in Escherichia coli. J Bacteriol 2001; 183:6017-6027.
134. Pernestig AK, Georgellis D, Romeo T et al. The Escherichia coli BarA-UvrY two-component system is needed for efficient switching between glycolytic and gluconeogenic carbon sources. J Bacteriol 2003; 185:843-853.
135. Frederick RD, Chiu J, Bennetzen JL et al. Identification of a pathogenicity locus, rpfA, in Erwinia carotovora subsp. carotovora that encodes a two-component sensor-regulator protein. Mol Plant Microbe Interact 1997; 10:407-415.
136. Eriksson AR, Andersson RA, Pirhonen M et al. Two-component regulators involved in the global control of virulence in Erwinia carotovora subsp. carotovora. Mol Plant Microbe Interact 1998; 11:743-752.
137. Cui Y, Chatterjee A, Chatterjee AK. Effects of the two-component system comprising GacA and GacS of Erwinia carotovora subsp. carotovora on the production of global regulatory rsmB RNA, extracellular enzymes and HarpinEcc. Mol Plant Microbe Interact 2001; 14:516-526.
138. Liu Y, Cui Y, Mukherjee A et al. Characterization of a novel RNA regulator of Erwinia carotovora ssp. carotovora that controls production of extracellular enzymes and secondary metabolites. Mol Microbiol 1998; 29:219-234.
139. Ma W, Cui Y, Liu Y et al. Molecular characterization of global regulatory RNA species that control pathogenicity factors in Erwinia amylovora and Erwinia herbicola pv. gypsophilae. J Bacteriol 2001; 183:1870-1880.
140. Cui Y, Mukherjee A, Dumenyo CK et al. rsmC of the soft-rotting bacterium Erwinia carotovora subsp. carotovora negatively controls extracellular enzyme and Harpin(Ecc) production and virulence by modulating levels of regulatory RNA (rsmB) and RNA-binding protein (RsmA). J Bacteriol 1999; 181:6042-6052.
141. Liu Y, Jiang G, Cui Y et al. kdgREcc negatively regulates genes for pectinases, cellulase, protease, HarpinEcc and a global RNA regulator in Erwinia carotovora subsp. carotovora. J Bacteriol 1999; 181:2411-2421.
142. Mukherjee A, Cui Y, Ma W et al. hexA of Erwinia carotovora ssp. carotovora strain Ecc71 negatively regulates production of RpoS and rsmB RNA, a global regulator of extracellular proteins, plant virulence and the quorum-sensing signal, N-(3-oxohexanoyl)-L-homoserine lactone. Environ Microbiol 2000; 2:203-215.
143. Cui Y, Chatterjee A, Hasegawa H et al. Erwinia carotovora subspecies produce duplicate variants of ExpR, LuxR homologs that activate rsmA transcription but differ in their interactions with N-acylhomoserine lactone signals. J Bacteriol 2006; 188:4715-4726.
144. Chatterjee A, Cui Y, Liu Y et al. Inactivation of rsmA leads to overproduction of extracellular pectinases, cellulases and proteases in Erwinia carotovora subsp. carotovora in the absence of the starvation/cell density-sensing signal, N-(3-oxohexanoyl)-L-homoserine lactone. Appl Environ Microbiol 1995; 61:1959-1967.
145. Cui Y, Chatterjee A, Liu Y et al. Identification of a global repressor gene, rsmA, of Erwinia carotovora subsp. carotovora that controls extracellular enzymes, N-(3-oxohexanoyl)-L-homoserine lactone and pathogenicity in soft-rotting Erwinia spp. J Bacteriol 1995; 177:5108-5115.
146. Chatterjee A, Cui Y, Chatterjee AK. RsmA and the quorum-sensing signal, N-(3-oxohexanoyl)-L-homoserine lactone, control the levels of rsmB RNA in Erwinia carotovora subsp. carotovora by affecting its stability. J Bacteriol 2002; 184:4089-4095.
147. Pemberton CL, Whitehead NA, Sebaihia M et al. Novel quorum-sensing-controlled genes in Erwinia carotovora subsp. carotovora: identification of a fungal elicitor homologue in a soft-rotting bacterium. Mol Plant Microbe Interact 2005; 18:343-353.
148. Murata H, Chatterjee A, Liu Y et al. Regulation of the production of extracellular pectinase, cellulase and protease in the soft rot bacterium Erwinia carotovora subsp. carotovora: evidence that aepH of E. carotovora subsp. carotovora 71 activates gene expression in E. carotovora subsp. carotovora, E. carotovora subsp. atroseptica and Escherichia coli. Appl Environ Microbiol 1994; 60:3150-3159.

149. Lenz DH, Mok KC, Lilley BN et al. The small RNA chaperone Hfq and multiple small RNAs control quorum sensing in Vibrio harveyi and Vibrio cholerae. Cell 2004; 118:69-82.
150. Wong SM, Carroll PA, Rahme LG et al. Modulation of expression of the ToxR regulon in Vibrio cholerae by a member of the two-component family of response regulators. Infect Immun 1998; 66:5854-5861.
151. Lenz DH, Bassler BL. The small nucleoid protein Fis is involved in Vibrio cholerae quorum sensing. Mol Microbiol 2007; 63:859-871.
152. Dorsey CW, Tomaras AP, Actis LA. Genetic and phenotypic analysis of Acinetobacter baumannii insertion derivatives generated with a transposome system. Appl Environ Microbiol 2002; 68:6353-6360.
153. Hammer BK, Tateda ES, Swanson MS. A two-component regulator induces the transmission phenotype of stationary-phase Legionella pneumophila. Mol Microbiol 2002; 44:107-118.
154. Molofsky AB, Swanson MS. Legionella pneumophila CsrA is a pivotal repressor of transmission traits and activator of replication. Mol Microbiol 2003; 50:445-461.
155. Kang BR, Cho BH, Anderson AJ et al. The global regulator GacS of a biocontrol bacterium Pseudomonas chlororaphis O6 regulates transcription from the rpoS gene encoding a stationary-phase sigma factor and affects survival in oxidative stress. Gene 2004; 325:137-143.
156. Han SH, Lee SJ, Moon JH et al. GacS-dependent production of 2R,3R-butanediol by Pseudomonas chlororaphis O6 is a major determinant for eliciting systemic resistance against Erwinia carotovora but not against Pseudomonas syringae pv. tabaci in tobacco. Mol Plant Microbe Interact 2006; 19:924-930.
157. Chin-A-Woeng TF, van den Broek D, Lugtenberg BJ et al. The Pseudomonas chlororaphis PCL1391 sigma regulator psrA represses the production of the antifungal metabolite phenazine-1-carboxamide. Mol Plant Microbe Interact 2005; 18:244-253.
158. Girard G, van Rij ET, Lugtenberg BJ et al. Regulatory roles of psrA and rpoS in phenazine-1-carboxamide synthesis by Pseudomonas chlororaphis PCL1391. Microbiology 2006; 152:43-58.
159. Dubern JF, Bloemberg GV. Influence of environmental conditions on putisolvins I and II production in Pseudomonas putida strain PCL1445. FEMS Microbiol Lett 2006; 263:169-175.
160. Schmidt-Eisenlohr H, Gast A, Baron C. Inactivation of gacS does not affect the competitiveness of Pseudomonas chlororaphis in the Arabidopsis thaliana rhizosphere. Appl Environ Microbiol 2003; 69:1817-1826.
161. Vodovar N, Vallenet D, Cruveiller S et al. Complete genome sequence of the entomopathogenic and metabolically versatile soil bacterium Pseudomonas entomophila. Nat Biotechnol 2006; 24:673-679.
162. Jousset A, Lara E, Wall LG et al. Secondary metabolites help biocontrol strain Pseudomonas fluorescens CHA0 to escape protozoan grazing. Appl Environ Microbiol 2006; 72:7083-7090.
163. Sánchez-Contreras M, Martín M, Villacieros M et al. Phenotypic selection and phase variation occur during alfalfa root colonization by Pseudomonas fluorescens F113. J Bacteriol 2002; 184:1587-1596.
164. Martinez-Granero F, Rivilla R, Martin M. Rhizosphere selection of highly motile phenotypic variants of Pseudomonas fluorescens with enhanced competitive colonization ability. Appl Environ Microbiol 2006; 72:3429-3434.
165. Ge Y, Huang X, Wang S et al. Phenazine-1-carboxylic acid is negatively regulated and pyoluteorin positively regulated by gacA in Pseudomonas sp. M18. FEMS Microbiol Lett 2004; 237:41-47.
166. van den Broek D, Chin-A-Woeng TF, Bloemberg GV et al. Molecular nature of spontaneous modifications in gacS which cause colony phase variation in Pseudomonas sp. strain PCL1171. J Bacteriol 2005; 187:593-600.
167. Kang BR, Yang KY, Cho BH et al. Production of indole-3-acetic acid in the plant-beneficial strain Pseudomonas chlororaphis O6 is negatively regulated by the global sensor kinase GacS. Curr Microbiol 2006; 52:473-476.
168. Dubern JF, Lagendijk EL, Lugtenberg BJ et al. The heat shock genes dnaK, dnaJ and grpE are involved in regulation of putisolvin biosynthesis in Pseudomonas putida PCL1445. J Bacteriol 2005; 187:5967-5976.
169. Kang H, Gross DC. Characterization of a resistance-nodulation-cell division transporter system associated with the syr-syp genomic island of Pseudomonas syringae pv. syringae. Appl Environ Microbiol 2005; 71:5056-5065.
170. Quinones B, Pujol CJ, Lindow SE. Regulation of AHL production and its contribution to epiphytic fitness in Pseudomonas syringae. Mol Plant Microbe Interact 2004; 17:521-531.
171. Ovadis M, Liu X, Gavriel S et al. The global regulator genes from biocontrol strain Serratia plymuthica IC1270: cloning, sequencing and functional studies. J Bacteriol 2004; 186:4986-4993.
172. Williamson NR, Fineran PC, Leeper FJ et al. The biosynthesis and regulation of bacterial prodiginines. Nat Rev Microbiol 2006; 4:887-899.
173. Whistler CA, Ruby EG. GacA regulates symbiotic colonization traits of Vibrio fischeri and facilitates a beneficial association with an animal host. J Bacteriol 2003; 185:7202-7212.

174. Majdalani N, Gottesman S. The Rcs phosphorelay: a complex signal transduction system. Annu Rev Microbiol 2005; 59:379-405.
175. Majdalani N, Hernandez D, Gottesman S. Regulation and mode of action of the second small RNA activator of RpoS translation, RprA. Mol Microbiol 2002; 46:813-826.
176. Shimizu T, Yaguchi H, Ohtani K et al. Clostridial VirR/VirS regulon involves a regulatory RNA molecule for expression of toxins. Mol Microbiol 2002; 43:257-265.
177. Ohtani K, Kawsar HI, Okumura K et al. The VirR/VirS regulatory cascade affects transcription of plasmid-encoded putative virulence genes in Clostridium perfringens strain 13. FEMS Microbiol Lett 2003; 222:137-141.
178. Okumura K, Kawsar HI, Shimizu T et al. Identification and characterization of a cell-wall anchored DNase gene in Clostridium perfringens. FEMS Microbiol Lett 2005; 242:281-285.
179. Silvaggi JM, Perkins JB, Losick R. Genes for small, noncoding RNAs under sporulation control in Bacillus subtilis. J Bacteriol 2006; 188:532-541.
180. Kreikemeyer B, Boyle MD, Buttaro BA et al. Group A streptococcal growth phase-associated virulence factor regulation by a novel operon (Fas) with homologies to two-component-type regulators requires a small RNA molecule. Mol Microbiol 2001; 39:392-406.

# CHAPTER 6

# Two-Component Signaling and Gram Negative Envelope Stress Response Systems

**Dawn M. MacRitchie, Daelynn R. Buelow, Nancy L. Price and Tracy L. Raivio***

## Introduction

Bacteria have evolved complex stress responses that allow them to respond to their surroundings. In Gram negative bacteria, these stress responses can be compartmentalized into the cytoplasmic and extracytoplasmic (envelope) stress responses. The extracytoplasmic stress response monitors the integrity of the envelope, which consists of the outer membrane (OM), inner membrane (IM) and the periplasm. The OM is an atypical membrane in that it is asymmetrical with lipopolysaccharides found only on the outer facet. Functions of the OM include acting as a permeability barrier, allowing for transport via porins and avoiding phagocytosis.[1] The IM is composed of phospholipids and proteins. The IM is involved in many activities including energy generation and conservation, biosynthetic and catabolic reactions, signal transduction and acting as a hydrophobic barrier to control and maintain the intracellular concentrations of cytoplasmic ions/molecules.[2] The periplasm, which lies between the IM and OM, contains the peptidogylcan layer that is involved in maintaining cell shape. Although it has been thought that the periplasm is extremely viscous due to its high protein content, it was recently shown that the viscosity of the periplasm is not that much different than the cytoplasm, with an average diffusion rate of $9.0 \pm 2.1\ \mu m^2 s^{-1}$ in the cytoplasm and $2.6 \pm 1.2\ \mu m^2 s^{-1}$ in the periplasm.[3] The periplasm is involved in processing essential nutrients for transport, biogenesis of major envelope components, detoxification and buffering the cytoplasmic environment from external stresses to maintain growth and viability.[4]

*Escherichia coli* has at least five known stress responses that allow it to monitor envelope homeostasis, the $\sigma^E$, Cpx (Conjugative Pilus Expression), Bae (Bacterial Adaptive Response), Psp (Phage-Shock-Protein) and Rcs (Regulator of Capsular Synthesis) responses. Of these five stress responses three of them are controlled by two-component systems and thus, two-component signal transduction is centrally involved in the bacterial response to envelope stress. Although there is some overlap, each of these pathways has been shown to respond to different stresses and regulate different processes. The $\sigma^E$ pathway senses and responds mainly to stresses that involve OM protein (OMP) maintenance and folding.[5] The Cpx pathway is induced and mediates adaptation to, misfolded proteins in the periplasm, specifically those associated with the periplasmic face of the inner membrane.[6] The Cpx response has also been shown to sense and facilitate adhesion to abiotic surfaces.[7] The Bae pathway is involved in detoxification by ridding the cell of toxic

*Corresponding Author: Tracy L. Raivio—Department of Biological Sciences, University of Alberta, Edmonton, Alberta T6G 2E9, Canada. Email: traivio@ualberta.ca

*Bacterial Signal Transduction: Networks and Drug Targets*, edited by Ryutaro Utsumi.

compounds.[8,9] The Phage-shock-protein response monitors both the proton motive force and the integrity of the inner membrane.[10] The Rcs pathway seems to be involved in the formation of surface structures.[11,12]

## The $\sigma^E$ Envelope Stress Response

### *Discovery of* $\sigma^E$

The use of ECF (ExtraCytoplasmic Function) sigma factors, which are divergent from $\sigma^{70}$, is one way in which bacteria sense and respond to their surroundings. Signal transduction events in response to alterations in the environment lead to an increased cellular concentration of a given ECF sigma factor, which can then associate with the core RNA polymerase (RNAP) to change gene expression.[13] To date, only two ECFs have been identified in *E. coli*, FecI (as reviewed by 14) and $\sigma^E$, which is essential.[15,16] The $\sigma^E$ ($\sigma^{24}$) subunit was first identified for its ability to transcribe the *rpoH* (encodes an alternative sigma factor that regulates heat shock response) and *degP* (*htrA*) (encodes a periplasmic serine endoprotease/chaperone) genes, under heat shock conditions[17,18] (Table 1). When they initially identified $\sigma^E$, Erickson and Gross[17] purified the sigma factor responsible for transcribing *rpoH* from the P3 promoter, however they did not identify the gene that encoded the sigma factor. Taking different approaches Rouviere et al[19] and Raina et al[20] both determined that an ORF at minute 55.5 on the *E. coli* genetic map encoded $\sigma^E$, *rpoE*. It was later determined that *rpoE* was the first gene in a four gene operon[21,22] and the *rpoE* promoter is driven by $\sigma^E$.[19,20]

### *Activating Cues*

From the discovery of $\sigma^E$, it was determined that one of the activating cues that stimulated $\sigma^E$-mediated transcription was heat shock conditions, an up shift in temperature above 42°C.[17]

***Table 1. Members of the $\sigma^E$ regulon that are involved in envelope homeostasis***

| Gene Name | Classification; Role | Essential |
|---|---|---|
| *rpoE* | ECF σ factor; role in envelope and oxidative stress responses | Yes |
| *rseA* | Regulatory protein; antisigma factor of $\sigma^E$ | No |
| *rseB* | Regulatory protein; modulates RseA activity | No |
| *rseC* | Regulatory protein; possible positive regulator of $\sigma^E$ | No |
| *rseP* (*yaeL*) | Zinc metalloprotease; degrades RseA | Yes |
| *rpoD* ($\sigma^{70}$) | Housekeeping σ | Yes |
| *rpoH* ($\sigma^H$) | Heat shock σ | Yes |
| *degP* (*htrA*) | Serine endoprotease with chaperone activity; Conditional; degradation of damaged periplasmic proteins | Temp[a] above 43°C |
| *skp* | Chaperone; OMP chaperone | No[b] |
| *fkpA* | PPI[c] and chaperone activity | No[b] |
| *surA* | PPI[c] and OM porin chaperone | No[b] |
| *dsbC* | disulfide bond isomerase with chaperone activity | No[b] |
| *yifO* | OM lipoprotein; OM biogenesis | Yes |
| *yraP* | OM lipoprotein; OMP assembly | No |
| *yaeT* | OMP assembly factor | Yes |
| *micA* | sRNA; reduces *ompA* RNA levels | No |
| *rybB* | sRNA; reduces *ompC* and *ompW* RNA levels | No |

a. Temp, temperature
b. *skp, fkpA,surA, dsbC* single or double mutants had no deleterious affects on bacterial growth, however triple mutants displayed growth defects.
c. PPI, peptidylprolyl *cis, trans* isomerase

However heat shock conditions encompass many different signals that could elicit many different responses. To better understand the role of $\sigma^E$ in the cell it was crucial that specific signals that activate $\sigma^E$ be identified. From the periplasmic location of DegP, it was suggested that $\sigma^E$ may be involved in monitoring the envelope.[23] To test this hypothesis, Mecsas et al[23] looked to see what affect overproduction and underproduction of OMPs would have on $\sigma^E$ activity. Overexpression of OMPs such as, OmpX, OmpT, OmpF and OmpC resulted in an increase in $\sigma^E$ activity, while limiting the amount of OMPs present reduced the activity of $\sigma^E$. To determine if the signal was generated in the cytoplasm or the envelope Mecsas et al[23] looked at what happened when OMP precursors were overexpressed and trapped in the cytoplasm in a *secB* mutant. They observed that the activity of $\sigma^E$ resembled that seen when OMPs were limited, suggesting that $\sigma^E$ is activated by a signal that originates in the envelope when OMPs are overproduced.[23] A few years later it was shown that ethanol, DTT (dithiothreitol) and puromycin, all compounds known to cause misfolded proteins, specifically increased $\sigma^E$ activity.[20] It was also shown that $\sigma^E$ senses and responds to the production of the misfolded P-pilus subunit PapG, when it is driven off-pathway because of the absence of its cognate chaperone, PapD.[24] All these observations suggested that $\sigma^E$ senses and responds to misfolded proteins in the envelope. If this is true one would think that defects in envelope protein folding factors would also affect pathway activity. Mutations to protein folding genes, such as the *dsb* (encoding disulfide bond oxidoreductases) genes, *degP*, *surA* (a peptidylproyl *cis, trans* isomerase (PPI)/chaperone) and *fkpA* (PPI/chaperone), all caused an increase in pathway activity,[20,25] which reaffirmed that $\sigma^E$ senses and responds to misfolded proteins, specifically to misfolded OMPs, in the envelope.

### *Regulation of* $\sigma^E$

When Raina et al[20] and Rouviere et al[19] were trying to clone and analyze the gene encoding $\sigma^E$, they noticed that in the absence of induction the activity of $\sigma^E$ was low, suggesting that it was negatively regulated by a possible antisigma factor. To try and identify a possible antisigma factor two approaches were taken. De Las Penas et al[21] took into account that the genes encoding sigma and antisigma factors are normally found in an operon. They looked at the sequence downstream from *rpoE* and identified three ORFs that were termed *rseABC* for regulator of sigma E. Missiakas et al[22] used transposon mutagenesis to identify genes that, when inactivated, led to an increase in $\sigma^E$ activity. Their approach identified two such insertional mutations that seemed to be involved in $\sigma^E$ regulation. These mutations localized to the second and third genes, *rseAB* respectively, in the *rpoErseABC* four-gene operon. Upon characterization of *rseABC*, it was determined that RseA is an inner membrane protein that transmits the signal generated in the envelope to $\sigma^E$.[21,22] RseA spans the membrane once with C-terminal periplasmic and N-terminal cytoplasmic domains. It is the cytoplasmic domain of RseA that interacts with $\sigma^E$ to inhibit its activity and this is all that is needed for inhibition[22,26-27] (Fig. 1). Specifically, the −35 binding region 4.2 of $\sigma^E$ has been shown to interact with RseA.[27] RseB, a periplasmic protein, is also a negative regulator of $\sigma^E$ activity; however its role is not pivotal in transmitting the signal to $\sigma^E$.[21,22] RseB has a role in stabilizing RseA and the RseA~$\sigma^E$ interaction[21,22,28,29] (Fig. 1). When RseB is absent there is a two-fold increase in activity compared to when it is present and RseA half-life is changed from $18.6 \pm 3.1$ mins to $44.3 \pm 6.1$ mins.[21,22,29] RseC, a cytoplasmic protein, does not seem to have a major role in the signaling cascade, although it has been suggested that it may be a positive regulator of $\sigma^E$.[22]

The effects of RseB on RseA and the observation that the relative RseA synthesis rate was the same under inducing and non-inducing conditions suggested that RseA is posttranslationally modified in response to inducing cues.[29] RseA is cleaved in a two step process by two inner membrane proteases, first by DegS, a serine protease and then RseP (YaeL), a zinc metalloprotease[29-30] (Fig. 1). Under inducing conditions, it is not RseA that senses the misfolded proteins, but rather the PDZ domain of DegS.[31] PDZ domains are found in diverse signaling proteins and are involved in protein:protein interactions.[32] Under non-inducing conditions the PDZ domain of DegS inhibits it proteolytic activity, preventing it from degrading RseA (Fig. 1). However, under inducing conditions, the PDZ domain binds the C-terminus of misfolded OMPs, which contain

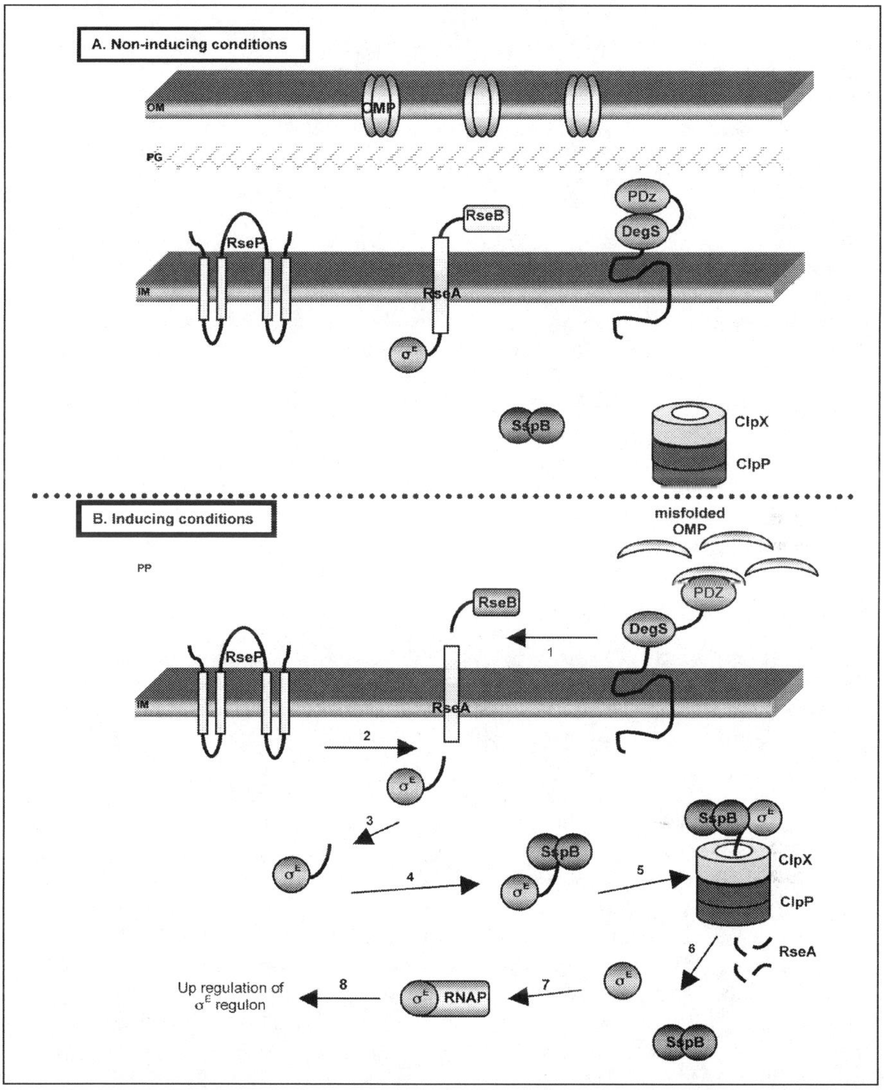


Figure 1. Model of the $\sigma^E$ pathway under induced and non-induced conditions. A. Under non-induced conditions, the PDZ domain of DegS inhibits degradation of RseA. RseB is bound to the periplasmic domain of RseA, stabilizing the RseA: $\sigma^E$ interaction. The presence of RseB and the periplasmic domain of RseA inhibit RseP from degrading RseA. B. Under inducing conditions, the PDZ domain of DegS interacts with the C-terminal domains of misfolded OMPs. The activated DegS then cleaves RseA near the transmembrane domain on the periplasmic side (1). This releases the periplasmic domain of RseA and RseB, which relieves the inhibition on RseP. RseP is then able to cleave RseA near the transmembrane domain in the cytoplasm (2). This releases the cytoplasmic domain of RseA with $\sigma^E$ still bound (3). SspB then delivers the cytoplasmic domain of RseA with $\sigma^E$ to the ClpXP complex (4, 5). ClpXP degrades the cytoplasmic domain of RseA freeing $\sigma^E$ (6). $\sigma^E$ is now free to bind with RNAP and up regulate the regulon (7, 8). The numbered arrows indicate the sequential order in which $\sigma^E$ is activated. OM, outer membrane, IM, inner membrane, PP, periplasm, PG, peptidoglycan, OMP, outer membrane protein, RNAP, RNA polymerase.

a YFF recognition peptide. This interaction activates DegS to cleave RseA adjacent to the transmembrane domain on the periplasmic side.[31] It has been determined that the cleavage by DegS is the rate determining step and that the DegS-cleaved RseA serves as a substrate for RseP proteolysis.[29,30,33] Once the periplasmic domain of RseA is removed then RseP is free to cleave RseA at the cytoplasmic cleavage site (HExxH), which is close to the transmembrane domain.[30] It is believed that the glutamine rich regions of RseA's periplasmic domain interact with the periplasmic PDZ domain of RseP, thus inhibiting it from degrading RseA until the proper stress is sensed.[30]

Grigorova et al[34] showed that RseB diminished RseP degradation of RseA. They speculated that the affect RseB had on RseP activity may have a role in sensing signals other than OMPs. It has been shown that RseB can interact with the misfolded periplasmic protein MalE31.[28] The hypothesis is that under inducing conditions that do not involve OMPs, RseB would be titrated away, where it may or may not interact with the misfolded proteins present. This removal of RseB would then allow RseP to cleave RseA independently of DegS.[34] If this does occur, one would speculate that the proteolysis is inefficient since the glutamine rich region in the periplasmic region of RseA that is important for RseP inhibition is still present. It has also been shown that the substrate for RseP is a DegS-cleaved RseA.[29,33] Thus the role of RseB in signal transduction remains mysterious. A better explanation for the inhibitory effect of RseB on RseP proteolysis of RseA may be that the glutamine rich region of RseA and RseB both contribute to the inhibition of RseP and that RseB may function to block any inappropriate degradation of RseA by DegS. In a similar fashion, RseB may also have a role in blocking the $\sigma^E$ pathway from normal OMP misfolding "noise". Another possibility is that RseB is important in shutting off the $\sigma^E$ pathway after OMP folding stress has been relieved.

After both cleavage events by DegS and RseP, the cytoplasmic domain of RseA is left intact which is still able to inhibit $\sigma^E$. This cytoplasmic domain of RseA must be degraded in order to free $\sigma^E$. After the cleavage by RseP, the cytoplasmic domain of RseA is targeted for degradation by ClpXP.[35] SspB, a ClpXP adaptor protein, recognizes the C-terminal VAA signal on RseA that is generated after RseP cleavage and delivers it to ClpXP, where it is degraded, releasing $\sigma^E$[35] (Fig. 1).

It was thought that RseA was the pivotal signal transduction protein of the $\sigma^E$ pathway. However, recently it has been shown that $\sigma^E$ activity increases upon entry into stationary phase independently of RseA.[36] This increased activity is due to an internal (cytoplasmic) signal related to starvation. Upon further analysis it was determined that the increased activity of the $\sigma^E$ pathway is correlated with increased levels of ppGpp (3',5'-bispyrophosphate).[36] ppGpp is a global regulator that is the general signal of starvation.[37] Along with activating $\sigma^E$, ppGpp also activates $\sigma^S$ and $\sigma^N$,[37] suggesting that it is used to coordinate a broad response to nutritional stress. The exact nature of how the signaling cascade between ppGpp and $\sigma^E$ works is still not understood. However this is the first time that a cytoplasmic regulator has been shown to influence $\sigma^E$ regulation. Understanding exactly how ppGpp regulates $\sigma^E$ will open the door to understanding how envelope stress responses are regulated by both external and internal signals.

### $\sigma^E$ *Regulon*

The $\sigma^E$ regulon has been extensively studied and it has been speculated that it consists of over 100 members. Originally, it was shown that the regulon consisted mainly of periplasmic protein folding and degrading factors, in keeping with a primary role in sensing and responding to misfolded proteins in the periplasm, especially OMPs. It is now known that the regulon contains members that are involved in multiple aspects of cell function including protein folding and degradation in the periplasm, assembly of outer membrane proteins, primary metabolism, transcription and translation, lipoproteins and lipid detoxification, DNA/RNA modification and repair, cell structure and division, regulation of small noncoding RNAs (sRNAs) and genes of unknown functions [17,25,38-43] (Table 1). For a comprehensive list of the members of the $\sigma^E$ regulon please refer to the microarray papers by Rhodius et al[43] and Kabir et al.[40] This section will only deal with the members of the regulon whose main purpose is in envelope protein biogenesis and degradation.

Initially, the main members of the $\sigma^E$ regulon identified were *rpoE* itself along with its regulators *rseABC*.[21,22] The first major characterization of the $\sigma^E$ regulon utilized promoterless *lacZ* elements to identify constructs that showed increased β-galactosidase activity when $\sigma^E$ was overexpressed.[38] *rseP*, another regulator of $\sigma^E$ was added to the regulon after this screen was carried out.[38] It was found that $\sigma^E$ not only regulates itself but also two other σ factors, *rpoD* ($\sigma^{70}$), the housekeeping sigma factor and *rpoH* ($\sigma^H$), the heat shock sigma factor.[17,18,38] Because of the regulation of *degP*, a periplasmic chaperone and protease and activation by OMPs, it was determined that $\sigma^E$ monitors the envelope, so it was not surprising when periplasmic chaperones and folding catalysts were added to the $\sigma^E$ regulon. These chaperones and folding catalysts include *skp*, a chaperone for OMPs, *fkpA*, a periplasmic peptidyl-prolyl-isomerase (PPI), *surA*, a periplasmic OMP chaperone, which also has PPI activity and *dsbC*, a disulfide bond isomerase which also has chaperone activity[38] (Table 1). Interestingly, prior to their identification as $\sigma^E$ regulon members, DsbC, SurA, FkpA and Skp were identified for their ability to modulate $\sigma^E$ activity when overexpressed or mutated.[25] Recent studies that characterized genes that were upregulated when $\sigma^E$ was overexpressed have expanded the regulon to include proteins involved in OM assembly, such as *yifO*, an essential lipoprotein required for OMP biogenesis, *yraP*, a non-essential lipoprotein that is important for cell envelope maintenance and *yaeT*, an essential OMP assembly factor[41,44,45] (Table 1).

Interestingly, it was recently shown that $\sigma^E$ alleviates OMP misfolding not only by inducing expression of OMP assembly and degradation factors but also by controlling expression of the OMPs themselves. When the $\sigma^E$ pathway is induced many OMPs, such as OmpA, OmpC, OmpF, OmpW and OmpX are down-regulated.[40,43] However, none of the OMP promoters contain the recognition sequence for $\sigma^E$.[43] It was discovered that $\sigma^E$ down-regulates OMPs upon pathway activation through the control of transcription of two sRNA, *micA* and *rybB*.[39] Upon induction of the $\sigma^E$ pathway there is an increase in MicA and RybB levels. MicA and RybB go on to bind with *ompA* mRNA and *ompC/W* mRNA, respectively, decreasing the steady-state levels of these mRNAs.[39] Thus, $\sigma^E$ mediated activation of these sRNA stops the accumulation of misfolded/mislocalized OMPs in the periplasm by controlling gene expression at the posttranscriptional level. This in turn, will allow the members of the $\sigma^E$ regulon to more efficiently clear the misfolded OMPs present and return the cell to homeostasis.

## $\sigma^E$ *and Virulence*

The $\sigma^E$ pathway is conserved throughout numerous Gram negative bacteria. While the $\sigma^E$ pathway has been studied extensively in *E.coli* with relation to monitoring and maintaining OMPs, most work in other Gram negative bacteria has linked $\sigma^E$ to virulence. $\sigma^E$ is essential in *E.coli*, but its homologues are not essential in other bacteria. Probably the most widely studied $\sigma^E$ homologues are in *Salmonella enterica* serovar Typhimurium and *Pseudomonas aeruginosa*. Studies of $\sigma^E$ in various pathogens showed that *rpoE* mutations attenuate virulence (Table 2) (for a more comprehensive review on $\sigma^E$ and pathogenesis please refer to the Raivio review ref. 46).

In *S. typhimurium*, *rpoE*/$\sigma^E$ mutants exhibit decreased survival and proliferation in macrophages and epithelial cell lines.[47] This decreased virulence is attributed to an increased sensitivity to reactive oxygen species, like $H_2O_2$ and superoxide.[47-49] It was also shown when *S. typhimurium* enters into macrophages there is an increase in $\sigma^E$ activity, allowing one to speculate that this increase in activity may help the bacteria survive against oxygen-dependent host defences.[49,50] Analysis of the $\sigma^E$ regulon in *S. typhimurium* also showed that it contains genes important for pathogenesis, which may also be why *rpoE* mutants are attenuated with a 50% lethal dose ($LD_{50}$) five orders of magnitude higher than wild type.[47,51] The phenotypes seen with *rpoE* mutants in *S. typhimurium* are common among other *rpoE* mutants in different pathogens (Table 2). For example *rpoE* mutants in *Vibrio cholerae* have a decreased ability to survive within the intestines and have a $LD_{50}$ three fold higher than wild type.[52] *Mycobacterium rpoE* mutants are unable to survive and grow within macrophages and when mice are infected with the mutants they have a delayed death.[53,54] $\sigma^E$/*rpoE* mutants in *Burkholderia pseudomallei* exhibit reduced survival in macrophages, which may be due to an increased sensitivity to $H_2O_2$.[55] As for the *P. aeruginosa rpoE* homologue, *algU*,

**Table 2. Affects that rpoE mutants have on pathogenesis**

| Pathogen | Affects on Virulence |
|---|---|
| *S. typhimurium* | Decreased survival and proliferation in macrophages and epithelial cell lines increased sensitivity to reactive oxygen species, $H_2O_2$ and superoxide $LD^{50a}$ is 5 fold higher than wild-type |
| *V. cholerae* | Decreased ability to survive in intestinal environments $LD^{50}$ is 3 fold higher than wild-type intestinal colonization is 30 fold reduced |
| *Mycobacterium* | Unable to survive and proliferate in macrophages results in delayed death in infected mice |
| *B. pseudomallei* | Reduced survival in macrophages increased sensitivity to $H_2O_2$ reduced ability to form biofilms, which help in resistance to antibiotics and host immune systems |
| *P. aeruginosa* | Increased sensitivity to phagocytic killing reduced ability to form biofilms, which are important in establishing infection |

a. $LD^{50}$, 50% lethal dose

mutants show increased sensitivity to phagocytic killing.[56] Thus, through the study of *rpoE* mutants in numerous pathogens it appears that $\sigma^E$ mediates adaptation to oxidative stresses, however the mechanism by which this occurs is not known.

## *Essential Nature of* $\sigma^E$

Not long after $\sigma^E$ was discovered and cloned it was noticed that it was essential for bacterial growth at high temperatures.[16] Initially, it was thought that the reason for the essential nature of $\sigma^E$ at high temperatures was due to the fact that it is needed to drive the temperature dependent promoters of *rpoH* and *degP*. However analysis of different *rpoE* mutants at 30° suggested that $\sigma^E$ might be essential at all temperatures, since no two *rpoE* mutants had similar growth patterns/phenotypes, suggesting the accumulation of suppressor mutations.[19,20] Upon analysis of cells lacking $\sigma^E$ it was determined they had acquired a suppressor in order to grow at low temperatures.[15] Understanding why and how $\sigma^E$ is essential is key to fully understanding the role of $\sigma^E$ within the bacterial cell. The $\sigma^E$ regulon includes numerous genes that are essential for viability such as *rpoE* itself, *rpoD, rpoH* and *rseP*[38] (Table 1). It was also noted that single and double deletions of *dsbC, skp, surA or fkpA* did not have deleterious effects, however deletions of more then two of these genes conferred growth defects[38,57] (Table 1). Thus, one reason why $\sigma^E$ is essential may be that it transcribes several genes that are essential for viability either alone or in combination. $\sigma^E$ may also contribute to other undefined essential cellular functions, since recently it has been shown through microarray analysis that $\sigma^E$ is involved in numerous cellular processes.[40,43]

As mentioned before the only way that $\sigma^E$ mutants are able to grow is in the presence of suppressors.[15] To try and understand in further depth the essential nature of $\sigma^E$, Douchin et al[58] looked at how suppressors alleviated the need for RseP in the normal activation of $\sigma^E$. There could be two reasons why RseP is needed in the normal activation of $\sigma^E$. The first reason is that without RseP, RseA would not be fully degraded and ultimately there would be no release of $\sigma^E$ and therefore no activation of the regulon. The second reason why RseP might be important is that *rseP* and *yaeT* are predicted to be in the same operon, so one could imagine that a mutation in *rseP* could affect *yaeT*, which is important in the insertion of OMPs into the OM.[44,45] It was found that overexpression of the sRNA, *rseX*, bypassed the RseP requirement for the activation of $\sigma^E$. RseX suppressed $\sigma^E$ lethality by down regulating levels of OmpC and OmpA,[58] thus decreasing the potential stress caused by these proteins and ultimately altering the requirement for $\sigma^E$. Thus, this study suggests that *rseP* (and therefore $\sigma^E$) is essential because it permits sufficient activation of the $\sigma^E$ pathway to allow the cell to deal with basal levels of stress conferred by normal/unstressed OMP elaboration.

In conclusion there are three potential reasons why $\sigma^E$ may be essential to the cell; the first being that the regulon encodes numerous genes with redundant functions. Thus if the bacteria loses an individual gene it still has other factors to overcome the deficit, however a loss of more than one would see that function eliminated resulting in reduced viability. The second reason that the $\sigma^E$ pathway may be essential is that it is involved in many aspects of cell physiology. The third reason $\sigma^E$ may be essential is that it controls the expression levels of OMPs through sRNAs, so when $\sigma^E$ is absent OMPs are expressed at higher levels but there is no pathway to upregulate the assembly and folding factors to facilitate their biogenesis, thus leading to an enormous stress that cannot be overcome.

### *Future Research Questions*

Although the $\sigma^E$ pathway has been studied in depth and it is the most characterized envelope stress response in *E.coli*, there are still questions that remain to be answered. Fully understanding the essentiality of $\sigma^E$ is probably the biggest question that still needs to be addressed. While initial studies of the $\sigma^E$ response addressed how the pathway is controlled by signal transduction that occurs across the inner membrane, recent studies indicate cytoplasmic signal transduction events, such as those involving generation of ppGpp also affect $\sigma^E$ regulated genes.[36] This begs the question of whether other cytoplasmic factors or signals exist that affect $\sigma^E$ regulated gene expression. Another unanswered question concerns the role RseB plays in signal transduction. Is RseB involved in shutting off the $\sigma^E$ pathway after the stress has been relieved, or is its role in preventing false signals from activating the pathway? Addressing these questions will shed light on how bacteria maintain homeostasis by integrating a variety of signal transduction mechanisms and pathways.

## The Cpx Two-Component System

### *Cpx History and Components of the Signal Transduction Pathway*

The discovery of the Cpx signal transduction pathway was originally not associated with studies of extracytoplasmic stress responses, but rather with a genetic screen for strains incapable of conjugation of the F plasmid.[59] Though non-essential for cell survival, gain-of-function mutations in the *cpx* locus yielded strains sensitive to elevated temperatures that possessed altered envelope protein composition, giving the first indication that the *cpx* genes may be involved with envelope stress.[60-63] Subsequent characterization of the *cpx* locus determined that it encoded two genes. Through analysis of fusion proteins and immunochemical experiments, the *cpxA* gene was determined to encode a transmembrane protein with sequence homology to the EnvZ two-component histidine kinase family.[64,65] Later analysis of the region upstream from *cpxA* uncovered its cognate response regulator, CpxR, which showed homology to the OmpR subfamily of DNA-binding proteins and together the CpxAR proteins were classified as a two-component signal transduction system.[66]

Subsequent studies uncovered a biological role for the Cpx two-component system. Using toxic, misfolded, mislocalized, secreted proteins, Silhavy and colleagues demonstrated that constitutively activated *cpx* mutants (*cpx**) endured high levels of misfolded and mislocalized envelope proteins.[67] These studies also demonstrated that the periplasmic protease, DegP, was an important component of the alleviation of toxicity.[67] Furthermore, the overexpression of a lipoprotein of unknown function, NlpE, was demonstrated to activate the Cpx pathway and also suppressed the toxicity of misfolded, mislocalized secreted proteins.[68] Since Cpx pathway activation rescued the cell from misfolded envelope proteins, the Cpx signal transduction system was proposed to control an envelope stress response in *E. coli*.[67,68]

Though CpxA and CpxR were initially classified as a traditional two-component signal transduction system, later work revealed a third signal transduction protein, CpxP.[69-71] CpxP was first identified as a small periplasmic protein that was induced in a Cpx-dependent manner in response to envelope stresses such as alkaline pH or overproduction of NlpE.[69] However in addition to being a downstream target of the Cpx system, the overproduction of CpxP resulted in a downregulation of Cpx-mediated gene expression.[70] Moreover, it was shown that tethering of CpxP to the inner membrane prevented the activation of the Cpx system after harsh envelope stress due to spheroplasting (i.e., removal of the outer membrane).[71] Although no direct interaction between CpxP and CpxA has been shown, it has been demonstrated that CpxP-mediated inhibition requires the periplasmic sensing domain of CpxA, which suggests that CpxP interacts with the sensor domain of CpxA to inhibit the Cpx pathway.[70] In support of this hypothesis, CpxP could inhibit CpxA enzymatic activity in reconstructed proteoliposomes.[72] Overall, early studies showed that the Cpx signal transduction pathway regulates an envelope stress response of which DegP is a component and consists of the membrane bound sensor kinase CpxA, the cognate response regulator CpxR and the periplasmic inhibitory protein CpxP (Fig. 2).

### *Cpx Pathway Activation*

The Cpx response is induced by envelope stressors that include alkaline pH, the overexpression of specific membrane-associated proteins and alterations in membrane structure.[6] In addition, adherence to abiotic surfaces has been shown to activate the Cpx response.[7] The first major

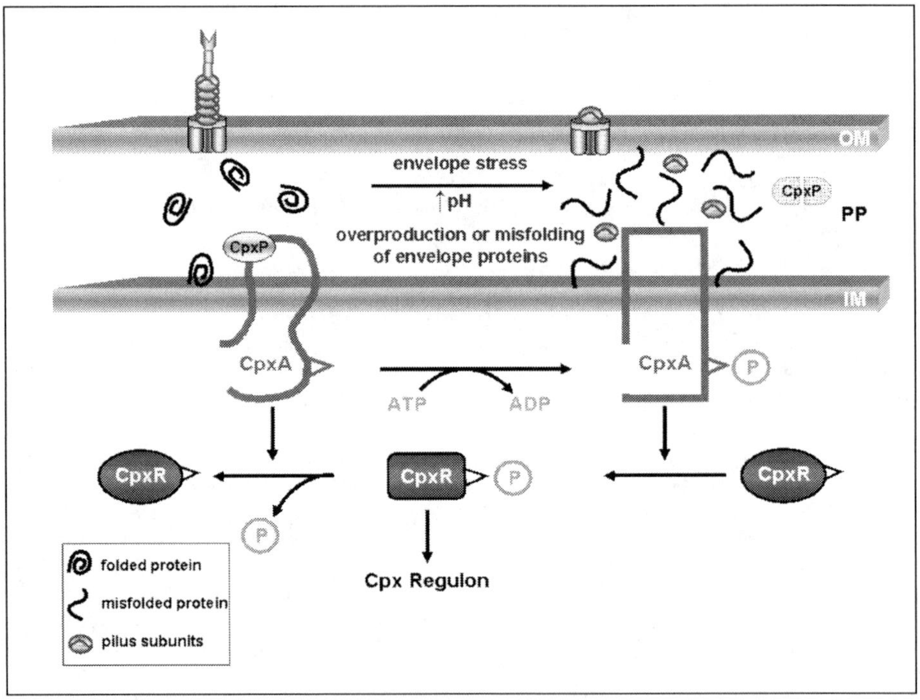


Figure 2. The Cpx envelope stress response. In non-inducing conditions, CpxA remains unphosphorylated, partly through interaction with the periplasmic protein, CpxP. Envelope stress brought about by either elevated pH or the presence of misfolded or overexpressed envelope proteins results in degradation of CpxP by DegP, thus relieving inhibition of CpxA. Inducing cues stimulate CpxA autokinase and kinase activities. The phosphorylated form of CpxA transfers the phosphate ion to CpxR which up-regulates transcription of several genes involved with protein folding, degradation and transcription regulation (e.g., *dsbA, degP, cpxRA, cpxP and* others). Once envelope stress is alleviated, CpxA acts as a phosphatase to remove the phosphate ion from CpxR thereby down-regulating the Cpx envelope stress response. Additional abbreviations: OM: outer membrane; PP: periplasmic space; IM: inner membrane. Figure adapted from Raivio et al, 1999, Raivio and Silhavy, 2001, Buelow and Raivio 2005.

inducing cue of the Cpx pathway was identified through a selection for multicopy suppressors of the toxicity exerted by secretion of a LamB-LacZ-PhoA fusion protein. This selection identified the overproduction of the outer membrane lipoprotein NlpE.[68] The mechanism of Cpx pathway activation by NlpE overexpression is not known, however NlpE overexpression results in protein mislocalization and likely misfolding, thus it is speculated that this may constitute part of the inducing cue.[6,68] In addition to NlpE, the overexpression of pilus subunits such as those of the uropathogenic *E. coli* (UPEC) P or Pap pilus (PapG and PapE) and the major subunit, BfpA, of the type IV bundle forming pilus of enteropathogenic *E. coli* (EPEC) has been shown to induce the Cpx system as well.[24,73] Since both Pap and Bfp subunits are misfolded and/or mislocalized when expressed in the absence of other pilus assembly components, these studies together with those on NlpE, provide strong evidence that misfolded proteins are a major Cpx pathway activating signal. It has been demonstrated that the Cpx system is induced during P pilus assembly and it is hypothesized that this is a result of occasional pilus subunits failing to follow the normal assembly pathway and becoming misfolded.[24,74] Thus, activation of the Cpx pathway by misfolded pilus subunits may have a physiological role during pilus assembly (discussed later); however, the molecular nature of the inducing cue that is generated by misfolded proteins is not known.

Interestingly, PapG was shown to activate the $\sigma^E$ and Bae envelope stress responses as well; however, overproduction of PapE appears to specifically induce the Cpx pathway.[24,75,76] The specificity of PapE to induce only the Cpx system has been attributed to its N-terminal extension, as deletion of this motif abolished the ability of PapE overexpression to activate the Cpx response.[76] It is hypothesized that in the absence of the Pap chaperone, PapE adopts a specific conformation that induces CpxA and that this conformation requires the N-terminal extension. Thus, further analysis of Pap subunit folding intermediates is likely to lead to a better understanding of how misfolded proteins activate the Cpx response.

As both Pap and bundle forming pili play an essential role in adherence to host cells, it is interesting that NlpE has also been implicated in cell adherence. Otto and Silhavy demonstrated the link between NlpE and adherence,[7] as bacterial attachment to abiotic surfaces was shown to induce the Cpx response in an NlpE-dependent fashion. Furthermore, both *cpxR* and *nlpE* mutants exhibited defects in adhesion.[7] Together, the induction of the Cpx pathway in response to assembly intermediates of bacterial pili and adherence to abiotic surfaces highlights a possible role for the Cpx system in bacterial pathogenesis and biofilm formation (see below).

Elevated pH is another inducing cue of the Cpx stress response that has been demonstrated in both *Shigella* and *E. coli* species.[69,77] Strains carrying *cpx* null mutations are hypersensitive to alkaline pH and the expression of Cpx-regulated genes has been demonstrated to be induced by elevated pH.[69] In *Shigella*, mutations in *cpxA* altered gene expression of the virulence regulator, *virF*, in a pH-dependent manner.[77] Although the nature of the Cpx inducing cue generated at alkaline pH remains unknown, it seems likely that it involves denatured proteins, given that misfolded proteins are known to activate CpxA.[24,73]

Finally, two separate research groups identified alterations in the cell envelope that activated the Cpx system. Milkeykovskaya and Dowhan[78] first reported that the lack of phosphatidylethanolamine (PE) in the membrane resulted in the increased Cpx-dependent expression of the periplasmic protease *degP*. In another study by Silhavy and colleagues,[79] the accumulation of an intermediate of enterobacterial common antigen (ECA) synthesis resulted in the activation of the Cpx system.[79] In both cases, the alteration of the membrane composition stimulated the Cpx response. Again, it is not clear how changes in membrane composition activate CpxA. One possibility is that these changes alter protein folding in the envelope.

### *Cpx Signal Transduction*

In the presence of the above mentioned envelope stresses, the activation of the Cpx response is thought to be mediated by classic two-component pathway signal transduction mechanisms. Typical two-component phosphotransfer reactions between the sensor kinase CpxA and the response regulator CpxR control the Cpx envelope stress response[72,80] (Fig. 2). Inducing cues result in autophosphorylation of CpxA, likely at the conserved histidine residue.[72,80] Once phosphorylated, CpxA transfers the phosphate ion likely to the conserved aspartate residue on the cognate response regulator, CpxR.[72,80] CpxR is most closely related to the OmpR subfamily of response regulators that are classified as winged-helix-turn-helix DNA binding proteins.[81] Based on this homology, it is expected that CpxR~P stimulates transcription through direct contacts with the carboxy-terminal domain of the $\alpha$ subunit of RNA polymerase. In addition, experimental evidence indicates that CpxA also acts as a phosphatase of phosphorylated CpxR.[72,80] Furthermore, since constitutively activated *cpx* mutations diminished the phosphatase activity of CpxA, this indicates that the relative activity of the pathway is controlled by the ratio of CpxA kinase to phosphatase activities.[80]

Efforts to elucidate the inhibitory role of CpxP on the Cpx system have proved to be challenging and have consisted of numerous experimental approaches. Investigations of various gain-of-function CpxA* mutants revealed that deletion or point mutations in the periplasmic domain of CpxA resulted in the inability of CpxA to sense Cpx inducing cues.[80] The authors speculated that this domain of CpxA interacted with an inhibitory protein which would be titrated away in the presence of activating signals.[80] Follow up studies revealed that overproduction of CpxP inhibited the Cpx response and this inhibition was dependent on the periplasmic domain of CpxA, which suggested that CpxP interacted with CpxA.[70] In support of this hypothesis it was demonstrated that tethering CpxP to the inner membrane prevented full activation of the Cpx response by spheroplasting.[71] Further, Fleischer et al[72] recently showed that the addition of purified CpxP to reconstructed proteoliposomes containing CpxA lead to an inhibition of CpxA autokinase activity. Together, the cumulative data suggest that CpxP associates with CpxA in the periplasm and the CpxP:CpxA interaction has a negative regulatory effect on the Cpx envelope stress response.

To examine the mechanisms of CpxP inhibition, Buelow and Raivio performed a mutagenesis screen that identified several *cpxP* mutants that no longer could inhibit the Cpx pathway.[82] Examination of these *cpxP* mutants revealed mutations in a highly conserved N-terminal domain.[82] Some of these mutations resulted in a large decrease in CpxP stability that was restored by mutations of the periplasmic protease DegP.[82] This observation suggested that DegP might normally be involved with CpxP stability and Cpx signal transduction. In fact, both the Silhavy and Raivio research groups have now demonstrated the involvement of the periplasmic protease DegP with CpxP degradation.[82,83] Buelow and Raivio[82] demonstrated that the inducing cue of alkaline pH led to a DegP-dependent degradation of CpxP, while Isaac et al [83] showed that CpxP was required for the degradation of misfolded pilus proteins and also that CpxP degradation was enhanced under these conditions.[82,83] These experiments suggest that misfolded periplasmic proteins may associate with CpxP thus titrating CpxP away from CpxA and relieving CpxP-mediated inhibition of the Cpx response. CpxP could then bring the misfolded protein to DegP where both CpxP and the misfolded protein would be degraded. Another possibility is that conditions that lead to envelope protein misfolding also result in misfolding of CpxP, causing it to become a substrate for DegP.

It is interesting to note that high levels of CpxP do not completely shut off the Cpx pathway, nor does deletion of *cpxP* prevent increased activation of the Cpx system.[70,84] Thus, it is proposed that the role of CpxP is to help "fine tune" the Cpx pathway during envelope stress and/or to prevent inappropriate activation of the Cpx response.[82] The elucidation of the molecular mechanism of CpxP-mediated inhibition, its relief and its precise cellular role await further study.

### *The Cpx Regulon*

Initial studies of genes regulated by CpxR identified three genes: *dsbA,* which encodes for the major periplasmic disulfide oxidoreductase, *degP,* coding for the periplasmic protease/chaperone and a gene encoding a peptidyl-prolyl isomerase, *ppiA*.[85,86] These findings were not surprising as each gene product is involved with proper envelope protein folding (*dsbA*, *degP* and *ppiA*) or degradation (*degP*). Through analysis of the upstream region of all three genes, Pogliano and colleagues[86] proposed a consensus binding site for CpxR (5′-GTAAN$_{(6\text{-}7)}$GTAA-3′) which has aided other research groups in identifying more putative Cpx-regulated genes.

There are currently 41 genes in 25 operons that are proposed to make up the Cpx regulon (Table 3). Examination of these genes reveals functional subcategories. One class consists of genes directly involved with envelope maintenance such as the aforementioned *dsbA*, *degP* and *ppiA* genes, as well as *psd*, *secA* and *spy*. Mutations in the *dsbA* locus resulted in misfolded envelope proteins due to the lack of disulfide bond formation.[87,88] Identification of DegP protease activity was shown through analysis of a *degP* mutant which was defective in breakdown of several misfolded periplasmic proteins, but not cytoplasmic proteins.[89] Though not essential for bacterial survival, proper protein folding in the periplasm is also mediated by peptidyl-prolyl isomerases such as *ppiA* and *ppiD*.[90,91] Identification of *ppiD* as a Cpx-regulated gene was based on the presence of putative CpxR binding sites in its upstream region,[91,92] the ability of a putative phosphatase that controls the Cpx response to down-regulate *ppiD* expression and its increased transcription in the presence of an uncharacterized *cpx** mutation.[91] However, *ppiD* has also been demonstrated to be under the regulatory control of the classical heat shock sigma factor, $\sigma^H$, which regulates the expression of

***Table 3. List of genes and operons proposed to be under Cpx regulation***

| Gene(s) | Function | Proposed Cpx-Regulation[a] |
|---|---|---|
| **Envelope Protein Maintenance** | | |
| *degP* | Periplasmic serine endoprotease | Positive |
| *yihEdsbA* | Disulfide oxidoreductase | Positive |
| *ppiA* | Periplasmic peptidyl-prolyl isomerase A | Positive |
| *ppiD* | Periplasmic peptidyl-prolyl isomerase D | Positive |
| *secA* | Secretion subunit A | Positive |
| *psd* | Phosphatidyl serine decarboxylase | Positive |
| *spy* | Spheroplast protein Y | Positive |
| **Envelope Components** | | |
| *ompC* | Outer membrane protein C | Positive |
| *ompF* | Outer membrane protein F | Negative |
| *nanC* | N-acetylneuraminic acid (NAN) channel | Positive |
| *acrD* | Component of efflux pump | Positive |
| *mdtABCD* | Multi drug transporter | Positive |
| **Signal Transduction** | | |
| *cpxP* | Periplasmic proteolytic adapter protein, modulator of CpxA | Positive |
| *cpxRA* | Cpx two-component regulators | Positive |
| *rpoErseABC* | Sigma E and regulators | Negative |
| **Bacterial appendages (flagella, frimbriae, pilins) and chemotaxis** | | |
| *motABcheAW* | Motility and chemotaxis signal transduction proteins | Negative |
| *tsr* | Serine chemotaxis | Negative |
| *aer* | Aerotaxis receptor | Negative |
| *csgDEFG, csgBAC* | Curlin adhesin regulation and assembly | Negative |
| *papBA* | UPEC P pilus assembly and regulation | Negative |
| **Unrelated to envelope components or stress** | | |
| *ung* | Uracil-DNA glycosylase | Positive or negative |
| *aroK* | Shikimate kinase I | Positive or negative |
| *mviM* | Putative virulence factor | Positive |
| *htpX* | Heat shock protease | Positive |

a. For references of the putative Cpx regulation of these genes, see text.

cytoplasmic chaperones and proteases.[91] Though no experimental evidence exists, the upstream regulatory region of $\sigma^H$ is speculated to contain a CpxR binding site, which suggests that $\sigma^H$ may be under Cpx regulatory control.[86,92] Thus, further work needs to be done to determine whether the proposed Cpx-regulation of *ppiD* is direct or the result of the Cpx system influencing $\sigma^H$ levels. The *psd* and *secA* genes encode for proteins that affect biogenesis of phospholipids and proteins destined for the bacterial envelope. These genes were proposed to be Cpx-regulated based on a bioinformatics screen for a CpxR consensus binding logo.[92] *psd* encodes a membrane-bound phosphatidylserine decarboxylase that is involved in phospholipid biosynthesis.[93] SecA is the ATPase subunit of the Sec bacterial protein translocation machinery which governs protein export across the inner membrane. The *spy* gene, of unknown function, may also be included in this category since its deletion causes induction of the $\sigma^E$ response, suggesting a role in outer OMP biogenesis.[71] A *spy-lacZ* reporter was shown to be up-regulated by several Cpx-inducing conditions, including NlpE overexpression, spheroplasting and the presence of *cpxA** mutations.[71]

A second class of genes proposed to be Cpx-regulated includes genes encoding envelope proteins involved in transport, such as outer membrane proteins OmpC, OmpF and NanC and membrane channels involved with active efflux, including AcrD and MdtABCD.[94-96] CpxR has been shown to bind upstream of the *ompC, ompF, acrD* and *mdtA* promoters.[94,96] Additionally, mutations and conditions that activate the Cpx response lead to increased *ompC* and decreased *ompF* expression.[94] Deletion of *cpxR* diminished *nanC* expression, suggesting that this putative N-acetylneuraminic acid channel may be positively Cpx-regulated.[95] Interestingly, all genes listed in this class are strongly regulated by other regulatory pathways. The porin genes *ompC* and *ompF* have been shown to be under the control of the EnvZ-OmpR signal transduction system.[97] Various regulators including NanR, NagC and OmpR regulate the expression of *nanC*.[95,98] The genes encoding the drug transporters AcrD and MdtABCD are also regulated by another envelope stress response, the Bae system (see next section).[8,99,100] Further, the CpxR binding site has been shown to overlap those of OmpR upstream of *ompC* and *ompF* and those of BaeR upstream of the *acrD* and *mdtABCD* loci.[94,96] At the *acrD* and *mdtABCD* gene clusters, CpxR appears to function to enhance or facilitate BaeR-mediated gene expression, since inducible transcription is still seen in *cpx* mutants, although at a diminished level.[96] These data suggest CpxR may function as an accessory regulator at these loci.

Another class of Cpx-regulated genes consists of regulatory genes, encoding the Cpx and $\sigma^E$ signal transduction proteins. Either constitutively activated *cpx** mutations or elevated pH levels resulted in increased levels of CpxA, CpxR and CpxP, indicating autoregulation of the Cpx system.[69,70,101] Autoregulation is noted in other stress response systems such as in $\sigma^W$ and $\sigma^X$ in *B. subtilus* and the *E. coli* $\sigma^E$ stress response. This probably reflects a need for tight control of these stress responses when responding to potentially harmful environments (reviewed in ref. 5, 13). In addition to regulating the Cpx system, the Cpx response also negatively regulates the $\sigma^E$ response.[92,102] CpxR~P binds upstream of the *rpoErseABC* locus which encodes $\sigma^E$ and its regulators and mutations that inactivate or induce the Cpx response cause increased or decreased expression of the *rpoErseABC* gene cluster, respectively.[92,102] Presumably, this cross regulation between envelope stress signal transduction pathways reflects the need for the cell to coordinate its response to different envelope stresses.

A growing number of proposed Cpx-regulated genes can be grouped together as genes involved with the production or function of extracellular bacterial structures such as flagella or pili. These include the motility and chemotaxis gene clusters *motABcheAW*, *tsr* and *aer*, the *csg* locus encoding the curli adhesin, its assembly machinery and regulators and the *pap* gene cluster encoding the UPEC P pilus. Experimental evidence for the link between the Cpx response and motility is limited as the motility and chemotaxis genes were reported to be downregulated only in a mutant *cpx** background, while the deletion of *cpxR* had minimal effect on the transcription of *motABcheAW* and *tsr.*[101] Nevertheless, the deletion of *cpxR* results in increased motility,[101] thus suggesting an inhibitory role of the Cpx response system on motility. As the regulation of motility, chemotaxis and flagellar components is governed by the alternative sigma factor, $\sigma^F$, it remains unclear whether

Cpx exerts its regulatory effect directly on these motility and chemotaxis genes, or acts on genes higher up in the flagellar regulation hierarchy. Alternatively, the effects of the Cpx response on motility may also be partly due to posttranscriptional effects. Curli are extracellular proteinaceous fibers important for biofilm formation and have also been proposed to be negatively regulated by the Cpx response, however only under certain growth conditions.[103] Using electrophoretic mobility shift assays, it was shown that CpxR, in concert with OmpR, binds to the upstream region of the *csgDEFG* operon which encodes the curli biosynthesis regulator and transport genes.[104] In a *cpxR* mutant grown in high salt, the expression of the *csgD* reporter is increased, suggesting that CpxR may be involved in repression of the *csgDEFG* operon under certain conditions.[104] However the Cpx regulatory effect was mild (~2 fold difference) and only noted in specific media conditions, thus it may be that the Cpx response regulates curli production only under specifically defined experimental conditions. A direct regulatory link between the Cpx response and P-pili has also been demonstrated.[74,105] Analysis of the upstream regions of the *pap* operons revealed a CpxR binding site which overlaps with that of another regulatory protein, Lrp. Phosphorylated CpxR competes with Lrp for binding to certain sites within this *pap* switch region that directs phase-variable expression of the *pap* operon, resulting in a decrease in in vivo production of P pili.[105] Interestingly, all of the genes encoding these extracellular appendages appear to be negatively regulated by the Cpx response. It is tempting to speculate that these genes have evolved to be under Cpx control because it is desirable for the cell to shut down production of extracellular envelope-localized structures when envelope stresses causing misfolding of proteins are present.

The remaining genes (*ung*, *aroK*, *htpX* and *mviM*) proposed to be part of the Cpx regulon do not involve envelope structure, maintenance, or regulation and thus the rationale for their regulation by the Cpx stress response is unclear. The majority of these genes (*ung*, *aroK* and *mviM*) were identified to be Cpx-regulated from a genome-wide bioinformatics screen for CpxR consensus binding sites.[92] The *ung* gene encodes uracil-DNA glycosylase and is involved in DNA repair. While DeWulf and colleagues[92] showed that *ung* transcript levels were reduced in a *cpxR* null background and elevated in the presence of a constitutively active *cpxA** mutation, Ogasawara et al[106] showed that CpxR~P bound upstream of the *ung* gene and apparently repressed transcription, leading to reduced Ung enzymatic activity. The experiments were performed in different strain backgrounds and so it may be that this is the explanation for the contradictory results. Differences in Cpx-regulation of genes have also been noted with the *aroK* gene that encodes a shikimate kinase that is involved in aromatic amino acid metabolism. DeWulf et al[92] have suggested *aroK* is positively regulated by the Cpx response, however using a different strain background, Price and Raivio[102] showed *aroK* to be negatively regulated by the Cpx response. *mviM* encodes a putative virulence factor and is proposed to be positively regulated by the Cpx response.[92] The cytosolic heat shock protease, HtpX, was suggested to be under Cpx-regulatory control by another research group which showed that a *htpX-lacZ* fusion was up-regulated in the presence of a *cpxA** mutation and down-regulated by ablation of *cpxR*.[107] Since these genes encode proteins with an array of functions including cytosolic proteolysis (*htpX*), DNA and amino acid metabolism (*ung*, *aroK*) and virulence (*mviM*), it is not clear what their cellular role is as part of the Cpx regulon. A detailed analysis of how and when the Cpx system exerts its regulatory effects on these genes is required to properly confirm the role of each gene in the Cpx regulon. It may be that the Cpx response plays cytoplasmic roles in metabolism and protein quality control that have yet to be fully described.

### *Cellular Role of the Cpx Response*

The Cpx system is proposed to regulate various genes with different functions not exclusively limited to envelope stress. In efforts to clarify the cellular role of the Cpx response, Price and Raivio[102] assayed the expression of the majority of the proposed Cpx regulon members in several *cpx* genetic backgrounds and in response to overexpression of NlpE, an inducer of the wild-type Cpx signal transduction pathway. The authors demonstrated a core group of genes that were strongly regulated by the Cpx response. The majority of these genes (*degP*, *dsbA*, *cpxP*, *spy*, *ompF*, *rpoErseABC* and *psd)* are directly involved with envelope protein folding and therefore suggest the primary role of the Cpx response is as an envelope stress response.[102]

Stemming from its role in envelope maintenance, the Cpx response is also likely to be involved in bacterial pathogenesis since it affects bacterial adherence via NlpE and also pilus assembly which is necessary for host-cell attachment.[7,24,73,74] In addition, the Cpx system itself has been shown to be important for bacterial pathogenesis in several pathogens. Mutations in *cpxA* homologues render pathogenic *Salmonellae* incapable of attaching to and invading host cells.[108,109] In *Shigella* spp., the Cpx signal transduction system plays an essential role in regulation of the *virF* and *invE* activators of virulence determinant expression.[110-112] Several studies have provided indirect evidence to support the theory that the Cpx pathway plays a crucial role in *E. coli* pathogenesis. Null mutations in *cpxR* resulted in a Pap pilus assembly defect in laboratory strains of *E. coli* carrying a plasmid encoding all of the P pilus assembly genes, resulting in shorter pili when compared to wild-type strains.[74] In addition, a *cpxR* null mutation in enteropathogenic *E. coli* (EPEC) resulted in a decrease in the production of bundle forming pili and localized adherence, the first step in EPEC pathogenesis.[73] Furthermore, both Cpx-regulated folding factor DsbA and protease DegP have been shown to be essential for bacterial virulence in several pathogenic species including *Pseudomonas*, EPEC, *Vibrio*, *Salmonella*, *Shigella*, *Yersinia*, and *Bordetella*.[113-121] DegP and DsbA are thought to help maintain proper function of various envelope-localized virulence determinants such as pili and type III secretion apparati. On whole, the Cpx system appears to play a role in bacterial pathogenesis, however, questions remain as to whether the Cpx response is directly involved with regulating virulence factor gene transcription, or rather indirectly involved through mediating the expression of folding factors and proteases that ensure proper function of these virulence determinants.

## The Bae Two-Component System

The BaeSR two-component system was the third signal transduction pathway to be officially classified as an envelope stress response in *Escherichia coli*. Based on the limited information available on the Bae pathway, it appears that its main function is to upregulate efflux pumps in response to toxic compounds. Here we will discuss what is presently known about Bae while highlighting important areas requiring further research.

### *Discovery and Classification as an Envelope Stress Response*

Characterization of Cpx regulon member *spy* revealed that induction of this gene during spheroplast formation was only partially dependent on a functional Cpx pathway.[71] In an effort to identify the coregulator of *spy*, Raffa and Raivio[75] performed transposon mutagenesis on a strain carrying a *spy::lacZ* fusion. This approach generated a single insertional mutant that dramatically increased *spy* expression. The mutant was determined to have a BaeS gain-of-function mutation and was capable of activating *spy* transcription independent of Cpx. The *baeS* gene had previously been identified for its ability to complement mutations in the histidine kinase genes *envZ*, *creC* and *phoR*.[122] Sequence analysis revealed that BaeS is itself a putative histidine kinase, encoded in an operon with a proposed response regulator, BaeR. Together they constitute the BaeSR (bacterial adaptive response) two-component system.[122] Further investigation into the link between the BaeSR two-component system and the Cpx envelope stress response revealed that both pathways were capable of activating *spy* expression in response to a variety of envelope stresses. In addition, a *baeR cpxR* double mutant was more sensitive to envelope perturbations than either single mutant.[75] While these findings suggest significant overlap between the pathways, it should be noted that BaeSR was shown to have no effect on any other Cpx regulated gene and that not every Cpx inducing cue tested was an effective signal for Bae pathway activation. Based on these findings the authors classified the Bae pathway as a distinct envelope stress response system.

### *Conquering Stress via Active Efflux*

In a screen performed by Baranova and Nikaido,[8] *baeR* overexpression was found to increase resistance of a hypersensitive *acrAB* mutant to novobiocin. The increased resistance was attributed to the fact that cells overexpressing BaeR failed to accumulate the drug. Upon closer inspection of the region upstream of the *baeSR* locus, the authors discovered that these genes were in fact part of a larger operon encoding the *yegMNOB* multi-drug efflux system.[8,123] After demonstrating that BaeR-mediated activation of *yegM* was responsible for the aforementioned phenotype, the operon was renamed *mdtABCD,* where *mdt* denotes multi-drug transporter.[8] The following year, the Yamaguchi group demonstrated that BaeR overexpression confers increased levels of resistance to β-lactam antibiotics, deoxycholate and low levels of SDS.[100,124] The BaeR-mediated resistance to β-lactams was due to an increase in expression of *mdtABC,* as well as an additional multidrug exporter gene, *acrD.*[100] Further investigation into *mdtA* and *acrD* regulation revealed that BaeR mediated activation of these genes is enhanced when CpxR binds multiple sites upstream of the BaeR binding sites in the *mdtA* and *acrD* promoter regions.[96] This modulation of Bae regulation by the Cpx pathway represents an example of how these two envelope stress response systems cooperatively combat envelope stress. Together, these findings suggest that one way in which the BaeSR two-component system fulfills its function as an envelope stress response system is through upregulation of genes involved in active efflux.

### *Present Model: Questions Still Remaining*

The BaeSR signal transduction pathway is thought to be a prototypical two-component system (Fig. 3). Phosphotransfer from the conserved H1 domain of BaeS to the conserved D1 domain of BaeR has been demonstrated in vitro in the presence of ATP.[125] Based on high homology between the periplasmic inhibitory protein, CpxP and Spy, there has been speculation as to whether Spy inhibits BaeS activity in a similar manner to how CpxP inhibits CpxA[71] (see previous section). This idea is attractive as it would reveal a common mechanism by which envelope stress response systems modulate activity in the absence of an inducing cue. To date, there is no experimental evidence in support of this model.

Activating cues that initiate BaeSR signal transduction include general envelope stresses such as spheroplast formation, overexpression of misfolded PapG pilus subunits and the presence of indole.[75] Increased sensitivity of *baeSR* mutants to myricetin, gallic acid, nickel chloride and sodium tungstate implies that the pathway may also respond to the presence of these antimicrobial compounds.[9] All of the aforementioned inducing cues elicit a cellular response that extends beyond the Bae pathway. For example, overexpression of PapG activates the $\sigma^E$, Cpx and Bae signaling pathways; while indole and spheroplast formation activate the Cpx and Bae pathways.[24,75] The identification of more specific inducers could reveal additional functions of the pathway and lead to the identification of new regulon members.

As it stands, the Bae regulon consists of a small but diverse group of genes not limited to those discussed above (*mdtABCD, acrD* and *spy*). Researchers have employed both genetic screens and expression profiling in an effort to complete the regulon. Baranova and Nikaido[8] screened a library of short genomic fragments for promoters that responded to BaeR overproduction. This approach led to the identification of two reputed regulon members, *ygcL* and *yicO,* encoding a protein of unknown function and a putative membrane transport protein respectively. Nishino et al[126] performed microarray analysis of an *E. coli* strain overexpressing BaeR, a *baeSR* deletion mutant and wild type *E. coli* cells exposed to indole. Using the three expression data sets the authors identified *mdtABCD, acrD, spy* and *ycaC* as direct targets of BaeR regulation. The only new regulon member to be uncovered, *ycaC,* encodes a putative membrane localized hydrolase.[126] The fact that the genetic screen and the microarray analyses exposed different genes as possible regulon members suggests that neither approach was exhaustive. This, combined with the knowledge that the more characterized envelope stress responses have extensive regulons, suggests that there are other BaeR regulated genes waiting to be identified.

## The Phage-Shock-Protein (Psp) Response

The Phage shock protein response was first discovered by Peter Model's group in *E. coli.*[127] The consensus within the field is that this unique pathway is upregulated in response to dissipation of the proton-motive force and the mislocalization of secretin proteins. In the following sections I

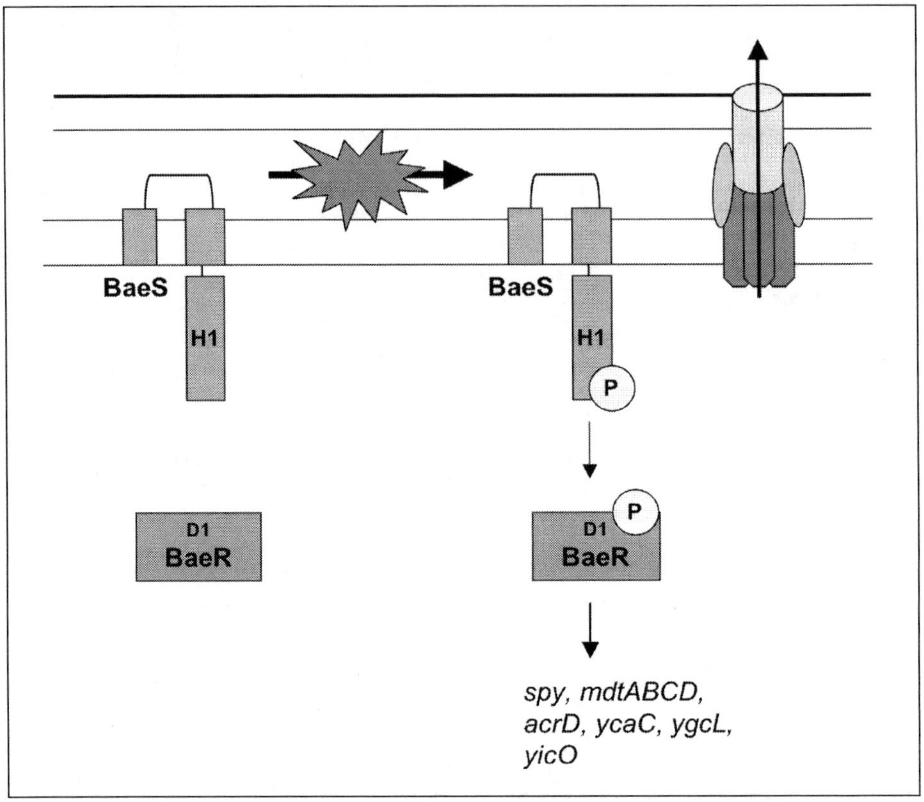


Figure 3. The BaeSR two-component system. Inducing cues such as general extracytoplasmic stresses and the presence of toxic compounds are sensed by the histidine sensor kinase, BaeS. This initiates autophosphorylation and subsequent phosphoryl transfer from BaeS to the cytoplasmic response regulator BaeR. BaeR~P functions as a transcriptional regulator activating expression of genes involved in active efflux and unknown functions. H1= conserved histidine, D1= conserved aspartate

will discuss the working model for Psp signal transduction with emphasis on its predicted activating signals.

## *Discovery of the Phage Shock Response*

The first phage shock protein (PspA) was identified during an examination of cellular proteins from *E. coli* cells infected with filamentous phage.[127] Researchers determined that the elevated level of PspA found in infected cells was a direct result of synthesis of the integral membrane protein encoded by phage gene IV. The phage gene IV protein (pIV) is required at high levels during infection to facilitate virus assembly.[128] Significant amounts of the mature form of this secretin protein become mislocalized during an infection, as the pilot protein that normally targets the secretin to the outer membrane is not encoded on the phage.[128] It is not clear why mislocalized secretin protein specifically induces PspA expression; however, generalized perturbation of the inner and outer membranes was ruled out as the signal.[127] Previous work with λ bacteriophage showed that many of the *E. coli* proteins synthesized during an infection are also induced by heat shock.[129] While exposure to extreme heat shock was found to increase PspA expression, it is not a classic heat shock protein as its expression is not regulated by $\sigma^{32}$.[127] Other extracytoplasmic stresses shown to induce PspA expression include exposure to ethanol and hyperosmotic shock.[127]

The observation that Psp expression is induced upon exposure to these general environmental stresses led researchers to categorize *pspA* as a stress responsive gene, in the same class as $\sigma^E$ and $\sigma^{32}$ regulated genes.[130]

## *Working Model of Psp Signal Transduction*

Examination of the newly identified locus encoding *pspA* in *E. coli* uncovered an operon encoding five genes, *pspA, B, C, D* and *E*.[131] Characterization of these genes revealed that *pspA* encodes a negative regulator of the *psp* operon, while the *pspB* and *pspC* gene products work together to activate *psp* expression.[130] The *pspE* gene is transcribed as part of the operon under inducing conditions and from its own promoter under non-inducing conditions.[131] For more than a decade, *pspD* was classified as a hypothetical gene.[131] Recently, it was demonstrated that *pspD* is in fact expressed in vivo and that it localizes to the inner membrane.[132] No function has been assigned to either PspD or PspE.

The entire *psp* operon is subject to positive regulation by $\sigma^{54}$, with a subset of inducers requiring the presence of IHF for full activation.[130,133] Constitutive activation of the *psp* operon in the absence of the negative regulator PspA and positive regulators PspB and PspC suggested the involvement of an additional positive regulator present in the cell under non-inducing conditions.[130] Transposon mutagenesis on a strain lacking the *pspA, B* and *C* genes led to the identification of the $\sigma^{54}$ transcriptional activator, PspF.[134] PspF belongs to the enhancer-binding family of proteins and is encoded immediately upstream of the *psp* operon in the opposite orientation.[134] In the absence of stress, PspA interacts with the ATPase domain of PspF and inhibits $\sigma^{54}$-mediated transcription.[135,136] Interaction between PspF and PspA has been demonstrated in vivo using the bacterial two-hybrid system.[137] Elderkin et al[137] went as far as to identify the exact residue in PspF that is required for PspA binding and subsequent inhibition.

PspA fractionates with both inner membrane and cytoplasmic proteins, suggesting it can exist as both a peripheral inner membrane protein and as a soluble protein.[131,138] Researchers investigating the ability of PspA to interact specifically with other phage shock proteins were able to crosslink PspA to PspB and PspC.[132] Interestingly, the interaction between PspA and PspB could only be demonstrated if the entire *psp* operon was overexpressed suggesting another Psp protein is needed. Based on the essentiality of *pspB* and *pspC* for the induction of the *E. coli* Psp response by certain inducers, it has been proposed that PspB and PspC interact upon presentation of an activating signal and that this interaction promotes binding to PspA.[130,139] While an interaction between PspB and PspC has not been demonstrated in *E. coli*, it has been shown for the *Yersinia enterocolitica* Psp response[140] (Fig. 4).

Protein-protein interactions appear to be the main mechanism by which Psp signal transduction is achieved. More work is required to determine the dynamics of these interactions under inducing and non-inducing conditions. Many questions remain concerning what takes place upon presentation of an inducing cue. As such, several aspects of the working model for Psp signal transduction are speculative and require further investigation (Fig. 4).

## *Determining Functionality by Means of Activating Signals*

As mentioned above, the original inducers of PspA expression identified were the mislocalization of a phage encoded secretin, pIV and several general extracytoplasmic stresses.[127] Shortly after the Psp response was identified, two separate groups studying protein export in *E. coli* isolated transport defective mutants capable of inducing PspA synthesis.[138,141] Mutation to the signal sequence of the outer membrane porin protein, LamB, causes aberrant localization of the protein. It was hypothesized that elevated levels of PspA functioned to counteract the deleterious effects of this toxic mutation. The authors tested the relevant phenotypes in a Δ*pspA-C::kan* background and found that this was not the case.[141] A separate study found that a collection of precursor PhoE mutants, for which normal biogenesis of this outer membrane protein is defective, activated PspA synthesis in a manner that was dependent on the Sec export system.[138] The requirement of the general secretory pathway for Psp induction suggested that entrance into the export system was generating the activating signal. Work showing that PspA synthesis was activated by addition of

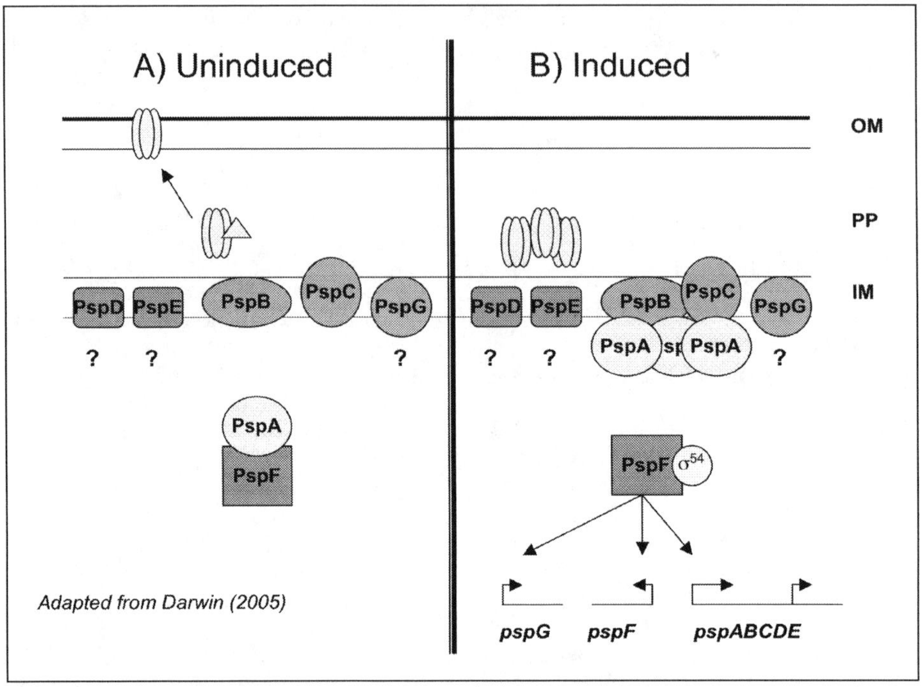


Figure 4. Proposed model for the *E. coli* Phage shock protein response. A) Under normal conditions PspA binds PspF in the cytoplasm and inhibits $\sigma^{54}$ mediated transcription of the Psp regulon. B) Inducing cues such as mislocalization of secretins or mutations that reduce membrane potential are sensed by PspB and/or PspC in the inner membrane. Interaction between PspB and PspC somehow enables or promotes PspA binding to PspB/C and the subsequent release of PspF. Free PspF initiates *psp* transcription. PspA accumulates at the cytoplasmic face of the inner membrane where it is thought to maintain the PMF. The functions of Psp regulon members PspG, PspD and PapE are unknown.

the proton ionophore, carbonylcyanide chorophenylhydrazone (CCCP),[142] led researchers to investigate the role of PspA in proton motive force (PMF) dependent translocation. They found that PspA was not required for efficient prePhoE transport by a PMF-independent mechanism but was required for PMF-dependent Sec transport, providing the first experimental evidence that the Psp system might respond to dissipation of the proton motive force.[139]

In line with the above observations, the Psp response has also been shown to respond to defects in the Sec secretion system and in YidC, a protein believed to be involved in clearing the Sec translocation channel.[143] YidC depletion was only found to affect secretion when Sec-dependent proteins were overexpressed.[143] This suggests that the Psp response is not required for efficient secretion unless the translocation apparatus becomes saturated. In support of this, PspA has been shown to enhance the transport efficiency of proteins targeted to the periplasm via the Tat pathway when the Tat system becomes saturated, but is not needed for normal Tat pathway function.[144] In both cases it has been proposed that saturation of the translocation apparatus somehow modulates the PMF, which in turn activates the Psp response. While direct activation of the Psp response by PMF dissipation has not been demonstrated experimentally, there are examples where mutations causing a decrease in membrane potential have detrimental effects in the absence of *pspA*.[145,146] It is not clear whether the Psp response is responding specifically to a reduction in membrane potential or simply sensing downstream effects of PMF dissipation.

A physiological role for the Phage shock protein response has been demonstrated in organisms other than *E. coli*. In *Salmonella enterica* serovar Typhimurium, Δ*rpoE* cells in stationary phase growth have elevated PspA levels and an *rpoE pspA* double mutant has reduced survival in stationary phase compared to a single mutant.[145] A decrease in PMF in the double mutant suggests that both genes are involved in PMF maintenance and that PspA can compensate for $\sigma^E$.[145] In *Yersinia enterocolitica* a *pspC* mutant is severely attenuated for survival in an animal model, due to YscC secretin production.[147,148] YscC forms the outer membrane channel through which type three secretion substrates are translocated. The Psp response is activated during assembly of a functional type three secretion system in this organism only when YscC is expressed.[148] Thus, in conjunction with, or as a result of, its role in monitoring membrane potential and secretin mislocalization the Psp response facilitates Gram-negative pathogenesis.

Transcriptional profiling of *E. coli* cells overexpressing the main effectors of the Psp response, PspA and the newest member of the pathway PspG,[149] revealed that upon induction of the Psp response cells adjust respiratory gene expression and reduce the expression of high energy, PMF consuming, processes such as motility.[150] This supports a model whereby the Phage shock protein response senses a variety of envelope stresses and responds by activating expression of PspA and PspG, which function to facilitate protein transport and restore membrane potential by downregulating energetically costly cellular events.

## The Rcs Phosphorelay System

The Rcs pathway was first identified during a screen for genes that regulate *E. coli* colanic acid capsular biosynthesis in the mid-1980s.[151] To date, this complex phosphorelay system has been shown to regulate genes associated with envelope integrity and structure and is believed to play a pivotal role in biofilm maturity and enteric pathogenesis. In the following sections we will review what is currently known about the Rcs pathway while clearly demonstrating why it has recently been classified as an extracytoplasmic stress response system.

### *Discovery and Early History of the Rcs Pathway*

The genes regulating capsular biosynthesis in *E. coli* K-12 are encoded by the *cps* operon and are expressed at relatively low levels under standard laboratory conditions. The observation that a mutation in the *lon* locus of *E. coli* results in the overproduction of the colanic acid capsular polysaccharide sent researchers on the hunt for genes involved in the biosynthesis and regulation of this cellular structure. Trisler and Gottesman[152] used insertional mutagenesis in an effort to identify genes responsible for the mucoidy phenotype seen in *lon*⁻ cells. In addition to identifying the six genes encoded in the *cps* operon, they also demonstrated that it is regulated by the ATP-dependent protease, Lon, at the transcriptional level.[152] One year later, the same group identified and characterized three new regulatory genes that affected the expression of several *cps::lac* fusions.[151] The genes were named *rcsA, B* and *C* (for regulator of capsular synthesis), where RcsA and RcsB were classified as positive regulators of *cps* and RcsC was classified (with Lon) as a negative regulator of *cps* expression.[151]

Genetic analysis of *rcsA* in the presence and absence of *lon* revealed that RcsA is targeted for proteolytic degradation by Lon and is a limiting factor in *cps* expression.[153] It was soon demonstrated that RcsB and RcsC can drive *cps* expression independent of RcsA revealing two modes of *cps* regulation.[154] Sequence homology of RcsB and RcsC to environmentally responsive two-component regulators led to the model whereby RcsC functions as a membrane bound sensor kinase and RcsB as a cytoplasmic response regulator.[155] Over the next decade researchers discovered the Rcs pathway was not a typical two-component system dedicated to capsule synthesis, but a unique phosphorelay encompassing numerous accessory proteins and an extensive regulon.

### *Rcs Phosphorelay Involves Auxiliary Regulators and Accessory Proteins*

The RcsC sensor kinase is unique in that it possesses an additional domain at its C-terminal end that shares homology to the N-terminal receiver domain of a response regulator.[155] It was speculated that this RcsC D1 domain facilitated RcsB D2 phosphorylation by mediating transfer of the phosphoryl group from the H1 domain of RcsC.[155] Further investigation into the proposed Rcs His-Asp-Asp phosphorelay led to the discovery of a unique phosphotransmitter encoded downstream of *rcsB*, YojN.[156] YojN, later renamed RcsD, contains an Hpt (histidine containing phosphotransfer) domain in its C-terminus that is required to transmit signals between RcsC and RcsB.[156] RcsD is a predicted integral membrane protein with a large periplasmic domain that shares limited homology to that of RcsC.[156,157] It has been suggested that RcsD receives signals independent of RcsC; however, no such signal has been uncovered.[12] It is now generally accepted that the RcsC-RcsD-RcsB pathway is a multi-step His-Asp-His-Asp phosphorelay concluding with RcsB~P binding DNA and altering transcription of target genes (Fig. 5).

RcsB is a member of the FixJ subfamily of response regulators and has been shown to function alone or in concert with auxiliary regulators. When acting alone, RcsB binds as a homodimer immediately upstream of the –35 region of the RNA polymerase binding site.[158,159] At a subset of promoters RcsB binds as a heterodimer with RcsA, at a conserved core sequence referred to as the RcsAB box (located 50-100 nucleotides upstream of the transcriptional start site).[160,161] RcsB and RcsA both contain a LuxR-type C-terminal DNA binding motif and function as positive regulators

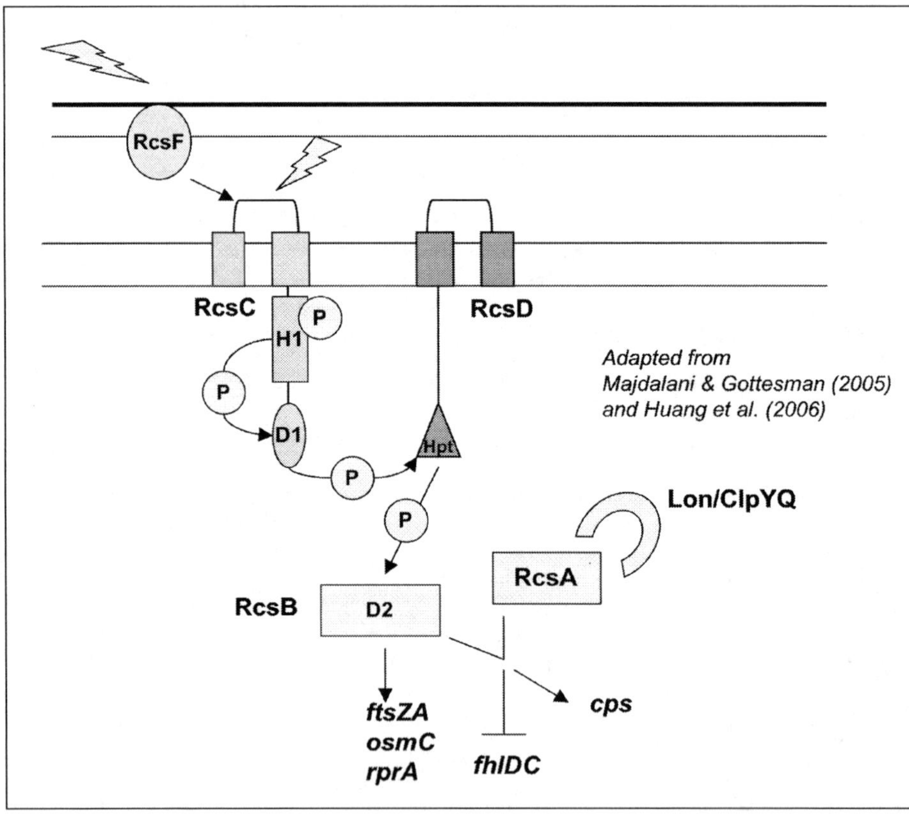


Figure 5. Proposed model for the *E. coli* Rcs phosphorelay system. Activating signals enter the Rcs pathway at the sensor kinase RcsC or at the outer membrane lipoprotein RcsF. RcsC autophosphorylates at an H1 domain and transfers the phosphate to a D1 domain in its C-terminus. The phosphate is then transferred to the Hpt domain of RcsD and finally to the D2 domain in the N-terminal end of the response regulator RcsB. RcsB~P works alone or in concert with RcsA to regulate transcription of the Rcs regulon. RcsA is an unstable protein that is proteolytically degraded by Lon and ClpYQ.

of colanic acid synthesis and *rcsA* autoregulation.[151,162] RcsA stability is controlled by the proteases Lon and ClpYQ.[153,163] RcsA-RcsB heterodimerization is not required for RcsB recognition of the RcsAB box, but significantly stabilizes the RcsB/DNA interaction.[164] It has been proposed that RcsB binding alone at these promoters allows for basal levels of activity, while the presence of the co-inducer RcsA enhances transcription in response to an inducing cue.[164] Another example of this type of cooperative action involves the TviA protein of Vi antigen synthesis in *Salmonella typhi.*[165,166] It is thought that the use of an auxiliary regulator to modulate RcsB activity allows the pathway to differentially control expression of regulon members in response to varied activating cues.

In addition to auxiliary regulators the Rcs phosphorelay also utilizes accessory proteins in order to carry out signal transduction. RcsF was first identified for its ability to activate exopolysaccharide synthesis when overproduced on a multi-copy plasmid.[167] Due to the loss of this function in an *rcsB* mutant, the authors proposed that RcsF was involved in RcsB phosphorylation. It was later determined through epitasis experiments that RcsF is positioned upstream of RcsC in the signaling cascade and that it is involved in signal transduction rather than direct phosphorylation.[168] Evidence showing that RcsF is an outer membrane lipoprotein orientated towards the periplasm supports a model whereby RcsF senses surface cues and transmits the signal to RcsC[168,169] (Fig. 5). It should be noted that not all Rcs inducing cues require RcsF. Overproduction of the DnaJ-like protein, DjlA, activates the Rcs phosphorelay independent of RcsF, suggesting there are two distinct signaling pathways towards RcsC.[169] The use of an outer membrane lipoprotein to transmit signals to a sensor kinase is not unique to this system. As mentioned earlier in this chapter, NlpE transmits a signal to the Cpx pathway upon adherence to hydrophobic surfaces.[7] It appears that lipoproteins may play a vital role in signal transduction.

Another accessory protein shown to influence the pathway is the putative inner membrane protein IgaA of *Salmonella enterica.* IgaA negatively regulates the RcsC-RcsD-RcsB system upon host colonization thereby facilitating *Salmonella* virulence.[170,171] This topic will be discussed further in a later section.

### *Activating Signals*

A commonality among the mutations and general environmental stresses known to activate the Rcs phosphorelay is an association with outer membrane structures and integrity. The majority of the mutations shown to activate Rcs, or the expression of Rcs-regulated genes, are involved in the production of cellular or extracellular polysaccharides. They include *rfa, mdo, tol, pmr* and *surA.* Collectively these genes play major roles in LPS biosynthesis (*rfa*), membrane derived oligosaccharide production (*mdo*), colanic acid synthesis (*tol, pmr*) and proper folding of envelope proteins (*surA*).[169,172-176] In addition to these mutations, the Rcs pathway is also activated by overproduction of certain envelope proteins. As previously mentioned, overproduction of accessory protein RcsF and the membrane anchored chaperone DjlA, activates *cps* expression through RcsC/B.[167,177] The exact signal generated by these proteins is not well understood, however, overproduction of envelope proteins seems to be a common inducer of envelope stress response systems. The environmental conditions shown to influence the Rcs regulon include osmotic shock, dessication, growth on solid media, as well as exposure to the cationic drug chlorpromazine and a subset of β-lactams.[178-183] Each of the aforementioned inducers cause serious stress to the bacterial cell envelope and elicit a response that extends beyond the Rcs pathway. For a more detailed review of this topic see references 11 and 12.

### *The Rcs Regulon: Determining the Physiological Role of the Pathway*

The Rcs regulon further supports the claim that this pathway is somehow involved in monitoring and maintaining cell surface structures. Historically, the Rcs phosphorelay is known to positively regulate colanic acid synthesis in *E. coli* K-12.[151] In line with this finding, the Rcs regulatory proteins RcsB and RcsA have been shown to positively regulate exopolysaccharide production in other organisms, including the plant pathogens *Erwinia amylovora* and *Pantoea stewartii,* as well as the enteric pathogen *Salmonella typhimurim.*[161,184,185] RcsB/RcsA also activate transcription of the *ugd* gene in *Salmonella.*[174,175] While this gene does not encode a component of a capsular structure it does encode UDP-glucose dehydrogenase, an enzyme required for production of a colanic acid sugar.[186] As previously mentioned, RcsB does not just work in concert with RcsA. In *Salmonella typhi,* RcsB interacts with another positive regulator, TviA, at the *tviA* promoter to initiate Vi antigen production.[165] More coregulation was found between Rcs and the PmrAB two-component system in *Salmonella.*[187] Researchers determined that both systems independently enhance transcription of the *wzz* gene from overlapping regions of its promoter.[187] Wzz is responsible for determining O-antigen chain length in *Salmonella enterica serovar* Typhimurium and so by affecting it expression, these pathways affect susceptibility to host immune responses.[187]

Other positively regulated genes in the Rcs regulon include *ftsA, ftsZ* and *osmC.*[158,188,189] The significance of the Rcs pathway regulating genes involved in cell division and osmoregulation is not clear at this point. A screen for *E. coli* promoters regulated by RcsB uncovered the first example of negative regulation by the Rcs response. Francez-Charlot et al[190] demonstrated that an RcsB-A heterodimer binds to an RcsAB box in the *flhDC* promoter and acts as a negative regulator of flagellin synthesis. A few years later the Dorel group showed that RcsB-A regulates the *csgDEFG* and *csgBA* operons.[191] The negative regulation of flagellin and curlin expression represents a clear overlap between Rcs, the Cpx pathway and the phage-shock-protein response regulons (refer to relevant sections in this chapter). It has also led to the suggestion that the Rcs pathway plays an important role in the transition from early to late biofilm formation, as it down-regulates motility genes and upregulates colanic acid, which functions in biofilm maturity.[11,182,192] The finding that an *rcsB* mutant elaborates premature flagella and an *rcsC* mutant takes longer to display flagella supports this idea.[193] This and other findings also provide support for RcsC phophatase activity.[168,193-196]

It has been demonstrated that the Rcs phosphorelay regulates the stationary phase sigma factor (RpoS) at the translational level.[197] This is accomplished by repressing the synthesis of LrhA, a repressor of the sRNA, RprA, which activates RpoS translation; as well as by directly stimulating RprA activity.[197,198] In addition to affecting RpoS expression, there is overlap between the Rcs and RpoS regulons. One such example is the newly identified periplasmic protein *ydel.*[157] It was observed that *rcsC, B, D* single mutants were significantly out competed by wild-type cells in a mouse model of infection; *ydel* was identified as the Rcs-regulated gene responsible for the persistence of *Salmonella* in mice.[157] The authors speculate that *ydel* somehow increases *Salmonella* resistance to antimicrobial peptides.[157] The Rcs pathway has also been shown to affect pathogenesis in other Gram-negative pathogens. The RcsC-RcsD-RcsB signaling system in enterohaemorrhagic *Escherichia coli* positively and negatively regulates type three secretion by acting through two opposing regulators, GrvA and PchA.[199] In *Yersinia enterocolitica* Biovar 1B, the Rcs phosphorelay positively regulates secretion through the Ysa type three secretion system by affecting expression of genes encoded on the Ysa pathogenicity island.[200]

Since the Rcs phosphorelay seems to play such an integral role in Gram-negative bacterial pathogenesis, it is not surprising that an organism would adapt a way to regulate Rcs activity. As mentioned previously, the IgaA membrane protein of *Salmonella* negatively regulates RcsBC upon host colonization.[170,171] IgaA is only required if the Rcs pathway is active and has proven to be required for *Salmonella* virulence.[170,171] It seems a common feature of envelope stress response systems is to regulate genes important for certain aspects of virulence, but when fully activated, to attenuate virulence.[109,111,201]

In summary, the Rcs regulon suggests that this complex signal transduction pathway is involved in envelope integrity and structure and suggests an important role for the Rcs system in biofilm maturity and Gram-negative pathogenesis.

### *Acknowledgements*

The authors gratefully acknowledge support from the Alberta Heritage Foundation for Medical Research, the Canadian Institutes of Health Research and the Natural Sciences and Engineering Research Foundation of Canada.

## References


1. Nikaido H, Vaara M. Outer membrane. In: Neidhardt FC et al, eds. Escherichia coli and Salmonella: cellular and molecular biology. Washington, DC: ASM Press, 1987:7-22.
2. Cronan JE, Gennis RB, Maloy SR. Cytoplasmic membrane. In: Neidhardt FC et al, eds. Escherichia coli and Salmonella: cellular and molecular biology. Washington, DC: ASM Press, 1987:31-55.
3. Mullineaux CW, Nenninger A, Ray N et al. Diffusion of green fluorescent protein in three cell environments in Escherichia coli. J Bacteriol 2006; 188:3442-3448.
4. Oliver DB. Periplasm and protein secretion. In: Neidhardt FC et al, eds. Escherichia coli and Salmonella: cellular and molecular biology. Washington, DC: ASM Press, 1987:56-69.
5. Alba BM, Gross CA. Regulation of the Escherichia coli sigma-dependent envelope stress response. Mol Microbiol 2004; 52:613-619.
6. Raivio TL, Silhavy TJ. Periplasmic stress and ECF sigma factors. Annu Rev Microbiol 2001; 55:591-624.
7. Otto K, Silhavy TJ. Surface sensing and adhesion of Escherichia coli controlled by the Cpx-signaling pathway. Proc Natl Acad Sci USA 2002; 99:2287-2292.
8. Baranova N, Nikaido H. The BaeSR two-component regulatory system activates transcription of the yegMNOB (mdtABCD) transporter gene cluster in Escherichia coli and increases its resistance to novobiocin and deoxycholate. J Bacteriol 2002; 184:4168-4176.
9. Zhou L, Lei XH, Bochner BR et al. Phenotype microarray analysis of Escherichia coli K-12 mutants with deletions of all two-component systems. J Bacteriol 2003; 185:4956-4972.
10. Darwin AJ. The phage-shock-protein response. Mol Microbiol 2005; 57:621-628.
11. Huang YH, Ferrieres L, Clarke DJ. The role of the Rcs phosphorelay in Enterobacteriaceae. Res Microbiol 2006; 157:206-212.
12. Majdalani N, Gottesman S. The Rcs phosphorelay: a complex signal transduction system. Annu Rev Microbiol 2005; 59:379-405.
13. Helmann JD. The extracytoplasmic function (ECF) sigma factors. Adv Microb Physiol 2002; 46:47-110.
14. Braun V, Mahren S, Ogierman M. Regulation of the FecI-type ECF sigma factor by transmembrane signalling. Curr Opin Microbiol 2003; 6:173-180.
15. De Las Penas A, Connolly L, Gross CA. $\sigma^E$ is an essential sigma factor in E.coli. J Bacteriol 1997; 179:6862-6864.
16. Hiratsu K, Amemura M, Nashimoto H et al. The rpoE gene of Escherichia coli, which encodes $\sigma^E$, is essential for bacterial growth at high temperature. J Bacteriol 1995; 177:2918-2922.
17. Erickson JW, Gross CA. Identification of the $\sigma^E$ subunit of E.coli RNA polymerase; a second alternate $\sigma$ factor involved in high-temperature gene expression. Genes Dev 1989; 3:1462-1471.
18. Wang QP, Kaguni JM. A novel sigma factor is involved in expression of the rpoH gene of Escherichia coli. J Bacteriol 1989; 171:4248-4253.
19. Rouviere PE, De Las Penas A, Mecsas J et al. rpoE, the gene encoding the second heat-shock sigma factor, $\sigma^E$ in E. coli. EMBO J 1995; 14:1032-1042.
20. Raina S, Missiakas D, Georgopoulos C. The rpoE gene encoding the $\sigma^E$ ($\sigma^{24}$) heat shock sigma factor E. coli. EMBO J 1995; 12:1043-1055.
21. De Las Penas A, Connolly L, Gross CA. The $\sigma^E$ -mediated response to extracytoplasmic stress in Escherichia coli is transduced by RseA and RseB, two negative regulators of $\sigma^E$ Mol Microbiol 1997; 24:373-385.
22. Missiakas D, Mayer MP, Lemaire M et al. Modulation of the Escherichia coli $\sigma^E$ (RpoE) heat-shock transcription-factor activity by the RseA, RseB and RseC proteins. Mol Microbiol 1997; 24:355-371.
23. Mecsas J, Rouviere PE, Erickson JW et al. The activity of $\sigma^E$, an Escherichia coli heat-inducible sigma-factor, is modulated by expression of outer membrane proteins. Genes Dev 1993; 7:2618-2628.
24. Jones CH, Danese PN, Pinkner JS et al. The chaperone-assisted membrane release and folding pathway is sensed by two signal transduction systems. EMBO J 1997; 16:6394-6406.
25. Missiakas D, Betton JM, Raina S. New components of protein folding in extracytoplasmic compartments of Escherichia coli SurA, FkpA and Skp/OmpH. Mol Microbiol 1996; 21:871-884.
26. Campbell EA, Tupy JL, Gruber TM et al. Crystal structure of Escherichia coli $\sigma^E$ with the cytoplasmic domain of its anti-sigma RseA. Mol Cell 2003; 11:1067-1078.
27. Tam C, Collinet B, Lau G et al. Interaction of the conserved region 4.2 of $\sigma^E$ with the RseA anti-sigma factor. J Biol Chem 2002; 277:27282-27287.
28. Collinet B, Yuzawa H, Chen T et al. RseB binding to the periplasmic domain of RseA modulates the RseA:$\sigma^E$ interaction in the cytoplasm and the availability of $\sigma^E$:RNA polymerase. J Biol Chem 2000; 275:33898-33904.
29. Ades SE, Connolly LE, Alba BM et al. The Escherichia coli $\sigma^E$-dependent extracytoplasmic stress response is controlled by the regulated proteolysis of an anti-sigma factor. Genes Dev 1999; 13:2449-2461.

30. Kanehara K, Ito K, Akiyama Y. YaeL (EcfE) activates the $\sigma^E$ pathway of stress response through a site-2 cleavage of anti-$\sigma^E$, RseA. Genes Dev 2002; 16:2147-2155.
31. Walsh NP, Alba BM, Bose B et al. OMP peptide signals initiate the envelope-stress response by activating DegS protease via relief of inhibition mediated by its PDZ domain. Cell 2003; 113:61-71.
32. Harris BZ, Lim WA. Mechanism and role of PDZ domains in signaling complex assembly. J Cell Sci 2001; 114:3219-3231.
33. Alba BM, Leeds JA, Onufryk C et al. DegS and YaeL participate sequentially in the cleavage of RseA to activate the $\sigma^E$ -dependent extracytoplasmic stress response. Genes Dev 2002; 16:2156-2168.
34. Grigorova IL, Chaba R, Zhong HJ et al. Fine-tuning of the Escherichia coli $\sigma^E$ envelope stress response relies on multiple mechanisms to inhibit signal-independent proteolysis of the transmembrane anti-sigma factor, RseA. Genes Dev 2004; 18:2686-2697.
35. Flynn JM, Levchenko I, Sauer RT et al. Modulating substrate choice: the SspB adaptor delivers a regulator of the extracytoplasmic-stress response to the AAA+ protease ClpXP for degradation. Genes Dev 2004; 18:2292-2301.
36. Costanzo A, Ades SE. Growth phase-dependent regulation of the extracytoplasmic stress factor, $\sigma^E$, by guanosine 3',5'-bispyrophosphate (ppGpp). J Bacteriol 2006; 188:4627-4634.
37. Magnusson LU, Farewell A, Nystrom T. ppGpp: a global regulator in Escherichia coli. Trends Microbiol 2005; 13:236-242.
38. Dartigalongue C, Missiakas D, Raina S. Characterization of the E. coli $\sigma^E$ regulon. J Biol Chem 2001; 276:20866-20875.
39. Johansen J, Rasmussen AA, Overgaard M et al. Conserved small noncoding RNAs that belong to the $\sigma^E$ regulon: role in down-regulation of outer membrane proteins. J Mol Biol 2006; 364:1-8.
40. Kabir MS, Yamashita D, Koyama S et al. Cell lysis directed by sigmaE in early stationary phase and effect of induction of the rpoE gene on global gene expression in Escherichia coli. Microbiology 2005; 151:2721-2735.
41. Onufryk C, Crouch ML, Fang FC et al. Characterization of six lipoproteins in the $\sigma^E$ regulon. J Bacteriol 2005; 187:4552-4561.
42. Rezuchova B, Miticka H, Homerova D et al. New members of the Escherichia coli sigmaE regulon identified by a two-plasmid system. FEMS Microbiol Lett 2003; 225:1-7.
43. Rhodius VA, Suh WC, Nonaka G et al. Conserved and variable functions of the $\sigma^E$ stress response in related genomes. PLoS Biol 2006; 4:e2.
44. Ruiz N, Falcone B, Kahne D et al. Chemical conditionality: a genetic strategy to probe organelle assembly. Cell 2005; 121:307-317.
45. Wu T, Malinverni J, Ruiz N et al. Identification of a multicomponent complex required for outer membrane biogenesis in E. coli. Cell 2005; 121:235-245.
46. Raivio TL. Envelope stress responses and Gram-negative bacterial pathogenesis. Mol Microbiol 2005; 56:1119-1128.
47. Humphreys S, Stevenson A, Bacon A et al. The alternative sigma factor, sigmaE, is critically important for the virulence of Salmonella typhimurium. Infect Immun 1999; 67:1560-1568.
48. Kenyon WJ, Sayers DG, Humphreys S et al. The starvation-stress response of Salmonella enterica serovar Typhimurium requires sigma(E)-, but not CpxR-regulated extracytoplasmic functions. Microbiology 2002; 148:113-122.
49. Testerman TL, Vazquez-Torres A, Xu Y et al. The alternative sigma factor sigmaE controls antioxidant defences required for Salmonella virulence and stationary-phase survival. Mol Microbiol 2002; 43:771-782.
50. Eriksson S, Lucchini S, Thompson A et al. Unravelling the biology of macrophage infection by gene expression profiling of intracellular Salmonella enterica. Mol Microbiol 2003; 47:103-118.
51. Skovierova H, Rowley G, Rezuchova B et al. Identification of the sigmaE regulon of Salmonella enterica serovar Typhimurium. Microbiology 2006; 152:1347-1359.
52. Kovacikova G, Skorupski K. The alternative sigma factor sigma(E) plays an important role in intestinal survival and virulence in Vibrio cholerae. Infect Immun 2002; 70:5355-5362.
53. Ando M, Yoshimatsu T, Ko C et al. Deletion of Mycobacterium tuberculosis sigma factor E results in delayed time to death with bacterial persistence in the lungs of aerosol-infected mice. Infect Immun 2003; 71:7170-7172.
54. Manganelli R, Fattorini L, Tan D et al. The extra cytoplasmic function sigma factor sigma(E) is essential for Mycobacterium tuberculosis virulence in mice. Infect Immun 2004; 72:3038-3041.
55. Korbsrisate S, Vanaporn M, Kerdsuk P et al. The Burkholderia pseudomallei RpoE (AlgU) operon is involved in environmental stress tolerance and biofilm formation. FEMS Microbiol Lett 2005; 252:243-249.

56. Deretic V, Schurr MJ, Boucher JC et al. Conversion of Pseudomonas aeruginosa to mucoidy in cystic fibrosis: environmental stress and regulation of bacterial virulence by alternative sigma factors. J Bacteriol 1994; 176:2773-2780.
57. Justice SS, Hunstad DA, Harper JR et al. Periplasmic peptidyl prolyl cis-trans isomerases are not essential for viability, but SurA is required for pilus biogenesis in Escherichia coli. J Bacteriol 2005; 187:7680-7686.
58. Douchin V, Bohn C, Bouloc P. Down-regulation of porins by a small RNA bypasses the essentiality of the regulated intramembrane proteolysis protease RseP in Escherichia coli. J Biol Chem 2006; 281:12253-12259.
59. McEwen J, Silverman P. Chromosomal mutations of Escherichia coli that alter expression of conjugative plasmid functions. Proc Natl Acad Sci USA 1980; 77:513-517.
60. McEwen J, Silverman P. Genetic analysis of Escherichia coli K-12 chromosomal mutants defective in expression of F-plasmid functions: identification of genes cpxA and cpxB. J Bacteriol 1980; 144:60-67.
61. McEwen J, Silverman P. Mutations in genes cpxA and cpxB of Escherichia coli K-12 cause a defect in isoleucine and valine syntheses. J Bacteriol 1980; 144:68-73.
62. McEwen J, Silverman PM. Mutations in genes cpxA and cpxB alter the protein composition of Escherichia coli inner and outer membranes. J Bacteriol 1982; 151:1553-1559.
63. McEwen J, Sambucetti L, Silverman PM. Synthesis of outer membrane proteins in cpxA cpxB mutants of Escherichia coli K-12. J Bacteriol 1983; 154:375-382.
64. Albin R, Weber R, Silverman PM. The Cpx proteins of Escherichia coli K12. Immunologic detection of the chromosomal cpxA gene product. J Biol Chem 1986; 261:4698-4705.
65. Weber RF, Silverman PM. The cpx proteins of Escherichia coli K12. Structure of the cpxA polypeptide as an inner membrane component. J Mol Biol 1988; 203:467-478.
66. Dong J, Iuchi S, Kwan HS et al. The deduced amino-acid sequence of the cloned cpxR gene suggests the protein is the cognate regulator for the membrane sensor, CpxA, in a two-component signal transduction system of Escherichia coli. Gene 1993; 136:227-230.
67. Cosma CL, Danese PN, Carlson JH et al. Mutational activation of the Cpx signal transduction pathway of Escherichia coli suppresses the toxicity conferred by certain envelope-associated stresses. Mol Microbiol 1995; 18:491-505.
68. Snyder WB, Davis LJ, Danese PN et al. Overproduction of NlpE, a new outer membrane lipoprotein, suppresses the toxicity of periplasmic LacZ by activation of the Cpx signal transduction pathway. J Bacteriol 1995; 177:4216-4223.
69. Danese PN, Silhavy TJ. CpxP, a stress-combative member of the Cpx regulon. J Bacteriol 1998; 180:831-839.
70. Raivio TL, Popkin DL, Silhavy TJ. The Cpx envelope stress response is controlled by amplification and feedback inhibition. J Bacteriol 1999; 181:5263-5272.
71. Raivio TL, Laird MW, Joly JC et al. Tethering of CpxP to the inner membrane prevents spheroplast induction of the Cpx envelope stress response. Mol Microbiol 2000; 37:1186-1197.
72. Fleischer R, Heermann R, Jung K et al. Purification, Reconstitution and Characterization of the CpxRAP Envelope Stress System of Escherichia coli. J Biol Chem 2007; 282:8583-8593.
73. Nevesinjac AZ, Raivio TL. The Cpx envelope stress response affects expression of the type IV bundle-forming pili of enteropathogenic Escherichia coli. J Bacteriol 2005; 187:672-686.
74. Hung DL, Raivio TL, Jones CH et al. Cpx signaling pathway monitors biogenesis and affects assembly and expression of P pili. EMBO J 2001; 20:1508-1518.
75. Raffa RG, Raivio TL. A third envelope stress signal transduction pathway in Escherichia coli. Mol Microbiol 2002; 45:1599-1611.
76. Lee YM, DiGiuseppe PA, Silhavy TJ et al. P pilus assembly motif necessary for activation of the CpxRA pathway by PapE in Escherichia coli. J Bacteriol 2004; 186:4326-4337.
77. Nakayama S, Watanabe H. Involvement of cpxA, a sensor of a two-component regulatory system, in the pH-dependent regulation of expression of Shigella sonnei virF gene. J Bacteriol 1995; 177:5062-5069.
78. Mileykovskaya E, Dowhan W. The Cpx two-component signal transduction pathway is activated in Escherichia coli mutant strains lacking phosphatidylethanolamine. J Bacteriol 1997; 179:1029-1034.
79. Danese PN, Oliver GR, Barr K et al. Accumulation of the enterobacterial common antigen lipid II biosynthetic intermediate stimulates degP transcription in Escherichia coli. J Bacteriol 1998; 180:5875-5884.
80. Raivio TL, Silhavy TJ. Transduction of envelope stress in Escherichia coli by the Cpx two-component system. J Bacteriol 1997; 179:7724-7733.
81. Kenney LJ. Structure/function relationships in OmpR and other winged-helix transcription factors. Curr Opin Microbiol 2002; 5:135-141.
82. Buelow DR, Raivio TL. Cpx signal transduction is influenced by a conserved N-terminal domain in the novel inhibitor CpxP and the periplasmic protease DegP. J Bacteriol 2005; 187:6622-6630.

83. Isaac DD, Pinkner JS, Hultgren SJ et al. The extracytoplasmic adaptor protein CpxP is degraded with substrate by DegP. Proc Natl Acad Sci USA 2005; 102:17775-17779.
84. DiGiuseppe PA, Silhavy TJ. Signal detection and target gene induction by the CpxRA two-component system. J Bacteriol 2003; 185:2432-2440.
85. Danese PN, Silhavy TJ. The $\sigma^E$ and the Cpx signal transduction systems control the synthesis of periplasmic protein-folding enzymes in Escherichia coli. Genes Dev 1997; 11:1183-1193.
86. Pogliano J, Lynch AS, Belin D et al. Regulation of Escherichia coli cell envelope proteins involved in protein folding and degradation by the Cpx two-component system. Genes Dev 1997; 11:1169-1182.
87. Kamitani S, Akiyama Y, Ito K. Identification and characterization of an Escherichia coli gene required for the formation of correctly folded alkaline phosphatase, a periplasmic enzyme. EMBO J 1992; 11:57-62.
88. Bardwell JC, McGovern K, Beckwith J. Identification of a protein required for disulfide bond formation in vivo. Cell 1991; 67:581-589.
89. Strauch KL, Beckwith J. An Escherichia coli mutation preventing degradation of abnormal periplasmic proteins. Proc Natl Acad Sci USA 1988; 85:1576-1580.
90. Kleerebezem M, Heutink M, Tommassen J. Characterization of an Escherichia coli rotA mutant, affected in periplasmic peptidyl-prolyl cis/trans isomerase. Mol Microbiol 1995; 18:313-320.
91. Dartigalongue C, Raina S. A new heat-shock gene, ppiD, encodes a peptidyl-prolyl isomerase required for folding of outer membrane proteins in Escherichia coli. EMBO J 1998; 17:3968-3980.
92. De Wulf P, McGuire AM, Liu X et al. Genome-wide profiling of promoter recognition by the two-component response regulator CpxR-P in Escherichia coli. J Biol Chem 2002; 277:26652-26661.
93. Hawrot E, Kennedy EP. Biogenesis of membrane lipids: mutants of Escherichia coli with temperature-sensitive phosphatidylserine decarboxylase. Proc Natl Acad Sci USA 1975; 72:1112-1116.
94. Batchelor E, Walthers D, Kenney LJ et al. The Escherichia coli CpxA-CpxR envelope stress response system regulates expression of the porins ompF and ompC. J Bacteriol 2005; 187:5723-5731.
95. Condemine G, Berrier C, Plumbridge J et al. Function and expression of an N-acetylneuraminic acid-inducible outer membrane channel in Escherichia coli. J Bacteriol 2005; 187:1959-1965.
96. Hirakawa H, Inazumi Y, Masaki T et al. Indole induces the expression of multidrug exporter genes in Escherichia coli. Mol Microbiol 2005; 55:1113-1126.
97. Hall MN, Silhavy TJ. Genetic analysis of the major outer membrane proteins of Escherichia coli. Annu Rev Genet 1981; 15:91-142.
98. Kalivoda KA, Steenbergen SM, Vimr ER et al. Regulation of sialic acid catabolism by the DNA binding protein NanR in Escherichia coli. J Bacteriol 2003; 185:4806-4815.
99. Nagakubo S, Nishino K, Hirata T et al. The putative response regulator BaeR stimulates multidrug resistance of Escherichia coli via a novel multidrug exporter system, MdtABC. J Bacteriol 2002; 184:4161-4167.
100. Hirakawa H, Nishino K, Yamada J et al. Beta-lactam resistance modulated by the overexpression of response regulators of two-component signal transduction systems in Escherichia coli. J Antimicrob Chemother 2003; 52:576-582.
101. De Wulf P, Kwon O, Lin EC. The CpxRA signal transduction system of Escherichia coli: growth-related autoactivation and control of unanticipated target operons. J Bacteriol 1999; 181:6772-6778.
102. Price NL, Raivio TL. submitted.
103. Dorel C, Vidal O, Prigent-Combaret C et al. Involvement of the Cpx signal transduction pathway of E. coli in biofilm formation. FEMS Microbiol Lett 1999; 178:169-175.
104. Jubelin G, Vianney A, Beloin C et al. CpxR/OmpR interplay regulates curli gene expression in response to osmolarity in Escherichia coli. J Bacteriol 2005; 187:2038-2049.
105. Hernday AD, Braaten BA, Broitman-Maduro G et al. Regulation of the pap epigenetic switch by CpxAR: phosphorylated CpxR inhibits transition to the phase ON state by competition with Lrp. Mol Cell 2004; 16:537-547.
106. Ogasawara H, Teramoto K, Hirao K et al. Negative regulation of DNA repair gene (ung) expression by the CpxR/CpxA two-component system in Escherichia coli K-12 and induction of mutations by increased expression of CpxR. J Bacteriol 2004; 186:8317-8325.
107. Shimohata N, Chiba S, Saikawa N et al. The Cpx stress response system of Escherichia coli senses plasma membrane proteins and controls HtpX, a membrane protease with a cytosolic active site. Genes Cells 2002; 7:653-662.
108. Leclerc GJ, Tartera C, Metcalf ES. Environmental regulation of Salmonella typhi invasion-defective mutants. Infect Immun 1998; 66:682-691.
109. Humphreys S, Rowley G, Stevenson A et al. Role of the two-component regulator CpxAR in the virulence of Salmonella enterica serotype Typhimurium. Infect Immun 2004; 72:4654-4661.
110. Nakayama S, Watanabe H. Identification of cpxR as a positive regulator essential for expression of the Shigella sonnei virF gene. J Bacteriol 1998; 180:3522-3528.

111. Nakayama S, Kushiro A, Asahara T et al. Activation of hilA expression at low pH requires the signal sensor CpxA, but not the cognate response regulator CpxR, in Salmonella enterica serovar Typhimurium. Microbiology 2003; 149:2809-2817.
112. Mitobe J, Arakawa E, Watanabe H. A sensor of the two-component system CpxA affects expression of the type III secretion system through posttranscriptional processing of InvE. J Bacteriol 2005; 187:107-113.
113. Ha UH, Wang Y, Jin S. DsbA of Pseudomonas aeruginosa is essential for multiple virulence factors. Infect Immun 2003; 71:1590-1595.
114. Stenson TH, Weiss AA. DsbA and DsbC are required for secretion of pertussis toxin by Bordetella pertussis. Infect Immun 2002; 70:2297-2303.
115. Yu J. Inactivation of DsbA, but not DsbC and DsbD, affects the intracellular survival and virulence of Shigella flexneri. Infect Immun 1998; 66:3909-3917.
116. Yu J, Edwards-Jones B, Neyrolles O et al. Key role for DsbA in cell-to-cell spread of Shigella flexneri, permitting secretion of Ipa proteins into interepithelial protrusions. Infect Immun 2000; 68:6449-6456.
117. Jackson MW, Plano GV. DsbA is required for stable expression of outer membrane protein YscC and for efficient Yop secretion in Yersinia pestis. J Bacteriol 1999; 181:5126-5130.
118. Zhang HZ, Donnenberg MS. DsbA is required for stability of the type IV pilin of enteropathogenic Escherichia coli. Mol Microbiol 1996; 21:787-797.
119. Johnson K, Charles I, Dougan G et al. The role of a stress-response protein in Salmonella typhimurium virulence. Mol Microbiol 1991; 5:401-407.
120. Peek JA, Taylor RK. Characterization of a periplasmic thiol:disulfide interchange protein required for the functional maturation of secreted virulence factors of Vibrio cholerae. Proc Natl Acad Sci USA 1992; 89:6210-6214.
121. Yu J, Webb H, Hirst TR. A homologue of the Escherichia coli DsbA protein involved in disulphide bond formation is required for enterotoxin biogenesis in Vibrio cholerae. Mol Microbiol 1992; 6:1949-1958.
122. Nagasawa S, Ishige K, Mizuno T. Novel members of the two-component signal transduction genes in Escherichia coli. J Biochem 1993; 114:350-357.
123. Nishino K, Yamaguchi A. Analysis of a complete library of putative drug transporter genes in Escherichia coli. J Bacteriol 2001; 183:5803-5812.
124. Hirakawa H, Nishino K, Hirata T et al. Comprehensive studies of drug resistance mediated by overexpression of response regulators of two-component signal transduction systems in Escherichia coli. J Bacteriol 2003; 185:1851-1856.
125. Yamamoto K, Hirao K, Oshima T et al. Functional characterization in vitro of all two-component signal transduction systems from Escherichia coli. J Biol Chem 2005; 280:1448-1456.
126. Nishino K, Honda T, Yamaguchi A. Genome-wide analyses of Escherichia coli gene expression responsive to the BaeSR two-component regulatory system. J Bacteriol 2005; 187:1763-1772.
127. Brissette JL, Russel M, Weiner L et al. Phage shock protein, a stress protein of Escherichia coli. Proc Natl Acad Sci USA 1990; 87:862-866.
128. Brissette JL, Russel M. Secretion and membrane integration of a filamentous phage-encoded morphogenetic protein. J Mol Biol 1990; 211:565-580.
129. Drahos DJ, Hendrix RW. Effect of bacteriophage lambda infection on synthesis of groE protein and other Escherichia coli proteins. J Bacteriol 1982; 149:1050-1063.
130. Weiner L, Brissette JL, Model P. Stress-induced expression of the Escherichia coli phage shock protein operon is dependent on sigma 54 and modulated by positive and negative feedback mechanisms. Genes Dev 1991; 5:1912-1923.
131. Brissette JL, Weiner L, Ripmaster TL et al. Characterization and sequence of the Escherichia coli stress-induced psp operon. J Mol Biol 1991; 220:35-48.
132. Adams H, Teertstra W, Demmers J et al. Interactions between phage-shock proteins in Escherichia coli. J Bacteriol 2003; 185:1174-1180.
133. Weiner L, Brissette JL, Ramani N et al. Analysis of the proteins and cis-acting elements regulating the stress-induced phage shock protein operon. Nucleic Acids Res 1995; 23:2030-2036.
134. Jovanovic G, Weiner L, Model P. Identification, nucleotide sequence and characterization of PspF, the transcriptional activator of the Escherichia coli stress-induced psp operon. J Bacteriol 1996; 178:1936-1945.
135. Dworkin J, Jovanovic G, Model P. The PspA protein of Escherichia coli is a negative regulator of sigma(54)-dependent transcription. J Bacteriol 2000; 182:311-319.
136. Elderkin S, Jones S, Schumacher J et al. Mechanism of action of the Escherichia coli phage shock protein PspA in repression of the AAA family transcription factor PspF. J Mol Biol 2002; 320:23-37.
137. Elderkin S, Bordes P, Jones S et al. Molecular determinants for PspA-mediated repression of the AAA transcriptional activator PspF. J Bacteriol 2005; 187:3238-3248.

138. Kleerebezem M, Tommassen J. Expression of the pspA gene stimulates efficient protein export in Escherichia coli. Mol Microbiol 1993; 7:947-956.
139. Kleerebezem M, Crielaard W, Tommassen J. Involvement of stress protein PspA (phage shock protein A) of Escherichia coli in maintenance of the protonmotive force under stress conditions. EMBO J 1996; 15:162-171.
140. Maxson ME, Darwin AJ. PspB and PspC of Yersinia enterocolitica are dual function proteins: regulators and effectors of the phage-shock-protein response. Mol Microbiol 2006; 59:1610-1623.
141. Carlson JH, Silhavy TJ. Signal sequence processing is required for the assembly of LamB trimers in the outer membrane of Escherichia coli. J Bacteriol 1993; 175:3327-3334.
142. Weiner L, Model P. Role of an Escherichia coli stress-response operon in stationary-phase survival. Proc Natl Acad Sci USA 1994; 91:2191-2195.
143. Jones SE, Lloyd LJ, Tan KK et al. Secretion defects that activate the phage shock response of Escherichia coli. J Bacteriol 2003; 185:6707-6711.
144. DeLisa MP, Lee P, Palmer T et al. Phage shock protein PspA of Escherichia coli relieves saturation of protein export via the Tat pathway. J Bacteriol 2004; 186:366-373.
145. Becker LA, Bang IS, Crouch ML et al. Compensatory role of PspA, a member of the phage shock protein operon, in rpoE mutant Salmonella enterica serovar Typhimurium. Mol Microbiol 2005; 56:1004-1016.
146. Guilvout I, Chami M, Engel A et al. Bacterial outer membrane secretin PulD assembles and inserts into the inner membrane in the absence of its pilotin. EMBO J 2006; 25:5241-5249.
147. Darwin AJ, Miller VL. Identification of Yersinia enterocolitica genes affecting survival in an animal host using signature-tagged transposon mutagenesis. Mol Microbiol 1999; 32:51-62.
148. Darwin AJ, Miller VL. The psp locus of Yersinia enterocolitica is required for virulence and for growth in vitro when the Ysc type III secretion system is produced. Mol Microbiol 2001; 39:429-444.
149. Lloyd LJ, Jones SE, Jovanovic G et al. Identification of a new member of the phage shock protein response in Escherichia coli, the phage shock protein G (PspG). J Biol Chem 2004; 279:55707-55714.
150. Jovanovic G, Lloyd LJ, Stumpf MP et al. Induction and function of the phage shock protein extracytoplasmic stress response in Escherichia coli. J Biol Chem 2006; 281:21147-21161.
151. Gottesman S, Trisler P, Torres-Cabassa A. Regulation of capsular polysaccharide synthesis in Escherichia coli K-12: characterization of three regulatory genes. J Bacteriol 1985; 162:1111-1119.
152. Trisler P, Gottesman S. lon transcriptional regulation of genes necessary for capsular polysaccharide synthesis in Escherichia coli K-12. J Bacteriol 1984; 160:184-191.
153. Torres-Cabassa AS, Gottesman S. Capsule synthesis in Escherichia coli K-12 is regulated by proteolysis. J Bacteriol 1987; 169:981-989.
154. Brill JA, Quinlan-Walshe C, Gottesman S. Fine-structure mapping and identification of two regulators of capsule synthesis in Escherichia coli K-12. J Bacteriol 1988; 170:2599-2611.
155. Stout V, Gottesman S. RcsB and RcsC: a two-component regulator of capsule synthesis in Escherichia coli. J Bacteriol 1990; 172:659-669.
156. Takeda S, Fujisawa Y, Matsubara M et al. A novel feature of the multistep phosphorelay in Escherichia coli: a revised model of the RcsC —> YojN —> RcsB signalling pathway implicated in capsular synthesis and swarming behaviour. Mol Microbiol 2001; 40:440-450.
157. Erickson KD, Detweiler CS. The Rcs phosphorelay system is specific to enteric pathogens/commensals and activates ydeI, a gene important for persistent Salmonella infection of mice. Mol Microbiol 2006; 62:883-894.
158. Davalos-Garcia M, Conter A, Toesca I et al. Regulation of osmC gene expression by the two-component system rcsB-rcsC in Escherichia coli. J Bacteriol 2001; 183:5870-5876.
159. Sturny R, Cam K, Gutierrez C et al. NhaR and RcsB independently regulate the osmCp1 promoter of Escherichia coli at overlapping regulatory sites. J Bacteriol 2003; 185:4298-4304.
160. Wehland M, Bernhard F. The RcsAB box. Characterization of a new operator essential for the regulation of exopolysaccharide biosynthesis in enteric bacteria. J Biol Chem 2000; 275:7013-7020.
161. Wehland M, Kiecker C, Coplin DL et al. Identification of an RcsA/RcsB recognition motif in the promoters of exopolysaccharide biosynthetic operons from Erwinia amylovora and Pantoea stewartii subspecies stewartii. J Biol Chem 1999; 274:3300-3307.
162. Ebel W, Trempy JE. Escherichia coli RcsA, a positive activator of colanic acid capsular polysaccharide synthesis, functions to activate its own expression. J Bacteriol 1999; 181:577-584.
163. Kuo MS, Chen KP, Wu WF. Regulation of RcsA by the ClpYQ (HslUV) protease in Escherichia coli. Microbiology 2004; 150:437-446.
164. Pristovsek P, Sengupta K, Lohr F et al. Structural analysis of the DNA-binding domain of the Erwinia amylovora RcsB protein and its interaction with the RcsAB box. J Biol Chem 2003; 278:17752-17759.

165. Virlogeux I, Waxin H, Ecobichon C et al. Characterization of the rcsA and rcsB genes from Salmonella typhi: rcsB through tviA is involved in regulation of Vi antigen synthesis. J Bacteriol 1996; 178:1691-1698.
166. Arricau N, Hermant D, Waxin H et al. The RcsB-RcsC regulatory system of Salmonella typhi differentially modulates the expression of invasion proteins, flagellin and Vi antigen in response to osmolarity. Mol Microbiol 1998; 29:835-850.
167. Gervais FG, Drapeau GR. Identification, cloning and characterization of rcsF, a new regulator gene for exopolysaccharide synthesis that suppresses the division mutation ftsZ84 in Escherichia coli K-12. J Bacteriol 1992; 174:8016-8022.
168. Majdalani N, Heck M, Stout V et al. Role of RcsF in signaling to the Rcs phosphorelay pathway in Escherichia coli. J Bacteriol 2005; 187:6770-6778.
169. Castanie-Cornet MP, Cam K, Jacq A. RcsF is an outer membrane lipoprotein involved in the RcsCDB phosphorelay signaling pathway in Escherichia coli. J Bacteriol 2006; 188:4264-4270.
170. Cano DA, Dominguez-Bernal G, Tierrez A et al. Regulation of capsule synthesis and cell motility in Salmonella enterica by the essential gene igaA. Genetics 2002; 162:1513-1523.
171. Dominguez-Bernal G, Pucciarelli MG, Ramos-Morales F et al. Repression of the RcsC-YojN-RcsB phosphorelay by the IgaA protein is a requisite for Salmonella virulence. Mol Microbiol 2004; 53:1437-1449.
172. Parker CT, Kloser AW, Schnaitman CA et al. Role of the rfaG and rfaP genes in determining the lipopolysaccharide core structure and cell surface properties of Escherichia coli K-12. J Bacteriol 1992; 174:2525-2538.
173. Clavel T, Lazzaroni JC, Vianney A et al. Expression of the tolQRA genes of Escherichia coli K-12 is controlled by the RcsC sensor protein involved in capsule synthesis. Mol Microbiol 1996; 19:19-25.
174. Mouslim C, Groisman EA. Control of the Salmonella ugd gene by three two-component regulatory systems. Mol Microbiol 2003; 47:335-344.
175. Mouslim C, Latifi T, Groisman EA. Signal-dependent requirement for the co-activator protein RcsA in transcription of the RcsB-regulated ugd gene. J Biol Chem 2003; 278:50588-50595.
176. Ebel W, Vaughn GJ, Peters HK et al. Inactivation of mdoH leads to increased expression of colanic acid capsular polysaccharide in Escherichia coli. J Bacteriol 1997; 179:6858-6861.
177. Kelley WL, Georgopoulos C. Positive control of the two-component RcsC/B signal transduction network by DjlA: a member of the DnaJ family of molecular chaperones in Escherichia coli. Mol Microbiol 1997; 25:913-931.
178. Gottesman S, Stout V. Regulation of capsular polysaccharide synthesis in Escherichia coli K12. Mol Microbiol 1991; 5:1599-1606.
179. Sledjeski DD, Gottesman S. Osmotic shock induction of capsule synthesis in Escherichia coli K-12. J Bacteriol 1996; 178:1204-1206.
180. Conter A, Sturny R, Gutierrez C et al. The RcsCB His-Asp phosphorelay system is essential to overcome chlorpromazine-induced stress in Escherichia coli. J Bacteriol 2002; 184:2850-2853.
181. Hagiwara D, Sugiura M, Oshima T et al. Genome-wide analyses revealing a signaling network of the RcsC-YojN-RcsB phosphorelay system in Escherichia coli. J Bacteriol 2003; 185:5735-5746.
182. Ferrieres L, Clarke DJ. The RcsC sensor kinase is required for normal biofilm formation in Escherichia coli K-12 and controls the expression of a regulon in response to growth on a solid surface. Mol Microbiol 2003; 50:1665-1682.
183. Sailer FC, Meberg BM, Young KD. beta-Lactam induction of colanic acid gene expression in Escherichia coli. FEMS Microbiol Lett 2003; 226:245-249.
184. Chatterjee A, Chun WA, Chatterjee AK. Isolation and characterization of rcsA like gene of Erwina amylovora that activates extracellular polysaccharide production in Erwinia species, Escherichia coli and Salmonella typhimurium. Mol Plant Microbe Interact 1990; 3:144-148.
185. Bereswill S, Geider K. Characterization of the rcsB gene from Erwinia amylovora and its influence on exoploysaccharide synthesis and virulence of the fire blight pathogen. J Bacteriol 1997; 179:1354-1361.
186. Stevenson G, Andrianopoulos K, Hobbs M et al. Organization of the Escherichia coli K-12 gene cluster responsible for production of the extracellular polysaccharide colanic acid. J Bacteriol 1996; 178:4885-4893.
187. Delgado MA, Mouslim C, Groisman EA. The PmrA/PmrB and RcsC/YojN/RcsB systems control expression of the Salmonella O-antigen chain length determinant. Mol Microbiol 2006; 60:39-50.
188. Gervais FG, Phoenix P, Drapeau GR. The rcsB gene, a positive regulator of colanic acid biosynthesis in Escherichia coli, is also an activator of ftsZ expression. J Bacteriol 1992; 174:3964-3971.
189. Carballes F, Bertrand C, Bouche JP et al. Regulation of Escherichia coli cell division genes ftsA and ftsZ by the two-component system rcsC-rcsB. Mol Microbiol 1999; 34:442-450.
190. Francez-Charlot A, Laugel B, Van Gemert A et al. RcsCDB His-Asp phosphorelay system negatively regulates the flhDC operon in Escherichia coli. Mol Microbiol 2003; 49:823-832.

191. Vianney A, Jubelin G, Renault S et al. Escherichia coli tol and rcs genes participate in the complex network affecting curli synthesis. Microbiology 2005; 151:2487-2497.
192. Prigent-Combaret C, Prensier G, Le Thi TT et al. Developmental pathway for biofilm formation in curli-producing Escherichia coli strains: role of flagella, curli and colanic acid. Environ Microbiol 2000; 2:450-464.
193. Fredericks CE, Shibata S, Aizawa S et al. Acetyl phosphate-sensitive regulation of flagellar biogenesis and capsular biosynthesis depends on the Rcs phosphorelay. Mol Microbiol 2006; 61:734-747.
194. Clarke DJ, Joyce SA, Toutain CM et al. Genetic analysis of the RcsC sensor kinase from Escherichia coli K-12. J Bacteriol 2002; 184:1204-1208.
195. Majdalani N, Hernandez D, Gottesman S. Regulation and mode of action of the second small RNA activator of RpoS translation, RprA. Mol Microbiol 2002; 46:813-826.
196. Garcia-Calderon CB, Garcia-Quintanilla M, Casadesus J et al. Virulence attenuation in Salmonella enterica rcsC mutants with constitutive activation of the Rcs system. Microbiology 2005; 151:579-588.
197. Majdalani N, Chen S, Murrow J et al. Regulation of RpoS by a novel small RNA: the characterization of RprA. Mol Microbiol 2001; 39:1382-1394.
198. Peterson CN, Carabetta VJ, Chowdhury T et al. LrhA regulates rpoS translation in response to the Rcs phosphorelay system in Escherichia coli. J Bacteriol 2006; 188:3175-3181.
199. Tobe T, Ando H, Ishikawa H et al. Dual regulatory pathways integrating the RcsC-RcsD-RcsB signalling system control enterohaemorrhagic Escherichia coli pathogenicity. Mol Microbiol 2005; 58:320-333.
200. Venecia K, Young GM. Environmental regulation and virulence attributes of the Ysa type III secretion system of Yersinia enterocolitica biovar 1B. Infect Immun 2005; 73:5961-5977.
201. Mouslim C, Delgado M, Groisman EA. Activation of the RcsC/YojN/RcsB phosphorelay system attenuates Salmonella virulence. Mol Microbiol 2004; 54:386-395.

# Chapter 7

# Dual Regulation with Ser/Thr Kinase Cascade and a His/Asp TCS in *Myxococcus xanthus*

**Sumiko Inouye* and Hirofumi Nariya**

## Abstract

Fruiting body development of *Myxococcus xanthus* is propelled by temporal gene expression directed via stage-specific intercellular signaling pathways. *M. xanthus* exhibits social behaviors during its complex life cycle and is a potential source for production of natural products such as secondary metabolites. The numerous signaling pathways of *M. xanthus* consist of not only the two-component His-Asp phosphorelay system (TCS) but also protein Ser/Thr kinases (PSTKs) that regulate gene expression, motility and multicellular development. Recent studies have uncovered the unique molecular regulatory mechanism of MrpC, a transcription factor essential for fruiting body development and sporulation. *mrpC* expression is activated early in development by MrpB, which belongs to the NtrC family of TCS. MrpC, is, in turn, a transcriptional activator of *fruA* that encodes another key transcription factor, FruA. FruA is essential for fruiting body development and sporulation and regulates positively and negatively the synthesis of many developmental proteins. In addition, expression of *mrpC* during vegetative growth is kept at a low level by the PSTK Pkn8-Pkn14 kinase cascade which negatively regulates MrpC-binding activity to its own promoter. Therefore, *M. xanthus* utilizes a novel dual system with both eukaryotic PSTK cascade and prokaryotic TCS signaling systems to tightly and precisely regulate MrpC levels, which activate timely *fruA* expression and propel fruiting body development and sporulation.

## Introduction

Multicellular development of *Myxococcus xanthus* is directed by a series of intercellular signaling pathways that regulate the stage-specific expression of a specific set of genes. The environmental signals are transmitted inside cells through receptor-type kinases that phosphorylate downstream regulatory proteins to modulate their activities. In prokaryotes, the major mechanism of signal transduction is a two-component His-Asp phosphorelay system (TCS) that consist of, in most cases, a histidine kinase (HK) and a response regulator (RR).[1] In contrast, protein Ser/Thr (PSTK) and Tyr kinases (PTK) play the major roles in eukaryotes by phosphorylating downstream PSTKs and regulatory proteins. There are two important differences between TCS and PSTK signaling systems. In the TCS system, HK auto-phosphorylates its His residue utilizing ATP and a high-energy phospho-group of the His residue of HK is then transferred to the Asp residue of a cognate response regulator. Furthermore, HK itself contains phosphatase activity that mediates dephosphorylation of response regulators.[2] In contrast, PSTKs use ATP to phosphorylate their

*Corresponding Author: Sumiko Inouye—Department of Biochemistry, Robert Wood Johnson Medical School, Piscataway, NJ 08854, USA. Email: inouyesu@umdnj.edu

*Bacterial Signal Transduction: Networks and Drug Targets*, edited by Ryutaro Utsumi.

own Ser/Thr residues for activation of their kinase activity. The activated PSTK phosphorylates its substrate by using ATP to modulate the function. The signal can be greatly amplified and stably maintained for long time periods since a specific phosphatase is required for the inactivation of PSTKs. *M. xanthus* genomics revealed 102 genes for putative PSTKs and 137 and 118 genes for putative HKs and RRs of TCS, respectively (http://cmr.tigr.org).

*M. xanthus* is a Gram-negative soil bacterium that exhibits a communal lifestyle during vegetative growth and fruiting body (FB) formation (Fig. 1). When *M. xanthus* cells are surrounded in a rich nutrient condition, they grow unicellularly, forming large swarms consisting of thousands of cells which coordinately secrete lytic enzymes that degrade macromolecules.[3] Upon nutrient depletion at high cell density, ~100,000 cells glide towards an aggregation center and form a multicellular FB. During the 30 h developmental process, approximately 10% of the initial cell population differentiates into spores inside the FB while others (80%) surrounding the FB undergo suicide autolysis, providing nutrients to the others that form the FB and spores.[4,5]

The cells sense nutrient limitation, in part, through the accumulation of guanosine penta- or tetra-phosphate, (p)ppGpp.[6,7] In addition, based on analysis of five extracellular complementation groups, five intercellular signals, A, B, C, D and E, are known to be involved and each is essential for multicellular FB formation and sporulation.[8,9] A-signal is a specific set of amino acids and peptides and appears to be used as quorum sensing to measure the cell density necessary for the initiation of successful multicellular development.[10] A-, B-, D- and E-signals are essential for the progression through the first 5 h of development, while C-signal is important for development after 6 h and is used repeatedly to achieve aggregation and later sporulation.[11] While the *csgA* gene is essential for C-signaling,[12] the precise nature of C-signal is controversial. One proposal is that CsgA generates the C-signal enzymatically and the other is that C-signal is a cell-surface-associated protein encoded by the *csgA* gene and is essential for aggregation, FB formation and sporulation. CsgA, a 25 kDa protein, with a high similarity to members of the short-chain alcohol dehydrogenase family,[13] is processed into C-signal, a 17 kDa protein, by a developmentally specific protease.[14]

The expression of many genes early in development is likely to be regulated by $\sigma^{54}$-RNA polymerase ($\sigma^{54}$-RNAP) recognizing specific promoters and initiating transcription with enhancer-binding proteins (EBPs).[15] Fifty two putative EBPs are found from genome analysis[16] and mutational analysis reveals that several EBPs are involved in motility and FB development.[16-21] MrpB, an EBP, is essential for *mrpC* expression that encodes a developmentally essential transcription factor, MrpC.[18] MrpC is a transcription factor of *fruA* that encodes another key transcription factor, FruA,[22] which is essential for FB development and sporulation.[23] The synthesis of many proteins during FB development is under the control of FruA in a C-signal-independent and dependent manner.[24]

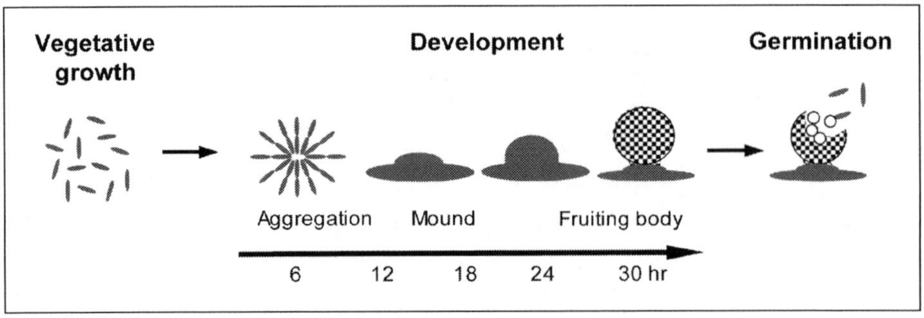


Figure 1. Life cycle of *Myxococcus xanthus*. See the text for details.

## FruA: A Key Developmental Transcription Factor

FruA is a key transcription factor that regulates a large number of gene expression early in development and is required for FB formation and sporulation. The *fruA* gene was identified by transposon mutagenesis as an essential locus for multicellular development and encodes a protein homologous to the response regulator family of TCS.[23] Later, the *fruA* gene was also identified as a *class II* gene, which is an essential component in the C-signal transduction pathway.[25] Although the putative phosphorylation site, the Asp residue at 59, is conserved in FruA and is important for in vivo function,[25] a kinase for FruA has not been identified yet. The expression of the *fruA* gene is induced depending on A- and E-signals at around 6 h of development and reaches the highest level at 12 h,[23] but is independent of C-signal.[25]

Analysis of protein expression patterns at 12 h of development revealed that 276 proteins are differentially expressed, of which the synthesis of 154 proteins is down-regulated, while the synthesis of 122 proteins is up-regulated.[24] Among the up-regulated proteins, 54 are dependent on both *fruA* and *csgA* and 7 are dependent on only *fruA*. Protein S, the product of the *tps* gene, is a spore coat protein[26] and DofA exhibits no similarity to known proteins.[27] Both proteins are nonessential for multicellular development and their expression is C-signal-independent.[24,28] In contrast, FrzCD methylation and *devRS* expression during development were found to be dependent on *fruA* as well as C-signal.[23,29] FrzCD is a component of the Frz system that controls developmental aggregation.[30] FruA regulates the expression of *frzF*, which encodes the methyltransferase for FrzCD and is in the same operon.[31] The *devRS* genes are essential for sporulation, but the function of the *devRS* gene products is unknown.[32] Another FB essential gene, *fdgA*, was identified by in vitro selection from the genomic library using the purified DNA-binding domain of FruA.[33] The *fdgA* gene encodes a protein homologous to the outer-membrane auxiliary (OMA) family protein involved in polysaccharide export system and is required for FB development and sporulation.

## MrpC: A Transcriptional Activator for *fruA* Expression

The *fruA* gene is expressed exclusively depending on A- and E-signals during FB development. The promoter region at–185 to–41 bp from the transcription start site is essential for the induction of developmental *fruA* expression based on *lacZ* reporter analysis.[22] In the upstream region of the *fruA* promoter, two types of complexes were observed with ammonium sulfate fractionated extracts prepared from cells at 12 h development. From footprint analysis, two binding sites were determined. Mutational analysis of the binding sites showed that they are indispensable for the induction of *fruA* expression during development. Thus, the DNA-binding site identified is an essential *cis*-acting element for *fruA* expression. Using the DNA affinity column specific to a *cis*-acting element, a *fruA* promoter-binding protein (FBP) was purified from developmental cell extracts. The N-terminal sequence of FBP was found to be identical to residues from 33 to 53 of the proposed MrpC sequence.[18] Since there is another ATG at the 26th codon in *mrpC*, a recombinant MrpC2 lacking the N-terminal 25 residues of MrpC was purified and found to bind to a *cis*-acting element of the *fruA* promoter as does FBP, indicating that MrpC/MrpC2 is a *trans*-acting factor for *fruA* expression.[22] Therefore, three essential transcription factors, MrpB, MrpC and FruA appear to form a transcriptional cascade, each of which sequentially activates downstream factors depending on developmental signaling pathways to propel FB development.

## Genetic Study of the *mrpC* Locus

*mrpC* is located downstream of *mrpA-mrpB* in the *mrp* locus which was identified via transposon mutagenesis as an essential locus for multicellular development (Fig. 2A).[18] Based on their predicted amino acid sequences, *mrpA* and *mrpB* encode a soluble His kinase and a response regulator of the NtrC family, respectively and *mrpC* encodes a homologue of the cAMP receptor protein (CRP) family of transcription factors. *mrpA* and *mrpB* are cotranscribed after A-signal and (p)ppGpp but before C-signal.[18,19] *mrpC*, located 274-bp downstream of *mrpB*, is transcribed independently and its expression is absolutely dependent on *mrpB* but not *mrpA*. Both *mrpB* and *mrpC* are required for aggregation and sporulation but *mrpA* is required only for sporulation.

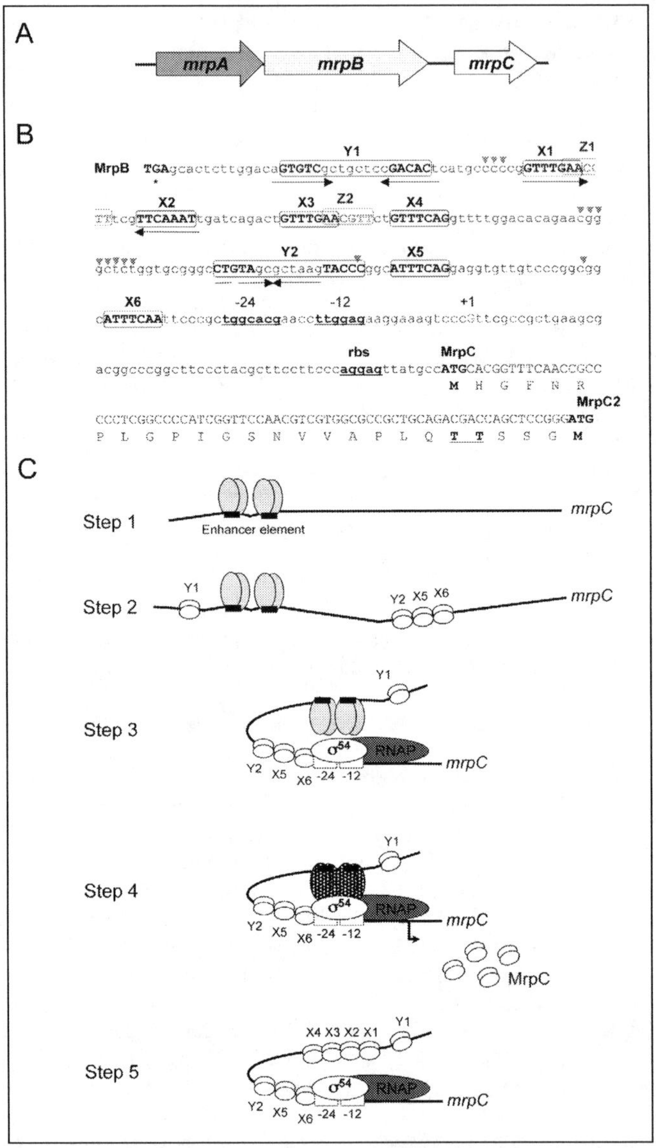


Figure 2. Transcriptional regulation of *mrpC*. A) Gene organization of the *mrp* locus. Arrows indicate transcriptional orientation. B) MrpB and MrpC binding sites on the *mrpC* promoter region. The boxes with the letter X and Y represent MrpC binding sites and boxes with the letter Z are MrpB binding sites. The translational stop codon TGA and the initiation codon ATG for the *mrpB* and *mrpC* genes are indicated by bold capitals, respectively. The codon ATG used for MrpC2 is also highlighted by bold capitals. The predicted phosphorylation site in the N-terminal 25 residues of MrpC is indicated with a bold letter. The sequences of the –12 and –24 boxes for the $\sigma^{54}$ promoter and a putative ribosome-binding site are represented by underlined bold bases. The transcriptional initiation site is indicated as +1 with uppercase G. The sites hypersensitive to DNase I treatment are indicated by grey arrowheads. C) Schematic diagram for the transcriptional initiation of *mrpC*. A gray oval and an empty circle represent MrpB and MrpC existing as dimers, respectively. See the text for details.

A *mrpA* mutant (Δ*mrpA*) is able to form translucent mounds (aggregation), but sporulation is delayed. Δ*mrpA* yields spores of 10 % after 3 days and 30 % after 7 days development relative to the parent strain, suggesting that *mrpA* is required for only myxospore formation. The defective phenotype of a *mrpB* mutant (Δ*mrpB*) was complemented by introducing the wild-type *mrpB* gene, but was not by a mutation at the putative phosphorylation site Asp58 to Alq, suggesting that activation of MrpB by phosphorylation with HK is essential for *mrpC* expression and mature FB development.

Therefore, a scenario proposed by Sun and Shi (2001)[18] early in development is that the expression of *mrpAB* is activated via a developmentally specific signaling pathway and produced MrpA and MrpB. MrpB phosphorylated by MrpA and/or a complement HK, activates downstream gene expressions particularly *mrpC* and then promotes cellular aggregation. MrpC might auto-regulate its own expression, because *mrpC* expression using the reporter gene fused at the N-terminal coding region of the MrpC was not observed in the *mrpC*-deletion strain. They speculated that MrpA serves as a phosphatase for MrpB in late development if starvation still persists, since MrpA is not required for *mrpC* expression or for cellular aggregation during FB development.

## MrpC Binding Sites in *mrpC* and *fruA* Promoter Regions

In the upstream region of the *mrpC* promoter, there is a characteristic sequence for the −12/−24 box of the $\sigma^{54}$-promoter, 13-bp upstream from the translation start site. The 637-bp upstream region is sufficient for the developmental expression of *mrpC*.[18] The transcriptional start site of *mrpC* is the G residue, 63 bases upstream from the initiation codon (Fig. 2B).[34] DNA-binding assay and footprint analysis using purified MrpC reveals that MrpC binds to at least eight sites in the upstream region of the *mrpC* promoter.[35] As described above, MrpC also binds to two sites in the upstream region of the *fruA* promoter.[22] Two types of consensus sequences, A/GTTTC/GAA/G (X) and GTGTC-$N_8$-GACAC (Y) (where N is any base) summarize the sequences for MrpC-binding sites based on analysis of binding sites in both the *mrpC* and *fruA* promoter regions, (Fig. 2).[35] In the *mrpC* promoter region, there are six X- and two Y-type sites, Y1-X1-X2-X3-X4-26-bp hypersensitive region-Y2-X4-X5, while there are two tandem repeats of Y-type sites in the *fruA* promoter region. The hypersensitive region between the X4 and Y2 sites of DNase I digestion suggests that MrpC-binding introduces DNA bending in the *mrpC* promoter region. Although MrpC belongs to the CRP family, purified MrpC does not require any cNMP for specific binding to the *fruA* and *mrpC* promoter regions. The N-terminal MrpC sequence exhibits similarity to the cyclic nucleotide binding domain, however, the residues which have been shown in *E. coli* CRP to contact the cyclic phosphate and purine ring of cAMP are not conserved in MrpC. Thus another type of nucleotide such as (p)ppGpp may serve as an effector for MrpC in vivo,[22] since (p)ppGpp is known to be essential early in development.[7]

A genomic survey using the consensus X site shows that several genes including developmental genes contain multiple tandem repeats and also a single X site in the putative promoter regions and at the predicted promoter sequences.[35] In the CRP family, such promoter/binding site arrangements can provide bipolar (activation and repression) regulation for gene expressions.[36] This suggests that MrpC is a global regulator for developmental gene expression, including *fruA* transcription.

## Regulation of *mrpC* Expression Mediated by MrpA, MrpB and MrpC

MrpB belongs to the NtrC family of enhancer-binding proteins (EBPs). EBPs bind to DNA enhancer elements typically located 100-150 bp upstream of the transcription initiation site and make contact with $\sigma^{54}$-RNA polymerase ($\sigma^{54}$-RNAP) to form a DNA loop.[37] Based on gel retardation analysis, purified MrpB was able to form two bands with the *mrpC* promoter region and these bands were super-shifted when MrpB concentration was increased. Footprint analysis shows that MrpB protects two regions which overlap with MrpC weak binding sites, X1-X2 and X3-X4, respectively and contains a common sequence, AACGTT, indicated by Z1 (−169 to −164) and Z2 (−135 to −132) in Figure 2B. Furthermore, MrpA is found to have autokinase activity, most likely phosphorylated at the His residue, judging by acid or alkaline treatment. The phosphoryl group of

MrpA is able to transfer to an Asp residue of MrpB, suggesting that MrpA may act as MrpB kinase in FB development and sporulation, although the genetic study of *mrpA* and *mrpB* suggests that *mrpA* is not essential for *mrpC* expression (Nariya and Inouye, manuscript in preparation).

The initial transcription of *mrpC* is carried out with MrpB and $\sigma^{54}$-RNAP, as described above. Once MrpC is produced, MrpC also plays a critical role for its own expression together with MrpB. Based on the MrpB and MrpC binding site analysis in the *mrpC* promoter region, a model for *mrpC* expression is depicted in Figure 2C. In early development, *mrpAB* transcript is activated via developmentally specific signaling pathways and produces MrpA and MrpB. MrpB dimers (EBPs are usually dimers), bind to two enhancer elements at Z1 and Z2. MrpC existing in a small amount during vegetative growth (see below) binds to MrpC strong binding sites, Y1, Y2, X5 and X6 (Step 2), inducing DNA bending at the hypersensitive region in DNase I digestion located between X4 and Y2, (indicated by a gray triangle in Fig. 2B). DNA looping can also be stabilized by intrinsic DNA curvature.[38] Interestingly, the probability of DNA curvature in this region (26 bp) is high, as predicted by the bend.it program (http://hydra.icgeb.trieste.it/~kristian/dna/bend_it.html). $\sigma^{54}$-RNAP binding to the −12/−24 box interacts with MrpB, forming a stable transcription competent complex (Step 3). MrpB binding to the *mrpC* promoter region induces DNA bending between the two binding sites, but not the region that forms DNA looping. In contrast, MrpC binding seems to have a large effect on DNA bending and the assembly of a stable transcription competent complex of the *mrpC* promoter. It is not clear at present whether MrpB and MrpC bind independently or synergistically to the *mrpC* promoter region. Oligomerization of MrpB may be induced by phosphorylation on its receiver domain with MrpA and/or a complement HK, yet identified and then the function of ATPase is activated. ATP hydrolysis by the conserved central domain of MrpB is activated to convert the $\sigma^{54}$-RNAP closed promoter complex to the transcription competent open complex and subsequently produces MrpC (Step 4). As the developmental process progresses, MrpC expression reaches a maximum at 18 h and then decreases.[35] Accumulation of MrpC to a high concentration may result in its binding to the low affinity binding sites overlapping with MrpB binding sites and may interfere with MrpB binding to repress *mrpC* expression (Step 5). Therefore, the expression of *mrpC* is controlled with a highly sophisticated bipolar feed back system with MrpA/other HK and MrpB in TCS together with MrpC that functions as an activator and a repressor for its expression under specific developmental conditions.

## Pkn8-Pkn14 Kinase Cascade

*M. xanthus* contains 102 PSTKs, several of which have been studied for their roles in multicellular FB development and sporulation.[5,39-44] Pkn4 is a 6-phosphofructokinase (PFK) kinase whose gene is located 18-bp downstream of the *pfk* gene forming operon.[44] Both *pkn4* and *pfk* are essential for glycogen consumption during early development and for an effective sporulation.[5,45] PFK interacts with the Pkn4 regulatory domain (Pkn4RD) in a phosphorylation-dependent manner and is activated by Pkn4-mediated phosphorylation at its allosteric effector site. PFK activation is inhibited by a new factor, MkapB, that was identified by the yeast two-hybrid screen using the *M. xanthus* genomic library with Pkn4 as bait.[46] MkapB interacts in a phosphorylation-dependent manner with Pkn4 and remains associated with Pkn4RD. As a result, Pkn4 is unable to interact with PFK.

In addition to MkapB, two other factors have been shown to associate with Pkn4 and other membrane PSTKs such as Pkn1, Pkn2, Pkn8 and Pkn9 by the yeast two-hybrid screen.[46] The three multi-kinase associated proteins, MkapA, MkapB and MkapC, contain well-known protein-protein interaction domains. MkapA is a small protein containing a $C_2H_2$-type zinc-finger like motif, MkapB contains eight tandem repeats of TPR (tetratrico peptide repeat) domain and MkapC has three tandem repeats of fibronectin type 3 (FN3)-like domain. Therefore, these PSTKs seem to form a complex network via Mkaps that also function as adapters for recruiting factors in the PSTK signaling pathway. Moreover, a membrane-associated PSTK, Pkn8 was found to interact with a cytoplasmic PSTK, Pkn14 and Pkn14 associates with MrpC.[46] This opens a new possibility for the regulation of the essential transcription factor, MrpC, of which expression early in development is, as described above, under the control of MrpA and MrpB which belong to the TCS.

## *mrpC* Expression in *pkn8* and *pkn14*-Deletion Strains during Vegetative Growth

Pkn8 and Pkn14 are active kinases and Pkn8 phosphorylates Pkn14 which phosphorylates MrpC at the Thr residue(s) based on the biochemical analysis.[34,46] Thus, Pkn8 is Pkn14 kinase, Pkn14 is MrpC kinase and Pkn8 and Pkn14 form an active kinase cascade. The C-terminal 41 residues of Pkn14 are required for tetramarization of Pkn14 that interacts with MrpC. In the *pkn8* (Δ*pkn8*) and *pkn14* (Δ*pkn14*) deletion strains, FB development progressed significantly faster with respect to that of the parental strain. In contrast to a low level of *mrpC* expression in the parent strain during vegetative growth, *mrpC* expression in both Δ*pkn8* and Δ*pkn14* is highly elevated, suggesting that the Pkn8-Pkn14 kinase cascade negatively regulates *mrpC* expression during vegetative growth.[34] Accumulation of MrpC in Δ*pkn8* and Δ*pkn14* induces *fruA* expression at early stationary phase, compared to the normal time of *fruA* expression at 6 h of development. Therefore, the developmental phenotype of Δ*pkn8* and Δ*pkn14* seems to be due to untimely FruA production mediated by elevated levels of MrpC during vegetative growth. Since *pkn14* is mainly expressed during vegetative growth, whereas the *pkn8* expression increases steadily during vegetative growth and development, *mrpC* expression seems to be released from the inhibition of the Pkn8-Pkn14 kinase cascade in early development. When developmental processes are initiated, *mrpC* expression is activated with MrpA and MrpB whose expression are highly induced under early developmental signals and accelerated further by auto-regulation of MrpC by itself. Therefore, *M. xanthus* appears to utilize both the eukaryotic PSTK kinase cascade and the prokaryotic two-component system to control the precise timing of *mrpC* expression in its life cycle, since untimely *mrpC* and *fruA* expressions result in failure to form normal mature FBs filled with a high number of spores.

## MrpC2, a Major Regulator for *mrpC* and *fruA* Expression

As described above, *mrpC* expression at a low level during vegetative growth is upregulated in early development and the Pkn8-Pkn14 kinase cascade appears to negatively regulate *mrpC* expression.[34] Furthermore, the protein that lacks the N-terminal 32 residues of MrpC has been isolated as *fruA*-binding protein (FBP) and MrpC2 lacking the N-terminal 25 residues has the same binding activity of FBP to the *fruA* promoter.[22] There are many important questions to be answered to understand *mrpC* and *fruA* expressions and successive FruA-mediated gene expressions in *M. xanthus* development: How does the Pkn8-Pkn14 kinase cascade inhibit *mrpC* expression by Pkn14-mediated phosphorylation and how and when is MrpC2 produced during development and what are the differences between MrpC and MrpC2. Through study of MrpC/MrpC2 profiles by Western blot analysis using anti-MrpC IgG, a low level of MrpC as a 30 kD band was observed during vegetative growth and greatly increased upon initiation of development reaching a maximum at 18 h development.[34] Interestingly, another band at 29 kD was detected relative to the increase in MrpC during developmental progression and reached the maximum amount at 18 h development, similar to MrpC. The 29 kD band migrates at the same position of FBP, suggesting that the 29 kD band might be a developmentally processed product of MrpC and may have a major role in *mrpC* and *fruA* expressions.[34] Indeed, MrpC2 has a 4- and 8-fold higher binding activity with respect to that of MrpC to the upstream region of the *mrpC* and *fruA* promoter, respectively. Therefore, if the full-length MrpC is a primary product of *mrpC* as predicted from the sequence analysis of *mrpC*,[18] the developmental proteomic processing of MrpC to MrpC2 is a critical event for progression of development.

The expression of *fruA* requires A- and E-signaling, but is also reduced in a B-signaling defective *bsgA* mutant early in development.[23] The *bsgA* (*lonD*) gene is essential for development and is responsible for the production of B-signal that is required at the earliest stage in development.[47] It encodes an ATP-dependent serine protease, a member of the Lon family (Gill et al 1993; Tojo et al 1993).[48,49] Based on the reporter analysis, *bsgA* expression is up-regulated during development but it is still expressed during vegetative growth at a low level.[49] MrpC was produced at a similar low level in the *lonD* mutant and the parent strains during vegetative growth, when MrpC/MrpC2 expression was examined in the *lonD* mutant strain.[49] MrpC expression was up-regulated, moreover MrpC2 was detected in the parent strain at 12 h of development, however its expression in the mutant strain was greatly reduced. Interestingly, MrpC2 was not detected and FruA expression was markedly reduced in the mutant strain when compared to that of the parental strain, suggesting that LonD is responsible for the proteolytic processing of MrpC into MrpC2. Therefore, LonD plays critical roles in the high level expression of *mrpC* and *fruA* that in turn propels the developmental progression of *M. xanthus*.

## Inhibition by Pkn8-Pkn14 Kinase Cascade on MrpC/MrpC2 Expression during Vegetative Growth

Pkn14 is able to phosphorylate MrpC and its phosphorylation site is a Thr residue(s).[34] Upon phosphorylation by Pkn14, MrpC binding activity to the *mrpC* and *fruA* promoters is decreased to 23 and 12%, respectively, of the binding activity of unphosphorylated MrpC. Furthermore, MrpC2 was found not to be phosphorylated by Pkn14, indicating that the phosphorylation site(s) of MrpC is(are) at Thr-21 and/or Thr-22 within the N-terminal 25 residues of MrpC (Fig. 2B).

Since *mrpC* expression is elevated in the Δ*pkn8* and Δ*pkn14* strains relative to the parental strain, based on the *lacZ*-reporter analysis, MrpC/MrpC2 in Δ*pkn8* and Δ*pkn14* cells during vegetative growth and early developmental stage were examined by western blot analysis. As expected, elevated levels of MrpC were detected during vegetative growth in both Δ*pkn8* and Δ*pkn14* relative to the parental strain and reached almost the same level of MrpC as the parental strain at 12 h of development. Interestingly, MrpC2 production was clearly observed in Δ*pkn8* and Δ*pkn14* while it was not detectable in the parental strain during vegetative growth. Therefore, the high level of *mrpC* expression in Δ*pkn8* and Δ*pkn14* during vegetative growth is most likely due to MrpC2 accumulation, because MrpC2 has a 4-fold higher binding activity to the *mrpC* promoter. Furthermore MrpC2 was undetectable in the parental strain during vegetative growth, whereas *lonD* expression has been reported during vegetative growth. This implies that MrpC phosphorylation by the Pkn8-Pkn14 cascade may also inhibit its proteolytic processing to MrpC2 by LonD in addition to reducing MrpC-binding activity. To avoid the untimely initiation of developmental processes, *mrpC* expression during vegetative growth is kept at the low level by inhibiting MrpC autoactivation activity by phosphorylation with the Pkn8-Pkn14 cascade.

## Overview

The regulatory mechanism of *mrpC* and *fruA* expression essential for *M. xanthus* development is summarized in Figure 3. The *mrpC* gene is located downstream of the *mrpAB* operon encoding MrpA (HK) and MrpB (RR) of the two-component system and its expression seems to be controlled under the $\sigma^{54}$ promoter with an enhancer binding protein, MrpB. Expression of *mrpC* is further amplified by MrpC. During vegetative growth, the *mrpC* gene is likely transcribed at a low level by $\sigma^{54}$-RNAP in the presence of basal levels of phosphorylated MrpB. The MrpC produced is phosphorylated by Pkn14, which is activated by membrane-associated Pkn8. Phosphorylated MrpC has a low binding affinity to the *mrpC* and *fruA* promoters and may not be proteolytically processed to MrpC2 by LonD (see below). Thus, the Pkn8-Pkn14 kinase cascade prevents FB development under unfavorable conditions by negatively regulating the DNA-binding activity of MrpC and by producing MrpC2, which has a higher affinity to the *mrpC* and *fruA* promoters.

As the cells sense nutrient limitation, the developmental processes are initiated. The expression of *pkn14* decreases and Pkn8 activity may be down-regulated by the multiple PSTK network systems via Mkaps, MkapA, MkapB and MkapC, while *mrpAB* expression is activated under the developmental signaling pathway. The developmental expression of *mrpC* is initiated by MrpB binding to two enhancer elements together with MrpC binding in the *mrpC* promoter region. MrpC binding at high affinity sites (Y1 and Y2-X5-X6) in the *mrpC* promoter region appears to introduce a DNA bend and to be essential in forming a stable transcription competent complex with

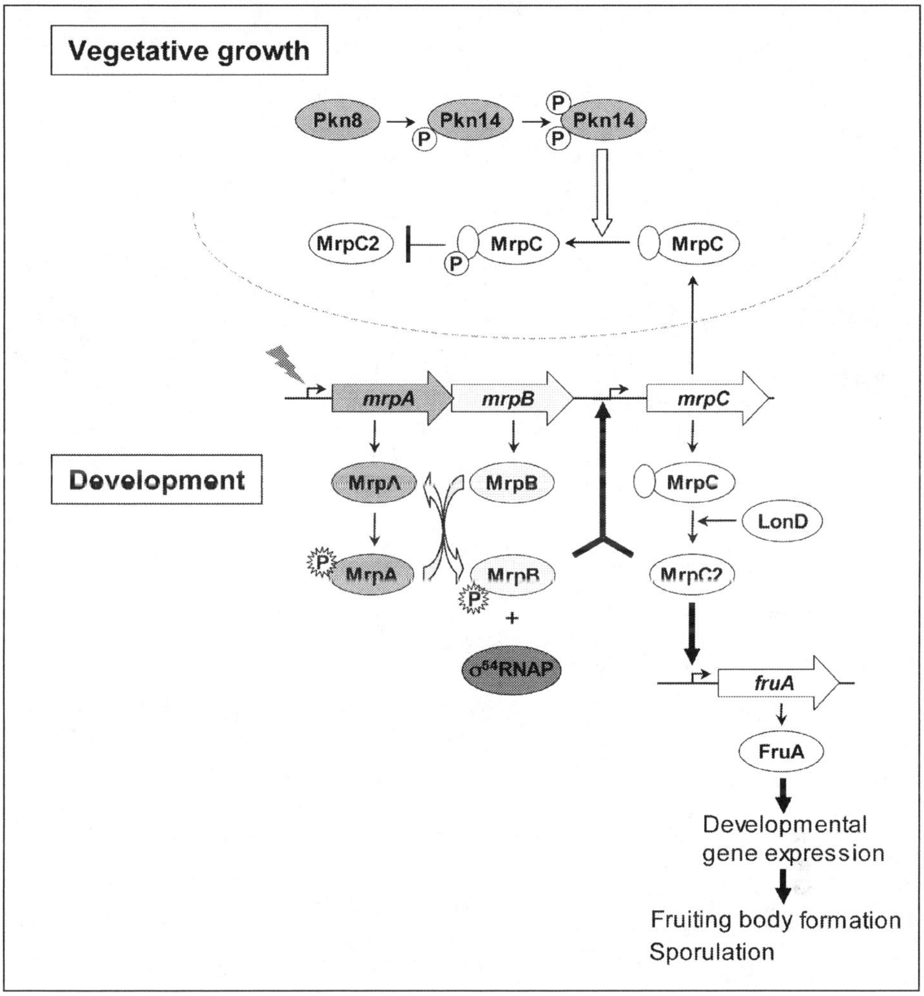


Figure 3. Proposed model for the regulation of *mrpC* and *fruA* expression regulated by PSTK and TCS signaling systems. See the text for details.

$\sigma^{54}$-RNAP (Fig. 2C). The developmentally synthesized MrpC is most likely not phosphorylated, but instead processed by LonD protease into MrpC2 that has a higher affinity for the *mrpC* and *fruA* promoter regions. The accumulation of MrpC2 activates the expression of *fruA* that encodes FruA, an essential transcription factor that regulates directly/indirectly a large number of gene and is essential for FB development and sporulation.

*M. xanthus* has numerous signal transduction systems with both the protein Ser/Thr kinase (PSTK) and the two-component His-Asp relay systems (TCS) that regulate the timely expression of genes under specific environmental signals. The molecular mechanisms guiding early developmental processes in *M. xanthus* are beginning to be understood and are different from those in other spore-forming bacteria. In contrast to *B. subtilis* endospore formation that is governed by sigma factor cascades, three essential transcription factors, MrpB, MrpC and FruA, most likely form a transcriptional cascade activating sequentially a downstream factor. Moreover, their expression is regulated by both a PSTK cascade and TCS. However, some key factors need to be elucidated.

Since phosphorylation likely controls the activity of MrpB and FruA, the most important factors are the HKs for FruA and MrpB. Particularly, the identification of FruA kinase and the characterization of its regulation on FruA activity are crucial to understand the expression of genes involved in developmental progression. Identifying the signals that affect *mrpC* expression via Pkn8-Pkn14 and MrpAB is a major challenge.

Among 102 PSTKs identified, over half of them are predicted to be receptor-type kinases and others are cytoplasmic kinases, implying that they may function for a variety of cellular processes by forming kinase cascades. Interestingly, one *M. xanthus* PSTK, containing HK domain (HK) in its regulatory domain, was found to interact with other PSTK containing a receiver domain (REC) via the HK and REC domains based on the genomic yeast two-hybrid screen. This suggests that TCS could be directly coupled to PSTK in the same signal transduction pathway (Nariya and Inouye, manuscript in preparation). While the social behavior and FB development of *M. xanthus* and the eukaryotic soil amoebae, *Dictyostelium discoideum*, are similar, their signal transduction systems appear to be different. The unique regulatory system that combines the PSTK and TCS discovered in *M. xanthus* may be utilized in other developmental and also pathogenic bacteria to maintain their life cycles.

## References

1. Parkinson JS. Signal transduction schemes of bacteria. Cell 1993; 73:857-871.
2. Stock JB, Ninfa AJ, Stock AM. Protein phosphorylation and regulation of adaptive responses in bacteria. Microbiol Rev 1989; 53:450-490.
3. Dworkin M. (1996) Recent advances in the social and developmental biology of the Myxobacteria. Microbiol Rev 1996; 60:70-102.
4. Dworkin M, Kaiser D, eds. Myxobacteria II. American Society for Microbiology. Washington, DC: ASM Press, 1993.
5. Nariya H, Inouye S. An effective sporulation of Myxococcus xanthus requires glycogen consumption via Pkn4-activated 6-phosphofructose kinase. Mol Microbiol 2003; 49:517-528.
6. Singer M, Kaiser D. Ectopic production of guanosine penta- and tetraphosphate can initiate early developmental gene expression in Myxococcus xanthus. Genes Dev 1995; 9:1633-1644.
7. Harris BZ, Kaiser D, Singer M. The guanosine nucleotide (p)ppGpp initiates development and A-factor production in Myxococcus xanthus. Genes Dev 1998; 12:1022-1035.
8. Shimkets LJ. Intercellular signaling during fruiting-body development of Myxococcus xanthus. Annu Rev Microbiol 1999; 53:525-549.
9. Kaiser D. Signaling in myxobacteria. Annu Rev Microbiol 2004; 58:75-98.
10. Kaplan HB, Plamann L. A Myxococcus xanthus cell density-sensing system required for multicellular development. FEMS Microbiol Lett 1996; 139:89-95.
11. Jelsbak L, Sogaard-Andersen L. Pattern formation by a cell surface-associated morphogen in Myxococcus xanthus. Proc Natl Acad Sci USA 2002; 99:2032-2037.
12. Hagen TJ, Shimkets LJ. Nucleotide sequence and transcriptional products of the csg locus of Myxococcus xanthus. J Bacteriol 1990; 172:15-23.
13. Lee BU, Lee K, Mendez J et al. A tactile sensory system of Myxococcus xanthus involves an extracellular NAD(P)(+)-containing protein. Genes Dev 1995; 9:2964-2973.
14. Lobedanz S, Søgaard-Andersen L. Identification of the C-signal, a contact-dependent morphogen coordinating multiple developmental responses in Myxococcus xanthus. Genes Dev 2003; 17:2151-2161.
15. Keseler IM, Kaiser D. σ54, a vital protein for Myxococcus xanthus. Proc Natl Acad Sci USA 1997; 94:1979-1984.
16. Jelsbak L, Givskov M, Kaiser D. Enhancer-binding proteins with a forkhead-associated domain and the sigma54 regulon in Myxococcus xanthus fruiting body development. Proc Natl Acad Sci USA 2005; 102:3010-3015.
17. Gorski L, Kaiser D. Targeted mutagenesis of sigma54 activator proteins in Myxococcus xanthus. J Bacteriol 1998; 180:5896-5905.
18. Sun H, Shi W. Genetic studies of mrp, a locus essential for cellular aggregation and sporulation of Myxococcus xanthus. J Bacteriol 2001; 183:4786-4795.
19. Sun H, Shi W. Analyses of mrp genes during Myxococcus xanthus development. J Bacteriol 2001; 183:6733-6739.
20. Caberoy NB, Welch RD, Jakobsen JS et al. Global mutational analysis of NtrC-like activators in Myxococcus xanthus: identifying activator mutants defective for motility and fruiting body development. J Bacteriol 2003; 185:6083-94.
21. Kirby JR, Zusman DR. Chemosensory regulation of developmental gene expression in Myxococcus xanthus. Proc Natl Acad Sci USA 2003; 100:2008-2013.

22. Ueki T, Inouye S. Identification of an activator protein required for the induction of fruA, a gene essential for fruiting body development in Myxococcus xanthus. Proc Natl Acad Sci USA 2003; 100:8782-8787.
23. Ogawa M, Fujitani S, Mao X et al. FruA, a putative transcriptional factor essential for the development of Myxococcus xanthus. Mol Microbiol 1996; 22: 757-767.
24. Horiuchi T, Taoka M, Isobe T et al. Role of fruA and csgA in gene expression during development of Myxococcus xanthus: Analysis by two-dimensional gel electrophoresis. J Biol Chem 2002; 277:26753-60.
25. Ellehauge E, Nørregaard-Madsen M, Søgaard-Andersen L. The FruA signal transduction protein provides a checkpoint for the temporal co-ordination of intercellular signals in Myxococcus xanthus development. Mol Microbiol 1998; 30:807-817.
26. Inouye M, Inouye S, Zusman DR. Biosynthesis and self-assembly of protein S, a development-specific protein of Myxococcus xanthus. Proc Natl Acad Sci USA 1979; 76:209-213.
27. Horiuchi T, Akiyama T, Inouye S et al. Analysis of dofA, a fruA-dependent developmental gene and its homologue, dofB, in Myxococcus xanthus. J Bacteriol 2002; 184:6803-6810.
28. Inouye M, Inouye S, Zusman DR. Gene expression during development of Myxococcus xanthus: pattern of protein synthesis. Dev Biol 1979; 68:579-591.
29. Søgaard-Andersen L, Kaiser D. C factor, a cell-surface-associated intercellular signaling protein, stimulates the Frz signal transduction system in Myxococcus xanthus. Proc Natl Acad Sci USA 1996; 93:2675-2679.
30. McCleary WR, McBride MJ, Zusman DR. Developmental sensory transduction in Myxococcus xanthus involves methylation and demethylation of FrzCD. J Bacteriol 1990; 172:4877-4887.
31. Ward MJ, Zusman DR. Motility in Myxococcus xanthus and its role in developmental aggregation. Curr Opin Microbiol 1999; 2:624-629.
32. Thony-Meyer L, Kaiser D. devRS, an autoregulated and essential genetic locus for fruiting body development in Myxococcus xanthus. J Bacteriol 1993; 175:7450-7462.
33. Ueki T, Inouye S. Identification of a gene involved in polysaccharide export as a transcription target of FruA, an essential factor for Myxococcus xanthus development. J Biol Chem 2005; 280:32279-32284.
34. Nariya H, Inouye S. Identification of a protein Ser/Thr kinase cascade that regulates essential transcriptional activators in Myxococcus xanthus. Mol Microbiol 2005; 58:367-379.
35. Nariya H, Inouye S. A protein Ser/Thr kinase cascade negatively regulates the DNA-binding activity of MrpC, a smaller form of which may be necessary for the Myxococcus xanthus development. Mol Microbiol 2006; 60:1205-1217.
36. Korner H, Sofia HJ, Zumft WG. Phylogeny of the bacterial superfamily of Crp-Fnr transcription regulators: exploiting the metabolic spectrum by controlling alternative gene programs. FEMS Microbiol Rev 2003; 27:559-592.
37. Buck M, Gallegos MT, Studholme DJ et al. The bacterial enhancer-dependent sigma(54) (sigma(N)) transcription factor. J Bacteriol 2000; 182:4129-4136.
38. Hoover TR, Santero E, Porter S et al. The integration host factor stimulates interaction of RNA polymerase with NIFA, the transcriptional activator for nitrogen fixation operons. Cell 1990; 63:11-22.
39. Muñoz-Dorado J, Inouye S, Inouye M. A gene encoding a protein serine/threonine kinase is required for normal development of M. xanthus, a gram-negative bacterium. Cell 1991; 67:996-1006.
40. Udo H, Muñoz-Dorado J, Inouye M et al. Myxococcus xanthus, a gram-negative bacterium, contains a transmembrane protein serine/threonine kinase that blocks the secretion of β-lactamase by phosphorylation. Genes Dev 1995; 9:972-983.
41. Zhang W, Inouye M, Inouye S. Reciprocal regulation of the differentiation of Myxococcus xanthus by Pkn5 and Pkn6, eukaryotic-like Ser/Thr protein kinases. Mol Microbiol 1996; 20:435-447.
42. Hanlon WA, Inouye M, Inouye S. Pkn9, a Ser/Thr protein kinase involved in the development of Myxococcus xanthus. Mol Microbiol 1997; 23:459-471.
43. Thomasson B, Link J, Stassinopoulos AG et al. MglA, a small GTPase, interacts with a tyrosine kinase to control type IV pili-mediated motility and development of Myxococcus xanthus. Mol Microbiol 2002; 46:1399-1413.
44. Nariya H, Inouye S. Activation of 6-phosphofructokinase via phosphorylation by Pkn4, a protein Ser/Thr kinase of Myxococcus xanthus. Mol Microbiol 2002; 46:1353-1366.
45. Nariya H, Inouye S. Factors that Modulate the Pkn4 Kinase Cascade in Myxococcus xanthus. J Mol Microbiol Biotechnol 2005; 9:147-153.
46. Nariya H, Inouye S. Modulating factors for the Pkn4 kinase cascade in regulating 6-phosphofructokinase in Myxococcus xanthus. Mol Microbiol 2005; 56:1314-1328.
47. Gill RE, Cull MG. Control of developmental gene expression by cell-to-cell interactions in Myxococcus xanthus. J Bacteriol 1986; 168:341-347.
48. Gill RE, Karlok M, Benton D. Myxococcus xanthus encodes an ATP-dependent protease which is required for developmental gene transcription and intercellular signaling. J Bacteriol 1993; 175:4538-4544.
49. Tojo N, Inouye S, Komano T. The lonD gene is homologous to the lon gene encoding an ATP-dependent protease and is essential for the development of Myxococcus xanthus. J Bacteriol 1993; 175:4545-4549.

# Chapter 8

# Two-Component Signaling Systems and Cell Cycle Control in *Caulobacter crescentus*

**Erin B. Purcell, Cara C. Boutte and Sean Crosson***

## Abstract

Recent work on the regulation of prokaryotic growth and development by two-component systems (TCS) has revealed unsuspected levels of complexity. In the dimorphic freshwater bacterium *Caulobacter crescentus,* TCS provide stringent temporal and spatial control of cellular development and cell-cycle progression. While the environmental signals modulating TCS regulatory networks are largely unknown, the components of the network and their interactions with each other are increasingly well-defined. Here, we present an overview of TCS regulation of cell-cycle control in *C. crescentus.*

## Introduction

The field of two-component signal transduction was primarily born out of early genetic and biochemical analyses of bacterial chemotaxis[1] and sporulation.[2,3] The realization that the regulators of *Bacillus subtilis* sporulation, Spo0A and Spo0F, share sequence homology with select regions of the *Escherichia coli* chemotaxis regulators CheY and CheB,[4] and the *E. coli* transcription factors ArcA and OmpR,[5] suggested that disparate physiological outputs were regulated by common protein structural motifs. The homologous domain shared by these proteins, now referred to as a receiver, is the defining domain of the downstream 'response regulator' component of a canonical TCS. The receiver domain is phosphorylated by an upstream sensor histidine kinase component in response to some regulatory signal.[6]

Since these early discoveries, the vast majority of characterized bacterial TCS have proven to be non-essential when grown under typical laboratory conditions and to regulate adaptive processes such as the control of cytoplasmic osmolarity, nutrient uptake and utilization, alternative cellular respiration and intracellular redox state in response to physical and chemical changes in the environment of the cell.[7] It can be argued that many such adaptive responses are likely to be essential in the context of the natural environment of the bacterium. Nevertheless, with one exception,[8] two-component genes in *E. coli* and *B. subtilis* have all been deleted in nutrient-rich medium without a significant loss in cell viability.[9,10] Our perspective on the functional roles of TCS proteins began to change when, over ten years ago, it was reported that that the gram-negative bacterium, *Caulobacter crescentus*, encoded at least two essential two-component response regulator proteins, DivK[11] and CtrA.[12] Subsequent work in *C. crescentus* identified two essential two-component

*Corresponding Author: Sean Crosson— Department of Biochemistry and Molecular Biology, and The Committee on Microbiology, The University of Chicago, Gordon Center for Integrative Science, 929 E. 57th Street, Chicago, IL 60637, USA. Email: scrosson@uchicago.edu

*Bacterial Signal Transduction: Networks and Drug Targets*, edited by Ryutaro Utsumi.

kinases, CckA[13] and DivL,[14] both of which are upstream regulators of CtrA phosphorylation[13,14] and act in concert with several other two-component proteins to regulate polar cell division and cell cycle progession.[15,16] All four of these TCS proteins exhibit dynamic subcellular localization across the cell cycle that relates to their function as cell-cycle regulators.

More recently, a systematic deletion of each predicted two-component gene in the genome of *C. crescentus* has identified 5 additional essential TCS proteins, two of which, CenK and CenR, act as a cognate pair and regulate cell envelope biogenesis and integrity.[17] Thus, *C. crescentus* encodes several essential TCS that function as regulators of cellular growth and development. This chapter will focus on the functional role of a defined set of essential and non-essential TCS that work in concert with several non-TCS regulatory proteins to control cell cycle progression in *C. crescentus.*

## *Caulobacter crescentus:* A Dimorphic Bacterial Model for Cell Cycle Regulation

Recent work in microbial cellular biology has decisively put an end to the long-held view of bacteria as largely unstructured cells devoid of spatial or temporal regulation. It is now widely recognized that prokaryotes exert complex and precise spatial control over many internal molecules and that this internal organization is dynamic and subject to regulation by diverse signals.[18] *C. crescentus* is one of the more intriguing model systems, as cellular replication involves a sequence of profound morphological transitions coordinated with cell-cycle specific changes in the expression, phosphorylation, sub-cellular localization and degradation of hundreds of proteins. Each *C. crescentus* cell cycle culminates in dimorphic cell division in which morphologically distinct swarmer and stalked cells are produced (Fig. 1B). The stalked cell is so named for a single membranous stalk that extends from the cell pole and has a structure known as the holdfast at its tip. The holdfast secretes a strongly adhesive polysaccharide which enables the bacterium to affix itself to surfaces and remain stationary.[19] After cell division is complete and the progeny swarmer cell is released, the stalked cell immediately recommences chromosome replication while the swarmer cell goes through an extended G1 phase in which DNA replication is blocked. Swarmer cells have a single polar flagellum and pili and are highly mobile, exhibiting chemotactic behavior. After a period of motility the swarmer cell sheds its flagellum and retracts its pili in response to some unknown cue(s), grows a polar stalk in their place and initiates chromosome replication and cell division (Fig. 1B). This dimorphic cell cycle allows the swarmer cells to seek advantageous new environments while the sessile stalked cells can strictly limit replication until local conditions are favorable; the pathways through which the bacteria evaluate their environment are generally unknown. Thus C. *crescentus* cell division requires coordinated regulation of metabolic, morphological and cell-division processes. Indeed, morphogenic and cell-cycle regulation is genetically linked by a series of regulatory circuits containing two-component proteins.[20,21]

## CtrA, GcrA and DnaA: Global Regulators of Cell Cycle Progression

At the top of the heirarchy of cell cycle control are three essential proteins: CtrA, a DNA-binding response regulator;[12] GcrA, a non-TCS regulatory protein conserved among the $\alpha$-proteobacteria;[22] and DnaA, a transcription factor which initiates chromosome replication.[23] CtrA, GcrA and DnaA have oscillating patterns of transcription across the cell cycle. Importantly, their oscillations are out of phase,[24] which allows them to regulate functionally complementary genetic modules at distinct temporal phases of the cell cycle (Fig. 1C).

CtrA acts as a transcription factor; like other DNA-binding response regulators, its activity is controlled by its phophorylation state. At least two upstream sensor histidine kinases, CckA and DivL, indirectly phosphorylate CtrA[25,26] (Fig. 2). When CtrA is phosphorylated, it binds to five target sequences in the chromosomal origin of replication and occludes the binding of DnaA, thus blocking the initiation of chromosomal replication.[27] CtrA~P is found in very high concentrations in swarmer cells, preventing them from entering S phase until the protein is cleared from the cell

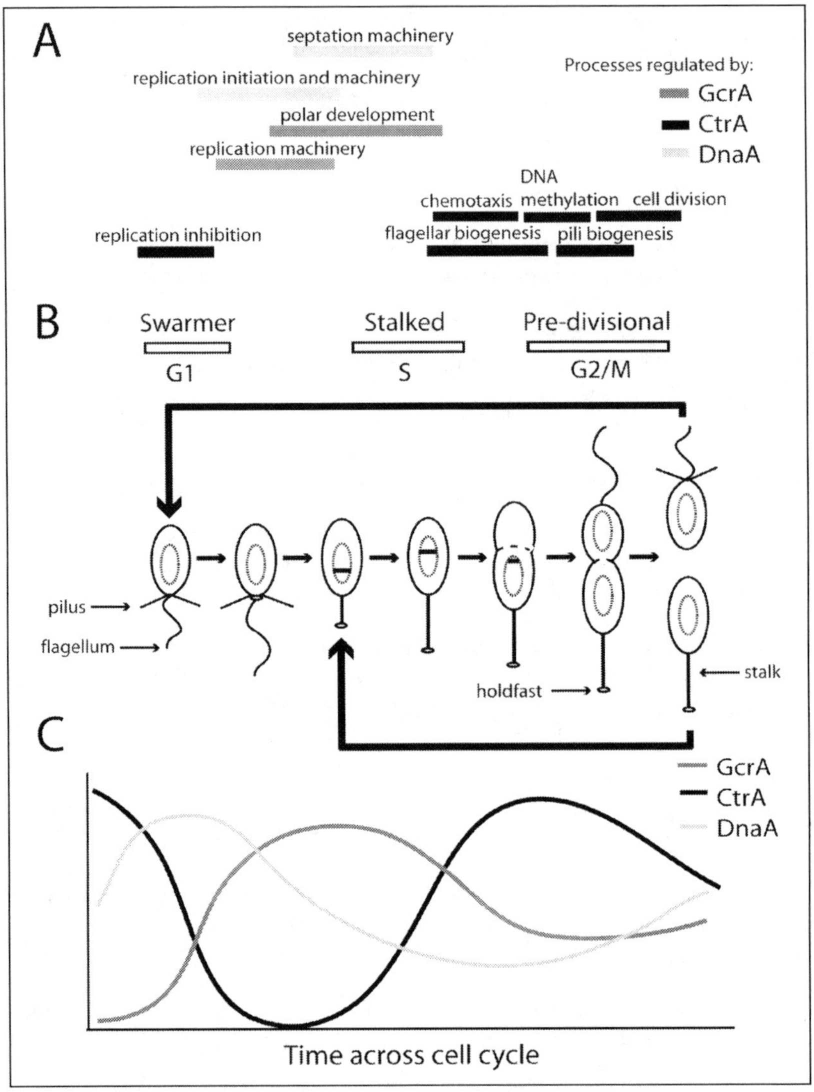


Figure 1. Coordination of cell cycle events by CtrA, GcrA and DnaA. The x-axis (time) is synchronous in the three panels. A) Approximate timing of cellular processes regulated by the three master regulators. B) Morphological events of the *C. crescentus* cell cycle. Pili, flagellum, stalk and holdfast are labeled. The dashed inner ring inside the *Caulobacter* cell represents the circular chromosome; the theta structure represents the replicating chromosome. C) Normalized cell-cycle protein levels of the three master regulators. The out-of-phase oscillations of CtrA, GcrA and DnaA allow "in-time" cueing of cell cycle events.

by proteolysis. This clearance occurs as the swarmer cell transitions to a stalked cell and coincides with attenuation of *ctrA* transcription, thus permitting chromosome replication to begin in the early stalked cell. Transcription of *ctrA* is upregulated again in late stalked and early predivisional cells, when it acts as a transcription factor for several classes of genes required for building a new swarmer cell.[12,28] In predivisional cells CtrA~P first activates transcription of genes required for flagellar biosynthesis and chemotaxis and later activates transcription of genes for pili biogenesis, DNA methylation and cell division[29] (Fig. 1A). Several other regulatory proteins including transcription factors, histidine kinases, response regulators and sigma factors are also regulated by CtrA~P at different times across the cell cycle.[29] Feedback regulation of *ctrA* by these downstream regulatory proteins will be discussed in a later section.

Although not a two-component protein, GcrA is conserved across a range of α-proteobacteria and acts as a key regulator of the two-component cell cycle circuitry of C. *crescentus.*[22] gcrA transcription is upregulated during the swarmer-to-stalked transition, just as CtrA is downregulated (Fig. 1C). Indeed, GcrA has been demonstrated to indirectly activate the transcription of *ctrA* in the stalked cell shortly after CtrA~P has been cleared from the cell during the G1 to S transition.[22] In addition to the negative regulation by CtrA~P, *gcrA* is subject to positive regulation by DnaA. While it is unclear if DnaA is the only transcriptional activator of *gcrA*, both CtrA~P clearance and DnaA activity are necessary for the accumulation of GcrA.[30] It has been shown that GcrA downregulates the transcription of DnaA, but it unclear whether this is regulated by a threshold level or by timing.[22] GcrA affects the transcription of many genes other than *ctrA*, many of which will be examined later in this chapter [22] (Fig. 1A). GcrA levels begin to decrease in predivisional cells, though they remain higher in the stalked compartment and the postdivisional stalked cell than in the swarmer cell.[24]

DnaA is essential for chromosome replication in C. *crescentus* and is also the transcription factor for multiple genes; some of these, such as ccrM and several genes necessary for flagellar biosynthesis, are redundantly regulated by CtrA[23] (Fig. 1A). DnaA is present throughout the cell cycle, but its level of accumulation is highest in cells undergoing the swamer-to-stalked transition and in early stalked cells (Fig. 1C). This is due to a combination of transcriptional and proteolytic regulation; transcription from the dnaA promoter increases roughly two-fold at this point in the cell cycle[31] and the half life of the unstable DnaA protein doubles at the swarmer-to-stalked transition.[32] DnaA proteolysis is dependent upon the same ClpP protease subunit responsible for CtrA degradation, although the ClpX chaperone required for CtrA degradation is unnecessary (Fig. 2). Intriguingly, the unknown regulatory mechanisms governing DnaA proteolysis allow for selective depletion of DnaA, but not CtrA, upon entry into stationary phase or upon starvation for carbon or nitrogen, although not phosphate.[32] Thus, DnaA acts as a positive regulator of *gcrA* transcription, GcrA acts both as a negative regulator of dnaA transcription and a positive regulator of *ctrA* transcription and CtrA acts as a negative transcriptional regulator of gcrA, forming an elegant, switch-like transcriptional feedback system which may be modulated by environmental conditions (Fig. 2).

## Transcriptional Control of CtrA

The out-of-phase oscillations of DnaA, GcrA and CtrA are tightly regulated by many factors including their own levels in the cell. There is an additional layer of complexity in the regulation of *ctrA* as two promoters, P1 and P2, control transcription of this critical response regulator.[33] Immediately after cell division the ctrA locus is fully methylated, which represses transcription.[34] After DnaA initiates replication and the replication fork passes the ctrA locus, two copies of the weaker P1 *ctrA* promoter are hemimethylated. It is some time before P1 is fully methylated by the essential CcrM DNA methyltransferase, because transcription of the ccrM gene is activated by CtrA~P.[34] CtrA begins to accumulate after the hemimethylated P1 promoter is indirectly activated by GcrA.[22] Once CtrA~P has accumulated to a threshold level, it binds to and represses its P1 promoter and activates its stronger P2 promoter, increasing the rate of its own transcription,[34] while repressing the *gcrA* promoter[22] (Fig. 2). It has been proposed that this step-wise increase of CtrA concentration within the cell may provide a mechanism for sequential activation of genes, as CtrA would temporally control regulated promoters in an order determined by their binding affinity to CtrA~P.[35]

## Proteolytic Control of Cell Cycle Regulatory Proteins

As previously mentioned, proteolysis of DnaA seems to depend upon both cell cycle progression and environmental input; the regulatory mechanisms governing this process are unknown. Both CtrA[36] and GcrA[24] are also selectively proteolyzed at different times across the cell cycle. While immunoblot analysis has shown that GcrA has a half-life four times longer in stalked cells than in swarmer cells, it is not yet known what proteins are involved in GcrA proteolysis.[24] The proteins involved in CtrA proteolysis have been comparatively well characterized. Specifically, CtrA is degraded by the ClpXP protease complex during the swarmer-to-stalked transition.[36] CtrA proteolysis is dependent on a bipartite signal within CtrA that is contained within both the N-terminal receiver domain and the final 15 amino acids at the C-terminus.[37] Clearance of CtrA from the cell and consequently from the CtrA binding sites on the origin of replication allows chromosome replication to begin.[38] In the predivisional cell CtrA is selectively proteolyzed in the stalked compartment while it accumulates in the swarmer compartment; this proteolysis helps to direct the different fates of these two cells.[39]

The ClpXP protease complex is dynamically localized across the cell cycle: It is found at the stalked pole of stalked cells, is transiently localized to the division plane of early predivisional cells and during the later stages of predivision it returns to the stalked pole where it remains until the next division.[40] This localization pattern is dependent upon the phosphorylation state of the single domain response regulator, CpdR.[28,41] During the swarmer-to-stalked transition, unphosphorylated CpdR accumulates and localizes to the stalked pole, where it is necessary for the localization of the ClpXP complex.[41] When CpdR is phosphorylated, ClpXP is delocalized and unable to efficiently degrade CtrA.[41] Notably, CpdR is phosphorylated by the same multi-component phosphorelay system that phosphorylates CtrA. The sensor histidine kinase CckA phosphorylates the histidine phosphotransferase ChpT, which in turn phosphorylates both CtrA and CpdR.[28] The activity of ChpT thus simultaneously actives CtrA and deactivates the machinery responsible for CtrA~P degredation (Fig. 2).

The small "adaptor" protein RcdA is expressed at low levels in swarmer cells but upregulated by CtrA~P during the swarmer-to-stalked transition.[40] Cells lacking RcdA do not localize CtrA to the ClpXP protease complex and cannot proteolyze CtrA[40]. The mechanistic and structural basis of polar ClpXP targeting by the TCS protein, CpdR and the adapter protein, RcdA, remain unexplored. No chaperones or adaptors have been reported to be necessary for degradation of DnaA by ClpXP,[32] and it is yet to be determined if other proteins are involved in this process.

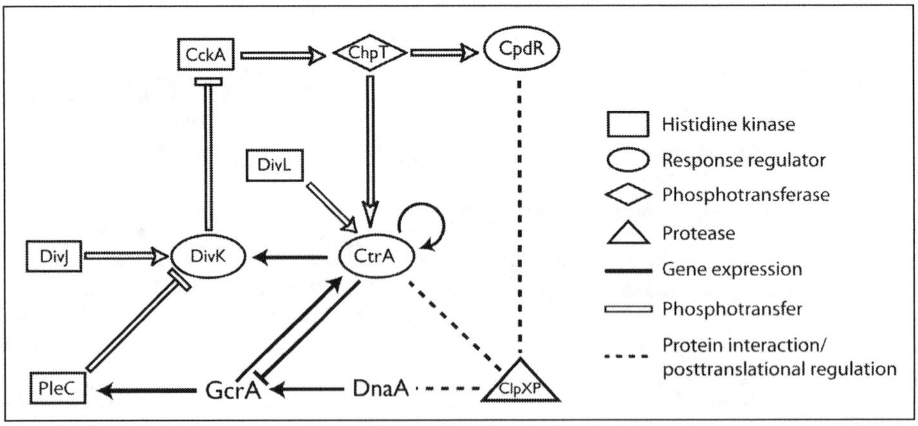


Figure 2. Schematic of *C. crescentus* regulatory components and their known interactions.

## Phosphorylative Control of Cell Cycle Regulatory Proteins

The phosphorylation profile of CtrA mirrors its transcriptional profile. CtrA~P is present in swarmer cells and degraded during the G1 to S transition (Fig. 1). After CtrA is translated in late stalked cells, it is phosphorylated and remains active in stalked and predivisional cells.[38] A network of signal transduction systems mediates the phosphorylative regulation of CtrA. The CckA hybrid histidine kinase autophosphorylates in response to an unknown signal[38] and then passes the phosphoryl group to the histidine phosphotransfer protein ChpT, which directly phosphorylates CtrA.[28] CckA is expressed at constant levels throughout the cell cycle, but is only active when localized to the cell poles. CckA localizes to swarmer poles while DNA replication is blocked, delocalizes during the G1 to S transition and then localizes to both poles of the predivisional cell.[25] The role played by the sensor domain in CckA autophosphorylation and/or localization is unknown. Both localization and activation of CckA are regulated by the response regulator DivK, which upon phosphorylation inhibits CckA activity and leads to its delocalization into the cytoplasm.[28] CtrA~P also directly controls the transcription of *divK*, creating a negative feedback regulatory circuit[29] (Fig. 2).

Like CckA, the DivL kinase is also genetically upstream of CtrA. Its purified catalytic domain phosphorylates purified CtrA in vitro, suggesting that DivL influences CtrA activity directly rather than through a phosphorelay or signal transduction pathway.[26] DivL is unusual among sensor kinases of two-component systems in that, although it has all the conserved motifs of regular histidine kinases, it is phosphorylated on a tyrosine residue.[26] DivL, like other histidine kinases, contains a conserved four-helix bundle[42] necessary for dimerization and catalysis. In a yeast two-hybrid assay, fragments of DivL were shown to weakly interact with DivK but the functional significance, if any, of this interaction remains unknown.[43] The transcription of DivL is not cell cycle regulated and localization is not required for DivL kinase activity,[44] although it does localize to the stalked cell pole in a manner dependent on another TCS sensor kinase, DivJ.[44] DivL is not involved in the regulation of chromosome replication and its exact role in the regulation of cell cycle progression and cell division remains unclear.[44]

PleC and DivJ are two additional TCS histidine kinases that regulate CtrA through the response regulator DivK. This additional two-component regulatory circuit is discussed below.

## Upstream Cell Cycle Control by Opposing Kinases

DivJ and PleC are non-essential two-component transmembrane sensor histidine kinases that, although similar in sequence, oppose each other to regulate the essential cytoplasmic response regulator DivK. Unidentified signals perceived by the kinase sensory domains regulate the kinase activity of both proteins, but the mechanisms by which this occur are unknown. In response to these unknown signals, DivJ autophosphorylates and transfers its phosphoryl group to DivK; PleC, also regulated by unknown signals, mainly acts to desphosphorylate DivK~P.[11] However, in cells devoid of both DivJ and PleC, DivK~P is still found, indicating that DivK phosphorylation is not exclusive to these kinases[45].

DivJ localizes to the stalked pole and PleC localizes to the flagellar pole in dividing and swarmer cells, while DivK 'shuttles' between them based on its phosphorylation state.[46] Mutant cells lacking divJ can be rescued by mutations in cckA and divL, which encode kinases that phosphorylate CtrA.[47] These genetic data provide evidence that DivJ and DivK~P ultimately inhibit CtrA activity, while PleC and DivK enhance CtrA activity.[47] DivJ is necessary for proper localization, length and morphology of stalks and holdfasts,[20,45] while PleC is necessary for the polar localization and assembly of pili components; loss of PleC activity leads to a loss of asymmetry in dividing cells, resulting in two flagellated poles.[48] While DivJ and PleC kinase activity on DivK oppose each other, proper DivJ localization to the stalked pole during division depends upon functional PleC.[45] PleC also controls the localization, at different time points, of a variety of proteins involved in pili formation and subsequent cell motility; its own localization is in turn controlled by PodJ, whose transcription and proteolytic processing is regulated by downstream targets of the PleC response regulator, DivK[49] (Fig. 2).

## Network Control of Histidine Kinase Localization and Polar Morphogenesis

The sensor-linked kinase PleC is a critical regulator of cellular morphology and development and its expression and localization, are subject to multiple levels of control by the core cell cycle regulatory network. Both PleC and PodJ, the transmembrane polarity factor which regulates PleC localization,[49] are subject to overlapping transcriptional activation by DnaA[30] and, indirectly, by GcrA.[22] Simultaneously, CtrA activates the transcription of PerP, which is one of two proteases that process PodJ in order to alter its activity[50]. PodJ is transcribed and translated in the stalked cell and present in the early predivisional cell. Full-length PodJ ($PodJ_L$) is a transmembrane protein with periplasmic and cytoplasmic domains, which localizes to the nascent swarmer pole during early cell division. Here, PodJ recruits PleC which is required for polar pili formation.[49] During cell division, PerP cleaves the periplasmic domain from PodJ while the remaining cytoplasmic domain and transmembrane anchor ($PodJ_S$) remain at the flagellar pole of the progeny swarmer cell. The membrane metalloprotease MmpA cleaves $PodJ_S$ during the swarmer-to-stalked transition and releases it from the membrane. The interval of time between PerP cleavage and MmpA cleavage appears to be determined by a regulatory mechanism distinct from the transcriptional regulation of PerP,[50] as MmpA is not localized to any region of the cell membrane and its transcription is not cell-cycle dependent [51].

## The Search for Regulatory Signals

Although there are almost certainly other unidentified components in the regulatory system that controls cell cycle progression, several major components of this control circuitry have been mapped and their interactions with each other have largely been defined (Fig. 2). However, the physical and chemical nature of signals that affect these regulatory pathways remain almost entirely undefined. For example, cell cycle arrest by carbon or nitrogen starvation does not affect cellular levels of CtrA but dramatically and reversibly enhances the rate of DnaA proteolysis.[32] This result indicates that environmental nutrient availability controls chromosome replication with high specificity.[32] The molecular mechanism(s) by which this process occurs remain a mystery, but are almost certainly tied into the core TCS control circuitry of *Caulobacter*. As the contents of prokaryotic cells become increasingly well-described, the next frontier in cellular biology is to determine not only which molecules interact, but how those interactions are governed in response to the intra- and extra-cellular environment. *Caulobacter crescentus,* which exhibits polar cell division and possesses an extensive suite of defined two-component proteins at the highest levels of its cell cycle control circuitry, is a uniquely suitable model organism in which to pursue answers to this next generation of questions in bacterial cell biology and signal transduction.

## References

1. Koshland Jr DE. A response regulator model in a simple sensory system. Science 1977; 196:1055-63.
2. Hoch JA. Genetic analysis of pleiotropic negative sporulation mutants in Bacillus subtilis. J Bacteriol 1971; 105:896-901.
3. Ollington JF, Haldenwang WG, Huynh TV et al. Developmentally regulated transcription in a cloned segment of the Bacillus subtilis chromosome. J Bacteriol 1981; 147:432-42.
4. Stock A, Koshland Jr DE, Stock J. Homologies between the Salmonella typhimurium CheY protein and proteins involved in the regulation of chemotaxis, membrane protein synthesis, and sporulation. Proc Natl Acad Sci USA 1985; 82:7989-93.
5. Trach KA, Chapman JW, Hoch JA. Deduced product of the stage 0 sporulation gene spo0F shares homology with the Spo0A, OmpR, and SfrA proteins. Proc Natl Acad Sci USA 1985; 82:7260-4.
6. Skerker JM, Prasol MS, Laub MT Two-Component Signal Transduction Pathways Regulating Growth and Cell Cycle Progression in a Bacterium: A System-Level Analysis. PLoS Biology 2005; 3:e.334.
7. Hoch JA, Silhavy TJ, eds. Two-Component signal transduction. Washington, DC: ASM; 1995.
8. Fabret C, Hoch JA. A two-component signal transduction system essential for growth of Bacillus subtilis: implications for anti-infective therapy. J Bacteriol 1998; 180:6375-83.
9. Kobayashi K, Ehrlich SD, Ogasawara N et al. Essential Bacillus subtilis genes. Proc Natl Acad Sci USA 2003; 100:4678-83.

10. Zhou L, Lei XH, Wanner BL et al. Phenotype microarray analysis of Escherichia coli K-12 mutants with deletions of all two-component systems. J Bacteriol 2003; 185:4956-72.
11. Hecht GB, Lane T, Ohta N et al. An essential single domain response regulator required for normal cell division and differentiation in Caulobacter crescentus. EMBO J 1995; 14:3915-24.
12. Quon KC, Marczynski GT, Shapiro L. Cell Cycle Control by an Essential Bacterial Two-Component Signal Transduction Protein. Cell 1996; 84:83-93.
13. Jacobs C, Domian IJ, Shapiro L et al. Cell cycle-dependent polar localization of an essential bacterial histidine kinase that controls DNA replication and cell division. Cell 1999; 97:111-20.
14. Wu J, Ohta N, Newton A et al. A novel bacterial tyrosine kinase essential for cell division and differentiation. Proc Natl Acad Sci USA 1999; 96:13068-73.
15. Jacobs-Wagner C. Regulatory proteins with a sense of direction: cell cycle signalling network in Caulobacter. Mol Microbiol 2004; 51:7-13.
16. Skerker JM, Laub MT. Cell-cycle progression and the generation of asymmetry in Caulobacter crescentus. Nat Rev Microbiol 2004; 2:325-337.
17. Skerker JM, Prasol MS, Laub MT et al. (2005) Two-component signal transduction pathways regulating growth and cell cycle progression in a bacterium: a system-level analysis. PLoS Biol 2005; 3:e.334.
18. Gitai Z. The New Bacterial Cell Biology: Moving Parts and Subcellular Architecture. Cell 2005; 120:577-586.
19. Poindexter JS. Biological Properties and Classification of the Caulobacter Group. Bacteriol Rev 1964; 28:231-295.
20. Sommer J, Newton A. Pseudoreversion Analysis Indicates a Direct Role of Cell Division Genes in Polar Morphogenesis and Differentiation in Caulobacter crescentus. Genetics 1991; 129:623-630.
21. McGrath P, Viollier P, McAdams H. Setting the pace: mechanisms tying Caulobacter cell-cycle progression to macroscopic cellular events. Curr Opin Microbiol 2004; 7:192-197.
22. Holtzendorff J, Hung D, Shapiro L et al. Oscillating Global Regulators Control the Genetic Circuit Driving a Bacterial Cell Cycle. Science 2004; 304:983-987.
23. Gorbatyuk B, Marczynski GT. Physiological consequences of blocked Caulobacter crescentus dnaA expression, an essential DNA replication gene. Molecular Microbiology 2001; 40:485-497.
24. Collier J, Murray SR, Shapiro L. DnaA couples DNA replication and the expression of two cell cycle master regulators. EMBO J 2006; 25:346-356.
25. Jacobs C, Domian IJ, Shapiro L et al. Cell Cycle-Dependent Polar Localization of an Essential Bacterial Histidine Kinase that Controls DNA Replication and Cell Division. Cell 1999; 97:111-120.
26. Wu J, Ohta N, Newton A et al. A novel bacterial tyrosine kinase essential for cell division and differentiation. Proc Natl Acad Sci USA 1999; 96:13068-13073.
27. Quon KC, Yang B, Marczynski GT et al. Negative control of bacterial DNA replication by a cell cycle regulatory protein that binds at the chromosome†origin. Proc Natl Acad Sci USA 1998; 95:120-125.
28. Biondi E, Reisinger S, Laub M et al. Regulation of the bacterial cell cycle by an integrated genetic circuit. Nature 2006; 444:899-404.
29. Laub M, Chen S, McAdams H et al. Genes directly controlled by CtrA, a master regulator of the Caulobacter cell cycle. Proc Natl Acad Sci USA 2002; 99:4632-4637.
30. Hottes A, Shapiro L, McAdams H. DnaA coordinates replication initiation and cell cycle transcription in Caulobacter crescentus. Mol Microbiol 2005; 58:1340-1353.
31. Zweiger G, Shapiro L. Expression of Caulobacter dnaA as a Function of the Cell Cycle. J Bacteriol 1994; 176:401-408.
32. Gorbatyuk B, Marczynski GT. Regulated degradation of chromosome replication proteins DnaA and CtrA in Caulobacter crescentus. Mol Microbiol 2005; 55:1233-1245.
33. Domian IJ, Reisenauer A, Shapiro L. Feedback control of a master bacterial cell-cycle regulator. Proc Natl Acad Sci USA 1999; 96:6648-6653.
34. Reisenauer A, Shapiro L. DNA methylation affects the cell cycle transcription of the CtrA global regulator in Caulobacter. EMBO J 2002; 21:4969-4977.
35. Reisenauer A, Quon K, Shapiro L. The CtrA Response Regulator Mediates Temporal Control of Gene Expression during the Caulobacter Cell Cycle. J Bacteriol 1999; 181:2430-2439.
36. Jenal U, Fuchs T. An essential protease involved in bacterial cell-cycle control. EMBO J 1998; 17:5658-5669.
37. Ryan KR, Judd EM, Shapiro L. J The CtrA Response Regulator Essential for Caulobacter crescentus Cell-cycle Progression Requires a Bipartite Degradation Signal for Temporally Controlled Proteolysis. Mol Biol 2002; 324:443-455.
38. Ausmees N, Jacobs-Wagner C. Spatial and Temporal Control of Differentiation and Cell Cycle Progression in Caulobacter Crescentus. Ann Rev Microbiol 2003; 57:225-247.
39. Domian I J, Quon KC, Shapiro L. Cell Type-Specific Phosphorylation and Proteolysis of a Transcriptional Regulator Controls the G1-to-S Transition in a Bacterial Cell Cycle. Cell 1997; 90:415-424.

40. McGrath P, Iniesta AA, McAdams H et al. A Dynamically Localized Protease Complex and a Polar Specificity Factor Control a Cell Cycle Master Regulator. Cell 2006; 124:535-547.
41. Iniesta A, McGrath P, Shapiro L et al. A phospho-signaling pathway control the localization and activity of a protease complex critical for bacterial cell cycle progression. Proc Natl Acad Sci USA 2006; 103:0935-10940.
42. Bylsma N, Drakenberg T. Prokaryotic calcium-binding protein of the calmodulin superfamily. FEBS Lett 1992; 299:44-47.
43. Ohta N, Newton A. The Core Dimerization Domains of Histidine Kinases Contain Recognition Specificity for the Cognate Response Regulator. J Bacteriol 2003; 185:4424-4431.
44. Sciochetti S, Ohta N, Newton A. The role of polar localization in the function of an essential Caulobacter crescentus tyrosine kinase. Mol Microbiol 2005; 56:1467-1480.
45. Wheeler RT, Shapiro L. Differential Localization of Two Histidine Kinases Controlling Bacterial Cell Differentiation. Mol Cell 1999; 4:683-694.
46. Matroule J-Y, Lam H, Jacobs-Wagner et al. Cytokinesis Monitoring during Development: Rapid Pole-to-Pole Shuttling of a Signaling Protein by Localized Kinase and Phosphatase in Caulobacter. Cell 2004; 118:579-590.
47. Pierce D, O'Donnol D, Brun Y et al. Mutations in DivL and CckA Rescue a divJ Null Mutant of Caulobacter crescentus by Reducing the Activity of CtrA. J Bacteriol 2006; 188:2473-2482.
48. Viollier PH, Sternheim N, Shapiro L. A dynamically localized histidine kinase controls the assymetric distribution of polar pili proteins. EMBO J 2002; 21:4420-4428.
49. Viollier PH, Sternheim N, Shapiro L. Identification of a localization factor for the polar positioning of bacterial structural and regulatory proteins. Proc Natl Acad Sci USA 2002; 99:13831-13836.
50. Chen JC, Hottes AK, Shapiro L et al. Cytokinesis signals truncation of the PodJ polarity factor by a cell cycle-regulated protease. EMBO J 2006; 25:377-386.
51. Chen JC, Viollier PH, Shapiro L. A membrane metalloprotease participates in the sequential degradation of a Caulobacter polarity determinant. Mol Microbiol 2005; 55:1085-1103.

# CHAPTER 9

# RegB/RegA, A Global Redox-Responding Two-Component System

**Jiang Wu and Carl E. Bauer***


## Abstract

The RegB-RegA regulon from *Rhodobacter capsulatus* and *Rhodobacter sphaeroides* encodes proteins involved in numerous energy-generating and energy-utilizing processes such as photosynthesis, carbon fixation, nitrogen fixation, hydrogen utilization, aerobic and anaerobic respiration, denitrification, electron transport and aerotaxis. The redox signal that is detected by the membrane-bound sensor kinase, RegB, has been identified to be the ubiquinone pool in the membrane. Regulation of RegB autophosphorylation also involves a redox-active cysteine that is present in the cytosolic region of RegB. Both phosphorylated and unphosphorylated forms of the cognate response regulator RegA are capable of activating or repressing a variety of genes in the regulon. Highly conserved homologues of RegB and RegA have been found in a wide number of photosynthetic and nonphotosynthetic bacteria with evidence suggesting that RegB/RegA have a fundamental role in the transcription of redox-regulated genes in many bacterial species.


## Introduction

Genetic screens initially identified RegB and RegA as regulators of the photosystem synthesis in *R. capsulatus*.[1,2] In this species, null mutations in *regB* and *regA* are defective in high-level expression of the photosystem which are normally only synthesized under conditions of low oxygen tension.[3] Expression of the *puh*, *puf* and *puc* operons that encode apoproteins for the light harvesting I, light harvesting II and reaction center complexes of the photosystem are significantly reduced in RegB and RegA mutants.[1,2] The similar phenotypes displayed by these mutants led to the hypothesis that they may be cognate *trans*-acting partners constituting a two-component regulatory system. This was confirmed by sequence analysis which demonstrated that RegB exhibits homology to histidine protein kinases[1,4-6] and RegA exhibits homology to DNA-binding response regulators.[2,4-6]

Subsequent to the discovery of RegB/RegA from *R. capsulatus*, homologous two-component regulatory systems were found and genetically characterized in many other species such as the RegB/RegA[7] homologs in *Rhodobacter sphaeroides* (also called PrrB/PrrA),[8,9] RegS/RegR system from *Bradyrhizobium japonicum*,[10] ActS/ActR from *Sinorhizobium meliloti*,[11] RoxS/RoxR from *Pseudomonas aeruginosa*[12] and RegB/RegA from *Rhodovulum sulfidophilum* and *Roseobacter denitrificans*[13] (Table 1). Genome sequence studies have also identified RegA and RegB homologues in many other photosynthetic as well as nonphotosynthetic α- and γ-proteobacterial species (Table 1).

Genetic shuttling studies have demonstrated that RegB and RegA homologues from different species are in vitro and in vivo interchangeable, which means phosphotransfer can be observed between different RegB and RegA homologues,[12,14] and that some RegA homologues can bind to promoters and regulate gene transcription in another species.[12,14] Indeed it is now well established

*Corresponding Author: Carl E. Bauer—Department of Biology, Indiana University, Myers Hall, 915 E. Third St., Bloomington, IN 47405-7170, USA. Email: bauer@indiana.edu

*Bacterial Signal Transduction: Networks and Drug Targets*, edited by Ryutaro Utsumi.

*Table 1.* ***RegB and RegA homologues identified by similarity based search***

| Organism | RegB | | | | | RegA | | | |
|---|---|---|---|---|---|---|---|---|---|
| | RegB | Identity | H-Box | Q-Binding Site | Redox-Active Cystein | RegA | Identity | Acid-Box | DNA-Binding Domain |
| *Rhodobacter capsulatus* | YES Q9L906 | | YES | YES | YES | YES P42508 | | YES | YES |
| *Rhodobacter sphaeroides* | YES Q3J6C1 | 58% | YES | YES | YES | YES Q53228 | 83% | YES | YES |
| *Silicibacter pomeroyi* | YES Q5LLQ5 | 54% | YES | YES | YES | YES Q1GCP6 | 82% | YES | YES |
| *Rhodovulum sulfidophilum* | YES O82866 | 54% | YES | YES | YES | YES O82868 | 83% | YES | YES |
| *Roseobacter denitrificans* | YES O82869 | 55% | YES | YES | YES | YES Q9ZNM4 | 81% | YES | YES |
| *Oceanicola granulosus* | YES Q2CJX1 | 53% | YES | YES | YES | YES Q2CJX3 | 84% | YES | YES |
| *Jannaschia sp.* | YES Q28JY5 | 51% | YES | YES | YES | YES Q28JX7 | 81% | YES | YES |
| *uncultured proteobacterium* | YES Q8KYV6 | 53% | YES | YES | YES | YES Q8KYV8 | 78% | YES | YES |
| *Rhizobium loti* | YES Q98C40 | 38% | YES | YES | YES | YES Q98C39 | 69% | YES | YES |
| *Brucella suis* | YES Q8G321 | 37% | YES | YES | YES | YES Q8G319 | 67% | YES | YES |
| *Brucella abortus* | YES Q2YP02 | 37% | YES | YES | YES | YES Q57FN7 | 67% | YES | YES |
| *Xanthobacter sp.* | YES Q26N86 | 38% | YES | YES | YES | YES Q26N85 | 69% | YES | YES |
| *Brucella melitensis* | YES Q8YER2 | 37% | YES | YES | YES | YES Q8YER6 | 67% | YES | YES |
| *Bradyrhizobium japonicum* | YES O86124 | 36% | YES | YES | YES | YES Q89VZ0 | 69% | YES | YES |
| *Aurantimonas sp.* | YES Q1YF90 | 35% | YES | YES | YES | YES Q1YF91 | 68% | YES | YES |
| *Nitrobacter winogradskyi* | YES Q3SWG3 | 35% | YES | YES | YES | YES Q3SWG2 | 70% | YES | YES |
| *Rhizobium meliloti* | YES Q92TA1 | 36% | YES | YES | YES | YES Q52913 | 70% | YES | YES |
| *Rhodopseudomonas palustris* | YES Q6NCA0 | 36% | YES | YES | YES | YES Q6NCA1 | 70% | YES | YES |
| *Rhizobium etli* | YES Q2KE47 | 36% | YES | YES | YES | YES Q2KE48 | 69% | YES | YES |
| *Sinorhizobium medicae* | YES Q52912 | 36% | YES | YES | YES | NO[1] | | | |
| *Nitrobacter hamburgensis* | YES Q1QRL7 | 34% | YES | YES | YES | YES Q1QRL6 | 70% | YES | YES |
| *Caulobacter crescentus* | YES Q9ABH9 | 38% | YES | YES | YES | YES Q9ABI0 | 68% | YES | YES |
| *Rhizobium leguminosarum* | YES Q1MNA6 | 36% | YES | YES | YES | YES Q1MNA7 | 69% | YES | YES |

**continued on next page**

***Table 1. Continued***

| | RegB | | | | | RegA | | | |
|---|---|---|---|---|---|---|---|---|---|
| **Organism** | **RegB** | **Identity** | **H-Box** | **Q-Binding Site** | **Redox-Active Cystein** | **RegA** | **Identity** | **Acid-Box** | **DNA-Binding Domain** |
| *Agrobacterium tumefaciens* | YES Q8UJ81 | 36% | YES | YES | YES | YES Q8UJ82 | 68% | YES | YES |
| *Nitrosospira multiformis* | YES Q2YD54 | 27% | YES | YES | YES | YES Q2YD55 | 59% | YES | YES |
| *Pseudomonas syringae* | YES Q4ZNL1 | 30% | YES | YES | YES | YES Q87WJ3 | 52% | YES | YES |
| *Chromohalobacter salexigens* | YES Q1QWH9 | 26% | YES | YES | YES | YES Q1QWI0 | 47% | YES | YES |
| *Pseudomonas fluorescens* | YES Q3KI43 | 29% | YES | YES | YES | YES Q3KI42 | 52% | YES | YES |
| *Azotobacter vinelandii* | YES Q4IY97 | 28% | YES | YES | YES | YES Q4IY98 | 50% | YES | YES |
| *Methylobacillus flagellatus* | YES Q1GZ69 | 27% | YES | YES | YES | YES Q1GZ68 | 51% | YES | YES |
| *Pseudomonas putida* | YES Q2XJG8 | 28% | YES | YES | YES | YES Q88PG2 | 50% | YES | YES |
| *Oceanospirillum sp* | YES Q2BQZ7 | 26% | YES | YES | YES | YES Q2BQZ8 | 49% | YES | YES |
| *Nitrosomonas europaea* | YES Q82V00 | 25% | YES | YES | YES | YES Q820M1 | 50% | YES | YES |
| *Pseudomonas aeruginosa* | YES Q9HVS7 | 28% | YES | YES | YES | YES Q9HVS8 | 50% | YES | YES |
| *Pseudomonas entomophila* | YES Q1IEE3 | 29% | YES | YES | YES | YES Q1IEE2 | 50% | YES | YES |
| *Sphingopyxis alaskensis* | *YES Q1GT06* | *26%* | *YES* | *YES* | *YES* | *YES Q1GT07* | *45%* | *YES* | *YES* |
| *Sphingomonas sp.* | YES Q1N6Y2 | 24% | YES | YES | YES | NO1 | | | |
| *Paracoccus denitrificans* | NO[2] | | | | | YES Q3PEU7 | 77% | YES | YES |
| *Uncultured Acidobacteria bacterium* | NO[2] | | | | | YES Q7X351 | 66% | YES | YES |
| *Pelagibacter ubique* | NO[2] | | | | | YES Q4FP64 | 67% | YES | YES |
| *Psychroflexus torquis* | NO | | | | | YES Q1VJU0 | 65% | YES | YES |
| *Nitrosomonas eutropha* | NO[2] | | | | | YES Q3N6K1 | 49% | YES | YES |
| *Thiobacillus denitrificans* | NO[2] | | | | | YES Q3SFG7 | 49% | YES | YES |
| *Alcaligenes eutrophus* | NO[2] | | | | | YES Q476X9 | 48% | YES | YES |
| *Anaeromyxobacter dehalogenans* | NO[2] | | | | | ES Q2II97 | 47% | YES | YES |

***continued on next page***

*Table 1. Continued*

| | RegB | | | | | RegA | | | |
|---|---|---|---|---|---|---|---|---|---|
| **Organism** | **RegB** | **Identity** | **H-Box** | **Q-Binding Site** | **Redox-Active Cystein** | **RegA** | **Identity** | **Acid-Box** | **DNA-Binding Domain** |
| *Anaeromyxobacter dehalogenans* | NO[2] | | | | | YES Q2II97 | 47% | YES | YES |
| *Ralstonia metallidurans* | NO[2] | | | | | YES Q1LS55 | 46% | YES | YES |
| *Saccharophagus degradans* | NO[2] | | | | | YES Q21GP2 | 39% | YES | YES |
| *Pseudoalteromonas haloplanktis* | NO[2] | | | | | YES Q3IBV7 | 42% | YES | YES |
| *Shewanella frigidimarina* | NO[2] | | | | | YES Q3NW53 | 42% | YES | YES |
| *Burkholderia cenocepacia* | NO[2] | | | | | YES Q44X66 | 43% | YES | YES |
| *Chromobacterium violaceum* | NO[2] | | | | | YES Q7NZM5 | 47% | YES | YES |
| *Burkholderia vietnamiensis* | NO[2] | | | | | YES Q4BNG8 | 43% | YES | YES |
| *Ralstonia solanacearum* | NO[2] | | | | | YES Q8Y3E0 | 45% | YES | YES |
| *Burkholderia pseudomallei* | NO[2] | | | | | YES Q3JXA5 | 43% | YES | YES |
| *Burkholderia thailandensis* | NO[2] | | | | | YES Q2T278 | 43% | YES | YES |
| *Burkholderia mallei* | NO[2] | | | | | YES Q62F01 | 51% | YES | YES |
| *Burkholderia ambifaria* | NO[2] | | | | | YES Q3FK41 | 43% | YES | YES |
| *Colwellia psychrerythraea* | NO[2] | | | | | YES Q47UR4 | 42% | YES | YES |
| *Methylococcus capsulatus* | NO[2] | | | | | YES Q602T5 | 46% | YES | YES |
| *Idiomarina loihiensis* | NO[2] | | | | | YES Q5QWI7 | 38% | YES | YES |
| *Shewanella amazonensis* | NO[2] | | | | | YES Q3QI67 | 42% | YES | YES |
| *Shewanella denitrificans* | NO[2] | | | | | YES Q3P3L2 | 37% | YES | YES |
| *Shewanella putrefaciens* | NO[2] | | | | | YES Q2ZVU1 | 38% | YES | YES |
| *Shewanella oneidensis* | NO[2] | | | | | YES Q8E9U1 | 38% | YES | YES |
| *Shewanella baltica* | NO[2] | | | | | YES Q3Q7V3 | 37% | YES | YES |
| *Dechloromonas aromatica* | NO[2] | | | | | YES Q47FP7 | 44% | YES | YES |
| *Hahella chejuensis* | NO[2] | | | | | YES Q2SM70 | 42% | YES | YES |

***continued on next page***

*Table 1. Continued*

| | RegB | | | | | RegA | | | |
|---|---|---|---|---|---|---|---|---|---|
| **Organism** | **RegB** | **Identity** | **H-Box** | **Q-Binding Site** | **Redox-Active Cystein** | **RegA** | **Identity** | **Acid-Box** | **DNA-Binding Domain** |
| *Marinobacter aquaeolei* | NO[2] | | | | | YES Q36SD4 | 38% | YES | YES |
| *Rhodopirellula baltica* | NO[2] | | | | | YES Q7UHV2 | 40% | YES | YES |
| *Bordetella pertussis* | NO[2] | | | | | YES Q7VUY1 | 42% | YES | YES |
| *Bordetella bronchiseptica* | NO[2] | | | | | YES Q7WMV8 | 42% | YES | YES |
| *Bordetella parapertussis* | NO[2] | | | | | YES Q7WBD8 | 42% | YES | YES |
| *Bordetella avium* | NO[2] | | | | | YES Q2KWQ0 | 42% | YES | YES |

[1]No homologues to RegA were found in Blast 2complete database. [2]No homologues to RegB were found in Blast2 complete database.

that RegB and RegA constitute a highly conserved global regulatory system that provides an overlying layer of redox-control on a variety of energy-generating and energy-utilizing biological processes in many diverse species of bacteria. Specifically, energy generating and energy utilizing metabolic and bioenergetic processes such as photosynthesis, tetrapyrrole synthesis, carbon fixation, nitrogen fixation, hydrogen oxidation, denitrification, aerobic/anaerobic respiration and electron transport are known members of the RegB/RegA regulon in *R. capsulatus* (Fig. 1).

## Members of the Reg Regulon

### *Photosynthesis*

The RegB/RegA system was initially discovered by selecting for mutations that exhibited reduced synthesis of the photosystem in *R. capsulatus*.[1,2] An intact copy of *regA* was subsequently shown to be required for *R. capsulatus* to grow photosynthetically under dim light.[2] As is the case for RegA, RegB is also necessary for anaerobic synthesis of the photosystem. Gene expression studies have indicated that expression of the *puc*, *puf* and *puh* operons that code for apoproteins of the light harvesting and reaction center complexes are significantly reduced when either RegB or RegA are disrupted.[1,2] Mutations in RegB and RegA homologues from *R. sphaeroides* also show similar effects with respect to the control of *puh, puf* and *puc* expression as was reported for *R. capsulatus*.[8]

In addition to controlling synthesis of light harvesting and reaction center apoproteins, RegA also affects tetrapyrrole synthesis whose branched pathways produce compounds such as bacteriochlorophyll and heme that are bound by photosystem and cytochrome apoproteins, respectively. For example, expression of the *bchE* in *R. sphaeroides* that encodes an enzyme in the bacteriochlorophyll biosynthesis pathway was reported to be regulated by PrrB/PrrA system.[15] There are also reports that RegA from *R. sphaeroides* controls expression of *hemA, hemZ* and *hemN*[15,16] that code for enzymes involved in the common branch used by both the heme and bacteriochlorophyll biosynthetic pathways. In *R. capsulatus* it has been demonstrated that *hemA, hemC, hemE, hemH* and *hemZ*, are also components of the RegB/RegA regulon.[17] Putative RegA-binding sites were also revealed in the promoter region of these *hem* genes and two PrrA-binding sites have been identified upstream of *hemA* in *R. sphaeroides*,[18] demonstrating that RegA directly regulates *hem* gene expression.[17]

### *Electron Transfer System*

The RegA homologue from *R. sphaeroides* (PrrA) was found to positively regulate the expression of *cycA*, that encodes cytochrome $c_2$.[8] In this species, cytochrome $c_2$ shuttles electrons from the cytochrome $bc_1$ complex to the photosystem reaction center as well as to respiratory component cytochrome oxidase. Primer extension and in vitro transcription studies indicated that PrrA directly activates *cycA* transcription.[19] Evidence that RegA directly controls *cycA* expression was provided by DNase I protection assays which showed that RegA* (constitutively active variant of RegA) from *R. capsulatus* binds to a region of the *cycA* P2 promoter centered -50 bp from the start site of transcription.[20]

Swem et al[21] demonstrated that RegB/RegA controls synthesis of cytochrome $c_2$ as well as cytochrome $c_y$ and the cytochrome $bc_1$ complex in *R. capsulatus*. It was shown that RegA activates biosynthesis of cytochromes $bc_1$ and $c_2$ under anaerobic, semi-aerobic and aerobic growth conditions, whereas it only activates cytochrome $c_y$ under semi-aerobic and anaerobic conditions. DNase I protection assays also demonstrated that RegA binds to 2 sites on the promoter of the *pet* ($bc_1$) operon and to 4 sites on the promoters of the *cycA* and *cycY* genes encoding cytochrome $c_2$ and cytochrome $c_y$, respectively.[21]

Like many bacterial species, *R. capsulatus* possesses a branched respiratory chain involving two different terminal oxidases. In one branch, the ubiquinol (ubihydroquinone) oxidase takes electrons directly from the quinone pool to reduce $O_2$ to $H_2O$. The second branch, which is similar to the mitochondrial electron transfer chain, is comprised of the cytochrome $bc_1$ complex, cytochromes $c_2$ or $c_y$ and a $cbb_3$-type cytochrome *c* oxidase.[22] RegA has been observed to activate cytochrome $cbb_3$ oxidase expression semi-aerobically and aerobically while repressing expression anaerobically.[21]

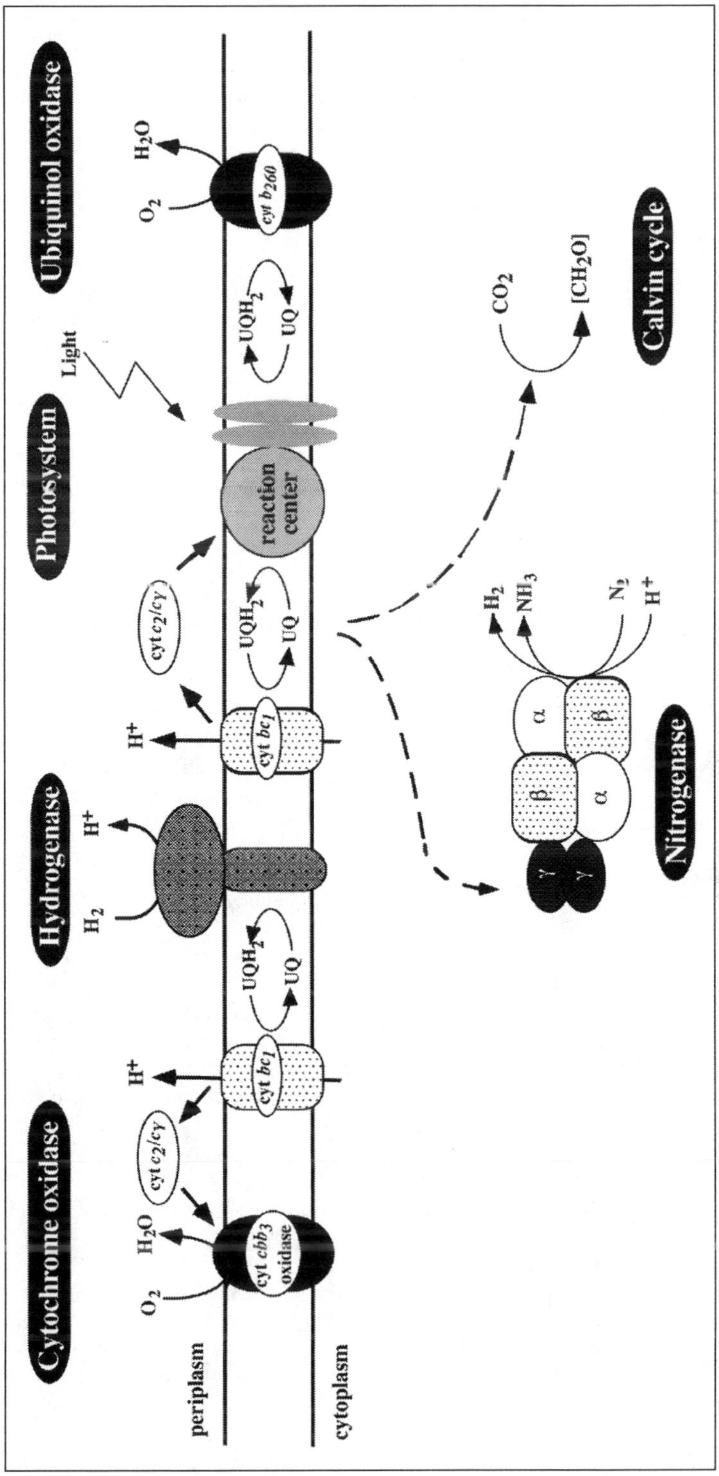


Figure 1. A diagram of various RegB/RegA controlled systems that have been identified in *Rhodobacter capsulatus*.

DNase I footprint analysis revealed that RegA directly controls synthesis of cytochrome *cbb*$_3$ oxidase by binding to a site on the *ccoNOQP* promoter located just upstream from the -35 sequence.[21]

As observed for *ccoNOQP* operon, RegB/RegA is also involved in the regulation of *cydAB* operon encoding ubiquinol oxidase, with RegA required for activation of *cydAB* transcription under all growth conditions tested. DNase I footprint assays indicate that RegA binds to two sites upstream from the -35 region of the promoter.[21]

Recently RegB/RegA homologues from *P. aeruginosa* (RoxS/RoxR) were reported to be involved in the control of aerobic respiration in this species.[12] More precisely, RoxS/RoxR controls the induction of the cyanide-insensitive oxidase, which is the terminal oxidase of one aerobic electron transport pathway, in the presence of cyanide. It is proposed that RoxR coregulates the *cioAB* promoter with another anaerobic regulator, ANR, thereby permitting the integration of different stimuli in the control of cyanide-insensitive oxidase expression.

*R. capsulatus* and *R. sphaeroides* are also both capable of anaerobic respiration using dimethylsulfoxide (DMSO) as a terminal electron acceptor.[23] The reduction of DMSO is catalyzed by a membrane-bound DMSO reductase enzyme that is encoded by the *dorCDA* operon. The *dor* operon is under the transcription control of a two-component signal transduction system, DorS/DorR, that responds to the availability of DMSO.[24-26] The sensor kinase, DorS is known to autophosphorylate in the presence of DMSO with the phosphate transferred to the response regulator, DorR, which then activates *dorCDA* expression. In addition to DorS/DorR, the *dorCDA* operon is also regulated by the RegB/RegA system with RegA acting as a repressor of the *dorCDA* operon during photoheterotrophic growth in the presence of malate as a carbon source.[27] However, RegA seems to lose control of the *dorCDA* operon if the cells are grown on pyruvate rather than malate. This indicates that another unidentified regulator can suppress the *regA* mutant phenotype in cells grown on pyruvate but not in cells grown on malate.

In addition to regulating synthesis of a larger number of cytochrome apoproteins, RegA also controls synthesis of the cytochrome cofactor heme. Smart et al[17] demonstrated that expression of *hemA, hemC, hemE, hemZ* and *hemH* genes that code for enzymes in the common heme/bacteriochlorophyll branch of the tetrapyrrole pathway are part of the RegB-RegA regulon. Thus, RegB-RegA are involved in regulating the stoicheometry of heme biosynthesis with synthesis of cytochrome apoproteins that bind this cofactor.

### *Carbon Fixation*

The Calvin-Benson-Bassham (CBB) reductive pentose phosphate pathway allows production of organic carbon via the assimilation of $CO_2$. Enzymes of the Calvin cycle are encoded by the *cbb*$_I$ and *cbb*$_{II}$ operons. Transcription of these operons is regulated in response to carbon by the transcriptional activator, CbbR.[28,29]

An involvement of the RegB/RegA system in the biosynthesis of Calvin cycle enzymes was first discovered in *R. sphaeroides* where it was demonstrated that RegB (PrrB) was required for positive regulation of the *cbb* operons, both anaerobically in the light and aerobically in the dark.[32] Using purified *R. capsulatus* RegA*, Dubbs et al[28,30] demonstrated that RegA directly controls *R. sphaeroides cbb* expression by binding to four sites in the *cbb*$_I$ promoter and to six sites on the *cbb*$_{II}$ promoter. The authors hypothesized that the locations of RegA binding could allow direct interactions with CbbR and/or with RNA polymerase. Furthermore, binding of RegA to the two sites located in the upstream activating sequence (UAS) in the *cbb*$_I$ promoter appears responsible for a RegA-mediated 41-fold enhancement in *cbb*$_I$ expression.[28]

Gibson et al[31] demonstrated that chemoautotrophically grown *regA* (*prrA*) mutants of *R. sphaeroides* differentially express the two *cbb* operons with expression of the *cbb*$_{II}$ promoter severely reduced and expression of the *cbb*$_I$ promoter enhanced in the *prrA* mutant strain. This result indicates that PrrA functions as an activator of *cbb*$_{II}$ and a repressor of *cbb*$_I$. Analysis of promoter mutants suggests that RegA may bind to distinct regions in *cbb*$_{II}$ and in *cbb*$_I$ during photoautotrophic and chemoautotrophic growth.

In *R. capsulatus*, the RegB/RegA system also controls expression of the two *cbb* operons that are present in this species.[32] Inactivation of *regA* and *regB* affects $cbb_I$ and $cbb_{II}$ expression with only 14 and 10% of wild-type levels, respectively, in a *regA*-disrupted strain under photoautotrophic growth condition. RegA* was also shown to bind to two DNA-binding sites in both the $cbb_I$ and $cbb_{II}$ promoter regions. There is a major high affinity RegA binding upstream of the $cbb_I$ transcription start site that is assumed to be involved in transcriptional activation in concert with $CbbR_I$. A low affinity RegA binding site overlapping a $CbbR_I$ DNA-binding site is proposed to has a negative role caused by RegA mediated occlusion of $CbbR_I$ binding to this region. On the $cbb_{II}$ promoter, there are 2 high affinity RegA binding sites. The upstream location of these binding sites suggests that they are involved in activation.

It has been reported that RegA homologues from *B. japonicum* (RegR) and from *S. meliloti* (ActR), also function as activators of *cbb* operons in concert with CbbR.[33,34] Thus, as was demonstrated for *R. capsulatus*, RegA homologues appear to control $CO_2$ fixation in number of photosynthetic and nonphotosynthetic bacteria.

## *Nitrogen Fixation*

Conditions of nitrogen and oxygen limitation are known to activate expression of *nif* genes that are required for biosynthesis of molybdenum nitrogenase (reviewed in ref. 35). Joshi and Tabita[36] observed that nitrogenase synthesis is de-repressed in the presence of excess ammonium in *R. sphaeroides* strains that lacked a functional $CO_2$ fixation pathway. They proposed that nitrogenase becomes de-repressed to serve as an alternative secondary electron sink in the absence of $CO_2$ fixation. Interestingly, a functional *regB* gene is required for de-repression of nitrogenase in the absence of carbon fixation, suggesting that RegB/RegA system is involved in the control of nitrogen fixation.

Elsen et al[37] shed light on the mechanism of de-repression of nitrogenase in *R. capsulatus* by showing that the RegB/RegA system indirectly controls expression of the *nifHDK* operon that encodes the molybdenum-containing nitrogenase complex. In *R. capsulatus* and in many other species, nitrogenase expression is regulated by nitrogen limitation through the NtrB/NtrC two-component system. Under nitrogen limiting conditions NtrB phosphorylates NtrC which then activates *nifA* transcription. NifA then activates expression of numerous *nif* genes including *nifHKD* (reviewed in ref. 35). In *R. capsulatus*, there are two functional copies of *nifA*, *nifA1* and *nifA2*, either of which can activate *nifHDK* expression. Elsen et al[37] demonstrated that RegA binds to the *nifA2* promoter and activate the transcription. Interestingly, RegA-mediated activation of *nifA2* transcription requires NtrC thereby indicating that RegA~P (phosphorylated RegA) alone is not sufficient to stimulate *nifA2* expression.[37] Thus, RegA appears to provide an overarching layer of redox control on top of the control of nitrogen availability that is provided by NtrC.

In *B. japonicum*, the RegB/RegA homologues (RegS/RegR) are required for the aerobic and anaerobic expression of the *fixRnifA* operon.[10] Interestingly a mutation that disrupts the response regulator RegR reduces *fixRnifA* expression and consequently nitrogen fixation activity. However, no related phenotype was observed upon disruption of the sensor kinase, RegS. RegR mutants of *B. japonicum* form nodules but the nodules are functionally incapable of fixing nitrogen (a *fix*$^-$ phenotype).

## *Denitrification*

Recently, the *R. sphaeroides* RegB/RegA system (PrrB/PrrA) was shown to control expression of nitrite reductase that is a terminal electron acceptor involved in denitrification.[38] Specifically, *regB* and *regA* disrupted strains reduced expression of the nitrite reductase structural gene, *nirK* which resulted in the inability to grow anaerobically on nitrite-containing medium. *nir* expression is also regulated by nitrite availability through the transcription factor NnrR. Thus, RegA presumably acts in concert with NnrR, to coordinate *nirK* expression.

## *Hydrogen Oxidation*

*R. capsulatus* possesses the *hupSLC* operon that codes for a membrane-bound uptake [NiFe]hydrogenase that catalyses $H_2$ oxidation. This enzyme allows the bacterium to grow autotrophically with $H_2$ as the sole electron source (reviewed in ref. 39).

$H_2$ regulation is mediated by the two-component regulatory system HupT/HupR with the response regulator HupR directly activating *hupSLC* transcription in the presence of $H_2$.[40] Maximal expression of *hupSLC* also requires the binding of IHF between the HupR and the RNA polymerase DNA-binding sites.[41]

Elsen et al[37] demonstrated that RegA is involved in repressing *hupSLC* expression under both aerobic and anaerobic heterotrophic growth conditions. A major DNA-binding site of RegA was shown to be located close to the −35 promoter recognition sequence with a second lower affinity RegA binding site overlapping the IHF DNA-binding region. At that location it is possible that RegA could prevent either the RNA polymerase, or the IHF protein, or both, from binding to the *hupSLC* promoter.

### *Dehydrogenases*

Glutathione-dependent formaldehyde dehydrogenase serves an important role in the detoxification of formaldehyde by conversion to formate. Analysis of expression of the glutathione-dependent formaldehyde dehydrogenase gene, *adhI*, demonstrated that *adhI* expression is under control of several effectors that respond to formaldehyde, methanol, or other formaldehyde adduct.[42] This enzyme is absolutely required for growth with carbon sources such as methanol that generates formaldehyde. Formaldehyde oxidation creates reducing power in the form of NADH thereby providing cellular energy as a product. Interestingly, in *R. sphaeroides* RegA (PrrA) was shown to be essential for normal aerobic expression of the *adhI* gene.[42] Analysis of RegA binding to the *adhI* promoter has not been undertaken, so it is not yet certain whether RegA directly or indirectly affects expression of formaldehyde dehydrogenase.

In *S. meliloti* the RegB/RegA homologues, ActS/ActR, control biosynthesis of three dehydrogenases: formaldehyde dehydrogenase, formate dehydrogenase and methanol dehydrogenase as well as $CO_2$ fixation.[34] The ActS/ActR system is also involved in acid tolerance.[11]

## The Sensor Kinase RegB

The *R. capsulatus regB* gene encodes a 460 amino acid (50.1 kDa) histidine protein kinase that is composed of two domains; a N-terminal trans-membrane domain containing six hydrophobic membrane-spanning regions[11,43-44] and a C-terminal cytoplasmic "transmitter" domain. A recent study identified the ubiquinone pool as the redox signal for RegB with a highly conserved quinone binding site found to be located in the trans-membrane domain thereby indicating that this region plays a role in redox-sensing (Fig. 2).[45] The transmembrane domain is followed by a cytosolic domain that contains an H-box site of autophosphorylation (His225) and the N, G1, F and G2 boxes that define the nucleotide binding cleft.[1,6,44] The transmitter domain also contains a conserved redox-active cysteine capable of regulating the activity of RegB through forming an intermolecular disulfide bond in response to the redox state that is located in a conserved "redox box" just downstream of the H-box (Fig. 2).[46]

### *Kinase Activity*

Initial kinase assays demonstrated that a His-tagged cytosolic domain of *R. capsulatus* RegB was capable of autophosphorylation in vitro as well as phosphotransfer to its cognate response regulator, RegA. The rate of autophosphorylation was initially reported to be low with half maximal phosphorylation observed after 45 minutes of incubation with $[\gamma{-}^{32}P]ATP$.[47,48] As the kinetics were not affected by the ATP concentration, it was suggested that the rate-limiting step of RegB autophosphorylation was phosphotransfer from bound ATP to the histidine residue or the dimerization of the protein, rather than binding of ATP.[47] Recent results from Swem et al[46] demonstrate that a nonHis-tagged (truncated) version of RegB exhibits a significantly faster autophosphorylation rate, reaching half maximal phosphorylation within 5 minutes. Similar results were seen with truncated cytosolic forms of RegB homologues from *R. sphaeroides*,[19] *B. japonicum*,[49] as well as with the full-length version from *R. sphaeroides*[50] and *R. capsulatus*.[45] Phosphorylated full-length RegB exhibits decreased stability of the phosphate compared to the truncated version of RegB (half-life of about 34 minutes versus 5.5 to 6 hours), pointing to a role of the transmembrane domain in

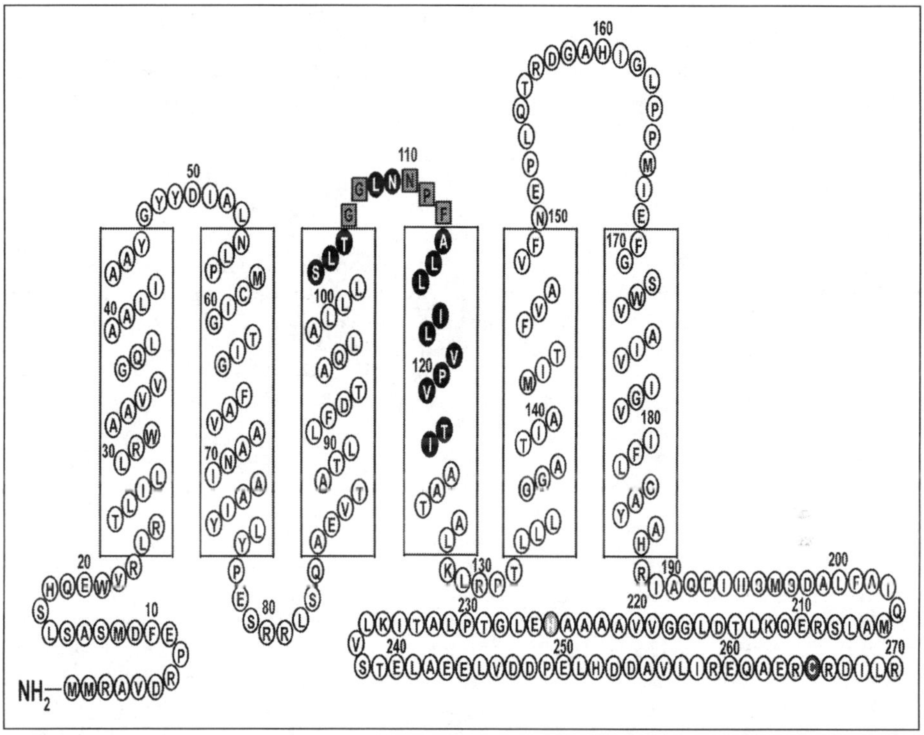


Figure 2. Membrane spanning, histidine autophosphorylation and cysteine regulatory portions of RegB. The photolytic portion of RegB that photo-affinity crosslinks to a quinone analog is indicated in inverse white lettering with universally conserved residues in this fragment shown with a red background. The histindine site of autophosphorylation is indicated in green and the redox active cystein is indicated in blue. A color version of this figure is available online at www.eurekah.com.

regulating the phosphorylation state of RegB.[48,50] As mutations in the N-terminal transmembrane domain led to RegB proteins with constitutive kinase activity in vivo, the "unsignaled" state of the protein has been proposed to be "autophosphorylation dominant".[9,51]

The first demonstration of phosphotransfer from *R. capsulatus* RegB~P (phosphorylated RegB) to RegA was reported by Inoue et al.[48] Phosphotransfer studies showed that the transfer of phosphate is rapid (<1 min) from the cytosolic domain of RegB to RegA in vitro.[47] Similar phosphotransfer kinetics has been reported using truncated cytosolic and full-length versions of *R. sphaeroides* RegB homologues and *B. japonicum* homologue ($RegS_C$).[49] No back-transfer of phosphate from RegA~P to RegB has been observed.[19,50] Since phosphotransfer is very rapid, autophosphorylation of RegB appears to be the rate-limiting step in the phosphorylation of RegA.

## *Phosphatase Activity*

Like many histidine kinases, RegB can modulate the level of phosphorylated RegA, not only through phosphorylation, but also by exerting phosphatase activity on RegA~P. Dephosphorylation of RegA~P in vitro has been shown to be dependent on the amount of unphosphorylated RegB, which is a good indication that RegB possesses phosphatase activity toward RegA.[47] Similar results were observed for the *B. japonicum* RegB homologue (RegS)[49] and for the full-length version of RegB from *R. sphaeroides.*[50] Phosphatase activity has also been characterized for a truncated soluble form of *R. sphaeroides* RegB by measuring the stability of the phosphate on RegA~P in the presence and absence of RegB. Collectively, these results showed that the presence of RegB resulted in > 16-fold reduction in the stability of the phosphate on RegA~P.[19] Because both truncated and full-length RegB exhibit the same phosphatase activity, it is assumed that modulation of phosphatase activity does not require the N-terminal domain of RegB.[50] However, further studies will be required to determine if phosphatase activity is redox-regulated.

In the *E. coli* sensor kinase EnvZ, it is apparent that a threonine residue positioned 4 residues downstream of the conserved phosphorylated histidine is important for phosphatase activity.[52] This threonine residue is positioned on the same α-helical face directly below the phosphorylated histidine. It is believed that the histidine residue may deprotonate the threonine hydroxyl group to form a good nucleophile that attacks the phosphoryl group bound to the aspartate residue of the cognate response regulator. Interestingly, RegB homologues also contain a 100% conserved threonine residue located four residues downstream of the phosphorylated histidine, suggesting that RegB may exhibit a similar phosphatase mechanism with RegA. Nonetheless, Potter et al[50] point out that a significant difference between RegB and EnvZ phosphatase activity exists given that EnvZ has a requirement for ATP, or a nonhydrolyzable analogue as a cofactor for phosphatase activity while RegB lacks this nucleotide requirement.

## Redox Signaling via Quinone

In vivo studies indicated that RegB/RegA-regulated photosynthesis genes was inhibited by growth under aerobic conditions. It was therefore initially presumed that RegB kinase activity was directly inhibited by oxygen. However, this possibility was subsequently excluded because *R. capsulatus* is fully capable of de-repressing pigment biosynthesis under chemiautotrophic growth conditions involving growth in the presence of oxygen, hydrogen and carbon dioxide.[53]

Another signal proposed to regulate RegB was the redox state of the respiratory electron transport chain.[8,36,54-57] This conclusion was based on the observation that mutations of *R. sphaeroides* and *R. capsulatus* cytochrome *cbb*$_3$ oxidase lead to elevated aerobic expression of RegB/RegA-regulated genes.[8,58] It was therefore suggested that cytochrome *cbb*$_3$ oxidase generates an "inhibitory" signal that represses the RegB/RegA two-component system. Recently Swem et al demonstrated that the redox state of the ubiquinone pool, which is known to be affected by respiration and photosynthesis is a direct signal controlling RegB autophosphorylaton.[45] In the Swem study, full-length RegB kinase activity was inhibited approximately 6-fold in vitro by the presence of oxidized coenzyme Q1 (Q1 is a derivative of ubiquinone) whereas kinase activity was not affected by the presence of reduced Q1. In purple photosynthetic bacteria, there is a large ubiquinone pool that functions as electron carrier of the electron transport chain in the membrane.[59] The oxidization/reduction state of the ubiquinone pool varies in response to changes in oxygen tension, being predominantly oxidized under aerobic conditions and predominantly reduced under anaerobic conditions.[60] Ubiquinone is a facile signal given that ubiquinones and RegB are both membrane-associated and the redox state of quinones reflects changes in environmental oxygen tension and changes in the redox state of these cells in general. Ubiquinone as redox signal for controlling RegB activity also correlates well with the observation that mutations in cytochrome *cbb*$_3$ oxidase lead to elevated RegB activity. In this case, a mutation in a terminal respiratory electron donor such as a *cbb*$_3$ oxidase would result in a more reduced ubiquinone pool, which would lead to elevation of RegB kinase activity.

With the use of $^{14}$C-azidoquinone photo affinity cross-linking, a ubiquinone-binding site was identified in a periplasmic loop between transmembrane helices three and four (Fig. 2).[45] In this region there is a heptapeptide sequence of GGXXNPF that is 100% conserved among all known RegB homologs.[45] It has been proposed that oxidized ubiquinone may bind to this heptapeptide through the π-π interactions between its para-hydroxybenzoate ring and the aromatic side group of the conserved phenylalanine (Phe112), as well as hydrogen bond interaction between ubiquinone and the conserved asparagine (Asp111).[45] The binding of oxidized ubiquinone may result in allosteric modification of RegB that leads to the inhibition of autophosphorylation. When the ubiquinone pool is shifted to protonated form under anaerobic conditions, the hydrogen bond between asparagine and ubiquinone could be disrupted, which could trigger structural changes facilitating the autophosphorylation of RegB. Subsequent in vivo mutational study on Phe112 resulted in elevated aerobic synthesis of photosystem, confirming that this heptapeptide is involved in the sensing of the ubiquinone pool redox state and regulating of the RegB activity.[45]

### *Additional Redox Signals*

In addition to the ubiquinone-binding site, there is a fully conserved cysteine that is also involved in redox-sensing.[46] This redox-active cysteine (Cys 265) is located in a "redox-box" that is harbored in the cytosolic dimerization interface located downstream of the H-box (Fig. 2). In vitro analysis using truncated RegB without the transmembrane domain indicates that an intermolecular disulfide bond forms between RegB dimmers under oxidizing conditions, converting active dimmers into inactive tetramers.[46] In vitro disulfide bond formation was shown to require the presence of a divalent metal ion which may help to fold RegB to the functional structure.[46] The involvement of an intermolecular disulphide bridge in the control of RegB activity is also supported by an increase in vitro full-length RegB phosphorylation in the presence of DTT.[50] Furthermore, Western blot has confirmed that the intermolecular disulfide bond formation under aerobic growth conditions regulates the RegB activity in vivo. However, the Western blot also showed only <20% Cys 265 forms disulfide bond in vivo. The reminder of Cys 265 has been proposed to form other derivatives, such as sulfenic acid (Cys-S-OH), to regulate the RegB activity.[46]

Various mutations have been constructed to probe the roles of the ubiquinone and Cys265 redox signals in regulating RegB activity. A Cys265 to alanine mutation (C265A) in full-length RegB led to attenuated, but not absent redox control by coenzyme Q1 in vitro and reduced redox control in vivo.[45,46] Similar results were observed with mutations in the ubiquinone binding domain which shows elevated expression which still harbors a level of redox control.[45] These results suggests that ubiquinone pool is a redox signal independent of the redox state of Cys265.[45] Given that ubiquinone binding site is located in transmembrane domain and Cys265 is located in cytosolic domain, they are not likely to directly interact. So, it seems ubiquinone pool and Cys265 function independently and that both contribute to the redox control of the RegB activity.

Genetic studies also implicate a role of SenC (also called PrrC) in transduction of a "redox signal" in *R. sphaeroides* and in *R. capsulatus*. Specifically, inactivation of *senC* (*prrC*), which is cotranscribed with *regA*, results in an oxygen-insensitive phenotype in *R. sphaeroides*.[8,36,54-57] Interestingly, SenC also has sequence similarity to a family of oxidoreductases that are involved in disulfide bond oxidation and reduction.[61] One possibility is that, SenC could be directly involved in modulation of the oxidation and reduction state of a redox-active cysteine residue within RegB.[46]

## The Response Regulator RegA

*R. capsulatus* RegA is a protein (184 amino acid residues, 20.4 kDa) containing conserved residues that are typically found in two-component response regulators including a phosphate accepting aspartate and an "acid pocket" containing two highly conserved aspartate residues in the N-terminal receiver domain. The receiver domain is linked by a four proline hinge to a 50 amino acid C-terminal output domain that contains a three helix bundle helix-turn-helix (H-T-H) DNA-binding motif.[2,62-63] RegA homologues have been found to be *highly* conserved among numerous α-proteobacterial species with an unprecedented complete conservation of the DNA-binding domain.

### *Effect of Phosphorylation*

DNA-binding activity of RegA was initially demonstrated using a constitutively active variant of RegA called RegA*.[62] Subsequent DNase I footprint analysis to the *puc* promoter region revealed that phosphorylated and unphosphorylated wild-type RegA, protect identical regions with varying affinities. This observation indicates that phosphorylation does not affect the points of protein-DNA interaction but rather the affinity for the binding site.[47] Indeed, phosphorylation is reported to increase the DNA-binding affinity of RegA by at least 16-fold.[47] A further 6-fold increase in the DNA binding activity is achieved upon phosphorylation of RegA*.[47] These results are consistent with what was reported for the *B. japonicum* RegA homologue (RegR), whose DNA-binding activity is increased by at least 8-fold upon phosphorylation.[49] Phosphorylation-induced stimulation of DNA-binding activity was also reported for *P. aeruginosa* RoxR.[12]

Mutational analysis has been undertaken at the site of phosphorylation (Asp63) in *R. capsulatus* RegA, as well as with RegA homologues from *B. japonicum* and *R. sphaeroides*. A D63N mutation in RegR of *B. japonicum* rendered the protein unable to be phosphorylated and also unable to bind DNA as demonstrated by gel retardation experiments with a *fixR-nifA* promoter probe.[49] In contrast, experiments involving a D63K RegA mutant from *R. capsulatus* showed that the mutant protein was capable of binding DNA despite an inability to be phosphorylated. This indicated that phosphorylation might not affect DNA-binding ability, but rather, facilitates a conformational change allowing appropriate interaction of RegA with RNA polymerase.[64] This theory was further supported by in vitro transcription assays by Comolli et al[19] involving wild-type and a D63A mutant of PrrA from *R. sphaeroides*. Their studies revealed that both unphosphorylated and phosphorylated wild type PrrA are able to activate in vitro transcription of the *cycA* P2 promoter with phosphorylated PrrA exhibiting greater activity than unphosphorylated PrrA. Interestingly, the D63A form of PrrA was unable to activate any detectable amounts of transcription. These data suggest that Asp63 is essential for function and that its presence in phosphorylated and unphosphorylated states may affect several steps of activation such as the DNA-binding of RegA and interaction with RNA polymerase.

Superimposed on the effect of phosphorylation are numerous in vivo observations that unphosphorylated RegA is also capable of affecting transcription. Specifically, mutational analysis indicates that phosphorylated RegA functions as an anaerobic repressor of cytochrome $cbb_3$ oxidase expression, while unphosphorylated RegA functions as an aerobic activator.[21,65] In addition, both phosphorylated and unphosphorylated RegA are involved in activation and repression of ubiquinol oxidase expression,[65] *cbb* (carbon fixation) operon expression,[32] and regulation of *hupSLC* expression.[37] Additional DNA-binding studies with both phosphorylated and unphosphorylated RegA are clearly needed to obtain an understanding of the mechanism of activation or repression by unphosphorylated RegA.

## DNA-Binding Sites

DNase I footprint analysis initially demonstrated that purified wild-type RegA and RegA* bind to identical specific sites in the *puf* and *puc* promoters.[62,67] Subsequent DNase I footprint analysis demonstrated that numerous other operons are under direct regulation of the RegB/RegA system, including *nifA2*,[37] *hupSLC*,[37] *regB*,[68] *senC-regA-hvrA*,[68] *petABC*,[21] *cycA*,[21] *cycY*,[21] *cydAB*,[21] *ccoNOPQ*,[21] *cbb*[32]*I and cbbII*,[32] and to *cheOp2*.[32] DNA-binding activity has also been reported for the phosphorylated forms of RegR in *B. japonicum*[49] and RoxR in *P. aeruginosa*[12] using gel mobility retardation experiments.

The number of RegA DNA-binding sites to target promoters ranges between 1 and 6, based on DNase I protection assays. Individual sites in the promoter regions have varying affinities as determined by the amount of RegA needed to obtain half maximal protection of individual sites.[21] These different binding locations may allow RegA to interact with RNA polymerase in more than one manner at these promoters.

The alignment of 21 RegA binding sites from *R. capsulatus* and *R. sphaeroides* revealed that RegA indeed binds to a consensus sequence of 5'-G (C/T)G(G/C)(G/C)(G/A) NN(T/A)(T/A) NNC(G/A)C-3'.[21] For RegR from *B. japonicum*, a related consensus "RegR box" 5'-GNG(A/G) C(A/G)TTNNGNCGC-3' on the *fixRnifA* promoter, was also identified.[69] More insights into the DNA recognition and binding ability of RegA was provided by an recently solved NMR structure of the DNA-binding output domain of RegA bound to DNA.[63] This structure demonstrated that the RegA DNA binding structure is comprised of a three-helix bundle encompassing a helix-turn-helix motif.[63] The NMR structure also confirmed that the consensus RegA recognition sequence consists of YGCGRCRx(T/A)(T/A)xGNCGC(x = a variable number of bases).[63] The three helices within the three-helix bundle can be labeled α6, α7 and α8, where α7 and α8 compose the predicted helix-turn-helix motif. α8 is the recognition helix, which binds specifically to the GCG inverted repeat sequence within the major DNA groove, while α6 seems to make nonspecific interaction with the phosphate backbone of the DNA.[63] A subsequent mutational analysis of the helix-turn-helix motif identified several residues critical for the function of PrrA, which support the proposed DNA-recognition site from the NMR structure.[70]

## Concluding Remarks

Genetic and biochemical analyses have revealed that the RegB/RegA system is a major global regulator of numerous energy-generating and energy-utilizing cellular processes. The systems controlled by RegB/RegA in *R. capsulatus* and in *R. sphaeroides* include such fundamental and diverse processes as photosynthesis, tetrapyrrole synthesis, $CO_2$ fixation, $N_2$ assimilation, hydrogen utilization, denitrification, dehydrogenases, electron transport and aerotaxis (Fig. 1). The Reg regulon is continuously growing and it is likely that there are many more, yet to be discovered, target genes under the control of RegB/RegA in these metabolically diverse bacteria.

Inspection of known members of the Reg regulon, as shown in Figure 1, exhibits an interesting interrelationship between the various regulated components. Specifically, photosynthesis, respiration (oxygen and DMSO mediated) and hydrogen oxidation all directly affect the oxidation/reduction state of the ubiquinone pool. Processes such as carbon fixation, nitrogen assimilation and formaldehyde dehydration all can function as electron sinks. In addition, formation of hydrogen by nitrogenase can be used as a substrate by the uptake hydrogenase system. Likewise, carbon generated by dehydration of formaldehyde can be used by the Calvin cycle during carbon fixation. Indeed evidence suggests that RegA can function as a "master controller" that is responsible for coordinating these various redox-responding systems.[71] For example, the carbon fixing Calvin cycle becomes de-repressed under photoheterotrophic conditions involving light plus organic compounds. Under this growth condition, carbon fixation is thought to function as an electron sink that bleeds off excess reducing power. If carbon fixation is incapacitated, such as when RubisCO is mutated, then nitrogenase becomes de-repressed, even in the presence of excess ammonium so that nitrogenase can take over their role of functioning as an electron sink in the absence of carbon fixation. Importantly, de-repression of carbon fixation and nitrogen fixation under conditions of excess carbon and ammonium are RegB/RegA-dependent events.[36] This clearly underscores the importance of RegB/RegA in controlling the overall cellular redox poise.

The observation that highly conserved RegB and RegA homologues exist in many other bacterial species also indicates that the RegB/RegA system constitutes an important redox control element that is not easily replaced by other regulators. Although many questions regarding the

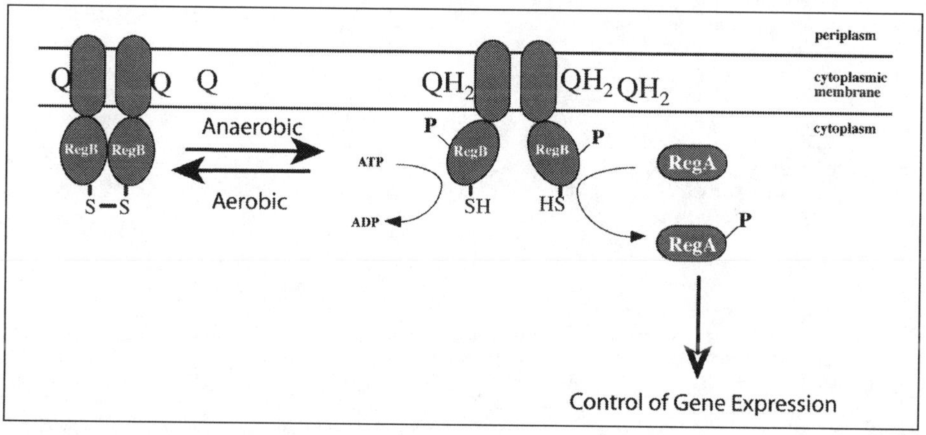


Figure 3. RegB kinase activity is controlled by at least two redox inputs. One is the redox state of the quinone pool (Q and $QH_2$) that interacts with the membrane spanning portion of RegB. A second input is the redox state of a conserved Cys (designated as SH or S-S) located downstream of the site of autophosphorylation (P).

function of RegB and RegA in other systems still need to be addressed, evidence is mounting that they control a similar set of target genes in a number of bacterial species.

Finally, there are several questions regarding the molecular mechanism of redox sensing by RegB. As depicted in Figure 3, there are at least two established redox sensing inputs in this membrane spanning kinase. One input is the redox state of the quinone pool that interacts with membrane spanning portion of RegB. The second input is the redox state of a conserved Cyst that is located downstream of the site of phosphorylation. In both cases, the inhibition of kinase activity by these different input signals remains to be established. There also remain many outstanding questions about the mechanism of transcription activation and repression by phosphorylated and dephosphorylated RegA as well as the nature of interactions that RegA may have with other transcription factors at target promoters.

## References

1. Mosley CS, Suzuki JY, Bauer CE. Identification and molecular genetic characterization of a sensor kinase responsible for coordinately regulating light harvesting and reaction center gene expression in response to anaerobiosis. J Bacteriol 1994; 176:7566-7573.
2. Sganga MW, Bauer CE. Regulatory factors controlling photosynthetic reaction center and light-harvesting gene expression in Rhodobacter capsulatus. Cell 1992; 68:945-954.
3. Cohen-Bazire G, Sistrom WR, Stanier RY. Kinetic studies of pigment synthesis by nonsulfur purple bacteria. J Cell Physiol 1957; 49:25-68.
4. Parkinson JS, Kofoid EC. Communication modules in bacterial signaling proteins. Annu Rev Genet 1992; 26:71-112.
5. Stock JB, Ninfa AJ, Stock AM. Protein phosphorylation and regulation of adaptive responses in bacteria. Microbiol Rev 1989; 53:450-490.
6. Stock AM, Robinson VL, Goudreau PN. Two-component signal transduction. Annu Rev Biochem 2000; 69:183-215.
7. Phillips-Jones MK, Hunter CN. Cloning and nucleotide sequence of regA, a putative response regulator gene of Rhodobacter sphaeroides. FEMS Microbiol Lett 1994; 116:269-275.
8. Eraso JM, Kaplan S. prrA, a putative response regulator involved in oxygen regulation of photosynthesis gene expression in Rhodobacter sphaeroides. J Bacteriol 1994; 176:32-43.
9. Eraso JM, Kaplan S. Oxygen-insensitive synthesis of the photosynthetic membranes of Rhodobacter sphaeroides: a mutant histidine kinase. J Bacteriol 1995; 177:2695-2706.
10. Bauer E, Kaspar T, Fischer HM et al. Expression of the fixR-nifA operon in Bradyrhizobium japonicum depends on a new response regulator, RegR. J Bacteriol 1998; 180:3853-3863.
11. Tiwari RP, Reeve WG, Dilworth MJ et al. Acid tolerance in Rhizobium meliloti strain WSM419 involves a two-component sensor-regulator system. Microbiology 1996; 142:1693-1704.
12. Comolli JC, Donohue TJ. Pseudomonas aeruginosa RoxR, a response regulator related to Rhodobacter sphaeroides PrrA, activates expression of the cyanide-insensitive terminal oxidase. Mol Microbiol 2002; 45:755-768.
13. Masuda S, Matsumoto Y, Nagashima KV et al. Structural and functional analyses of photosynthetic regulatory genes regA and regB from Rhodovulum sulfidophilum, Roseobacter denitrificans and Rhodobacter capsulatus. J Bacteriol 1999; 181:4205-4215.
14. Emmerich R, Hennecke H, Fischer HM. Evidence for a functional similarity between the two-component regulatory systems RegSR, ActSR and RegBA (PrrBA) in alpha-Proteobacteria. Arch Microbiol 2000; 174:307-313.
15. Oh JI, Eraso JM, Kaplan S. Interacting regulatory circuits involved in orderly control of photosynthesis gene expression in Rhodobacter sphaeroides 2.4.1. J Bacteriol 2000; 182:3081-3087.
16. Oh JI, Kaplan S. Generalized approach to the regulation and integration of gene expression. Mol Microbiol 2001; 39:1116-1123.
17. Smart JL, Willett JW, Bauer CE. Regulation of hem gene expression in Rhodobacter capsulatus by redox and photosystem regulators RegA, CrtJ, FnrL and AerR. J Mol Biol 2004; 342:1171-1186.
18. Ranson-Olson B, Jones DF, Donohue TJ et al. In vitro and in vivo analysis of the role of PrrA in Rhodobacter sphaeroides 2.4.1 hemA gene expression. J Bacteriol 2006; 188:3208-3218.
19. Comolli JC, Carl AJ, Hall C et al. Transcriptional activation of the Rhodobacter sphaeroides cytochrome c(2) gene P2 promoter by the response regulator PrrA. J Bacteriol 2002; 184:390-399.
20. Karls RK, Wolf JR, Donohue TJ. Activation of the cycA P2 promoter for the Rhodobacter sphaeroides cytochrome c2 gene by the photosynthesis response regulator. Mol Microbiol 1999; 34:822-835.
21. Swem LR, Elsen S, Bird TH et al. The RegB/RegA two-component regulatory system controls synthesis of photosynthesis and respiratory electron transfer components in Rhodobacter capsulatus. Mol Microbiol 2001; 309:121-138.

22. Gray KAaD, F. Mutational studies of the cytochrome $bc_1$ complexes. In: Blankenship RE, Madigan MT, Bauer CE, eds. Anoxygenic Photosynthetic Bacteria. Dordrecht: Kluwer Academic Publishers, 1995:747-774.
23. Zannoni D. Aerobic and anaerobic transport chains in anoxygenic phototrophic bacteria. In: Blankenship RE, Madigan MT, Bauer CE, eds. Anoxygenic Photosynthetic Bacteria. Dordrecht: Kluwer Academic Publishers, 1995:949-971.
24. Mouncey NJ, Kaplan S. Redox-dependent gene regulation in Rhodobacter sphaeroides 2.4.1(T): effects on dimethyl sulfoxide reductase (dor) gene expression. J Bacteriol 1998; 180:5612-5618.
25. Mouncey NJ, Choudhary M, Kaplan S. Characterization of genes encoding dimethyl sulfoxide reductase of Rhodobacter sphaeroides 2.4.1T: an essential metabolic gene function encoded on chromosome II. J Bacteriol 1997; 179:7617-7624.
26. Shaw AL, Leimkuhler S, Klipp W et al. Mutational analysis of the dimethylsulfoxide respiratory (dor) operon of Rhodobacter capsulatus. Microbiology 1999; 145:1409-1420.
27. Kappler U, Huston WM, McEwan AG. Control of dimethylsulfoxide reductase expression in Rhodobacter capsulatus: the role of carbon metabolites and the response regulators DorR and RegA. Microbiology 2002; 148:605-614.
28. Dubbs JM, Bird TH, Bauer CE et al. Interaction of CbbR and RegA* transcription regulators with the Rhodobacter sphaeroides cbbI promoter-operator region. J Biol Chem 2000; 275:19224-19230.
29. Dubbs JM, Tabita FR. Two functionally distinct regions upstream of the cbbI operon of Rhodobacter sphaeroides regulate gene expression. J Bacteriol 1998; 180:4903-4911.
30. Dubbs JM, Tabita FR. Interactions of the cbbII promoter-operator region with CbbR and RegA (PrrA) regulators indicate distinct mechanisms to control expression of the two cbb operons of Rhodobacter sphaeroides. J Biol Chem 2003; 278:16443-16450.
31. Gibson JL, Dubbs JM, Tabita FR. Differential expression of the $CO_2$ fixation operons of Rhodobacter sphaeroides by the Prr/Reg two-component system during chemoautotrophic growth. J Bacteriol 2002; 184:6654-6664.
32. Vichivanives P, Bird TH, Bauer CE et al. Multiple regulators and their interactions in vivo and in vitro with the cbb regulons of Rhodobacter capsulatus. J Mol Biol 2000; 300:1079-1099.
33. Emmerich R, Strehler P, Bauer E et al. Functional analysis of the Bradyrhizobium japonicum RegSR two-component regulatory system. In: Pedrosa FO, Hungria M, Yates G, Newton WE, eds. Nitrogen Fixation: From Molecules to Crop Productivity. London: Kluwer, 2000:89-90.
34. Fenner BJ, Tiwari RP, Reeve WG et al. ActR is a global genetic regulator in Sinorhizobium meliloti. In: Pedrosa FO, Hungria M, Yates G, Newton WE, eds. Nitrogen Fixation: From Molecules to Crop Productivity. London: Kluwer, 2000:488.
35. Masepohl BKW. Organization and regulation of genes encoding the molybdenum nitrogenase and the alternative nitrogenase in Rhodobacter capsulatus. Arch. Microbiol 1996; 165:80-90.
36. Joshi HM, Tabita FR. A global two component signal transduction system that integrates the control of photosynthesis, carbon dioxide assimilation and nitrogen fixation. Proc Natl Acad Sci USA. 1996; 93:14515-14520.
37. Elsen S, Dischert W, Colbeau A et al. Expression of uptake hydrogenase and molybdenum nitrogenase in Rhodobacter capsulatus is coregulated by the RegB-RegA two-component regulatory system. J Bacteriol 2000; 182:2831-2837.
38. Laratta WP, Choi PS, Tosques IE et al. Involvement of the PrrB/PrrA two-component system in nitrite respiration in Rhodobacter sphaeroides 2.4.3: evidence for transcriptional regulation. J Bacteriol 2002; 184:3521-3529.
39. Vignais PM, Toussaint B, Colbeau A. Regulation of hydrogenase gene expression. In: Blankenship RE, Madigan MT, Bauer CE, eds. Anoxygenic Photosynthetic Bacteria. Dordrecht: Kluwer Academic Publishers, 1995:1175-1190.
40. Dischert W, Vignais PM, Colbeau A. The synthesis of Rhodobacter capsulatus HupSL hydrogenase is regulated by the two-component HupT/HupR system. Mol Microbiol 1999; 34:995-1006.
41. Toussaint B, de Sury d'Aspremont R, Delic-Attree I et al. The Rhodobacter capsulatus hupSLC promoter: identification of cis-regulatory elements and of trans-activating factors involved in H2 activation of hupSLC transcription. Mol Microbiol 1997; 26:927-937.
42. Barber RD, Donohue TJ. Pathways for transcriptional activation of a glutathione-dependent formaldehyde dehydrogenase gene. J Mol Biol 1998; 280:775-784.
43. Chen W, Jager A, Klug G. Correction of the DNA sequence of the regB gene of Rhodobacter capsulatus with implications for the membrane topology of the sensor kinase RegB. J Bacteriol 2000; 182:818-820.
44. Ouchane S, Kaplan S. Topological analysis of the membrane-localized redox-responsive sensor kinase PrrB from Rhodobacter sphaeroides 2.4.1. J Biol Chem 1999; 274:17290-17296.
45. Swem LR, Gong X, Yu CA et al. Identification of a ubiquinone-binding site that affects autophosphorylation of the sensor kinase RegB. J Biol Chem 2006; 281:6768-6775.

46. Swem LR, Kraft BJ, Swem DL et al. Signal transduction by the global regulator RegB is mediated by a redox-active cysteine. EMBO J 2003; 22:4699-4708.
47. Bird TH, Du S, Bauer CE. Autophosphorylation, phosphotransfer and DNA-binding properties of the RegB/RegA two-component regulatory system in Rhodobacter capsulatus. J Biol Chem 1999; 274:16343-16348.
48. Inoue K, Kouadio JL, Mosley CS et al. Isolation and in vitro phosphorylation of sensory transduction components controlling anaerobic induction of light harvesting and reaction center gene expression in Rhodobacter capsulatus. Biochemistry 1995; 34:391-396.
49. Emmerich R, Panglungtshang K, Strehler P et al. Phosphorylation, dephosphorylation and DNA-binding of the Bradyrhizobium japonicum RegSR two-component regulatory proteins. Eur J Biochem 1999; 263:455-463.
50. Potter CA, Ward A, Laguri C et al. Expression, purification and characterisation of full-length histidine protein kinase RegB from Rhodobacter sphaeroides. J Mol Biol 2002; 320:201-213.
51. Oh JI, Ko IJ, Kaplan S. The default state of the membrane-localized histidine kinase PrrB of Rhodobacter sphaeroides 2.4.1 is in the kinase-positive mode. J Bacteriol 2001; 183:6807-6814.
52. Inouye M, Dutta R, Zhu Y. Regulation of porins in Escherichia coli by the osmosensing histidine kinase/Phosphatase EnvZ. In: M. Inouye, ed. Histidine Kinases in Signal Tranduction. Burlington: Academic Press, 2003:25-46.
53. Madigan MT, Gest H. Growth of the photosynthetic bacterium Rhodopseudomonas capsulata chemoautotrophically in darkness with $H_2$ as the energy source. J Bacteriol 1979; 137:524-530.
54. Eraso JM, Kaplan S. From redox flow to gene regulation: role of the PrrC protein of Rhodobacter sphaeroides 2.4.1. Biochemistry 2000; 39:2052-2062.
55. O'Gara JP, Kaplan S. Evidence for the role of redox carriers in photosynthesis gene expression and carotenoid biosynthesis in Rhodobacter sphaeroides 2.4.1. J Bacteriol 1997; 179:1951-1961.
56. Oh JI, Kaplan S. Redox signaling: globalization of gene expression. EMBO J 2000; 19:4237-4247.
57. Roh JH, Kaplan S. Genetic and phenotypic analyses of the rdx locus of Rhodobacter sphaeroides 2.4.1. J Bacteriol 2000; 182:3475-3481.
58. Buggy J, Bauer CE. Cloning and characterization of senC, a gene involved in both aerobic respiration and photosynthesis gene expression in Rhodobacter capsulatus. J Bacteriol 1995; 177:6958-6965.
59. Bolton JR. In: Clayton RK, Sistrom WR, eds. The Photosynthetic Bacteria. New York: Plenum Press, 1978:419-429.
60. Parson W. In: Clayton RK, Sistrom WR, eds. The Photosynthetic Bacteria. New York: Plenum Press, 1978:455-469.
61. McEwan AG, Lewin A, Davy SL et al. PrrC from Rhodobacter sphaeroides, a homologue of eukaryotic Sco proteins, is a copper-binding protein and may have a thiol-disulfide oxidoreductase activity. FEBS Lett 2002; 518:10-16.
62. Du S, Bird TH, Bauer CE. DNA binding characteristics of RegA. A constitutively active anaerobic activator of photosynthesis gene expression in Rhodobacter capsulatus. J Biol Chem 1998; 273:18509-18513.
63. Laguri C, Phillips-Jones MK, Williamson MP. Solution structure and DNA binding of the effector domain from the global regulator PrrA (RegA) from Rhodobacter sphaeroides: insights into DNA binding specificity. Nucleic Acids Res 2003; 31:6778-6787.
64. Hemschemeier SK, Ebel U, Jager A et al. In vivo and in vitro analysis of RegA response regulator mutants of Rhodobacter capsulatus. J Mol Microbiol Biotechnol 2000; 2:291-300.
65. Swem DL, Bauer CE. Coordination of ubiquinol oxidase and cytochrome cbb(3) oxidase expression by multiple regulators in Rhodobacter capsulatus. J Bacteriol 2002; 184:2815-2820.
66. Qian Y, Tabita FR. A global signal transduction system regulates aerobic and anaerobic $CO_2$ fixation in Rhodobacter sphaeroides. J Bacteriol 1996; 178:12-18.
67. Hemschemeier SK, Kirndorfer M, Hebermehl M et al. DNA binding of wild type RegA protein and its differential effect on the expression of pigment binding proteins in Rhodobacter capsulatus. J Mol Microbiol Biotechnol 2000; 2:235-243.
68. Du S, Kouadio JL, Bauer CE. Regulated expression of a highly conserved regulatory gene cluster is necessary for controlling photosynthesis gene expression in response to anaerobiosis in Rhodobacter capsulatus. J Bacteriol 1999; 181:4334-4341.
69. Emmerich R, Strehler P, Hennecke H et al. An imperfect inverted repeat is critical for DNA binding of the response regulator RegR of Bradyrhizobium japonicum. Nucleic Acids Res 2000; 28:4166-4171.
70. Jones DF, Stenzel RA, Donohue TJ. Mutational analysis of the C-terminal domain of the Rhodobacter sphaeroides response regulator PrrA. Microbiology 2005; 151:4103-4110.
71. Tichi MA, Tabita FR. Interactive control of Rhodobacter capsulatus redox-balancing systems during phototrophic metabolism. J Bacteriol 2001; 183:6344-6354.

# Chapter 10

# The BvgS/BvgA Phosphorelay System of Pathogenic *Bordetellae*

## Structure, Function and Evolution

**Dagmar Beier and Roy Gross***

## Abstract

In the genus *Bordetella* several important human and animal pathogens are classified with *B. pertussis*, the etiological agent of whooping cough, being medically the most relevant. In these bacteria expression of the most important virulence factors including several toxins, adhesins and colonization factors is controlled by a single master regulatory two-component system, the BvgS/BvgA system. This system represents a paradigm of a complex phosphorelay system that mediates a fine-tuned transcriptional response resulting in different expression levels of virulence factors during different stages of the infection process. In this chapter the current knowledge about signal perception and the molecular basis of differential gene expression controlled by a single two-component system is discussed.

## Introduction

In many bacterial pathogens two-component systems (TCS) play a dominant role in the expression of their virulence phenotype.[1,2] This is particularly true for bacteria which encounter substantially different environmental conditions during their life cycle such as facultatively pathogenic bacteria which have environmental phases but have also acquired the ability to colonize host organisms. In addition, during infection the invasion into different host niches and the defence mechanisms mounted by the host as a response to infection may require a quick and efficient adaptation to such changes as exemplified by facultatively intracellular bacteria which have to express different genetic programs during their extracellular and intracellular phases. Accordingly, in many cases complex regulatory networks consisting of several TCS but also of additional regulatory proteins or small RNAs are required for efficient colonization of a host organism. In contrast, *B. pertussis* is an intriguing example of a pathogenic bacterium which apparently requires the activity of a single master regulatory TCS, the BvgS/BvgA system, to modulate expression of virtually all virulence traits.[1,3] *B. pertussis* is the etiological agent of whooping cough, an acute disease which despite the availability of effective vaccines still is a major cause of morbidity and mortality in the world particularly in children.[4] A successful infection by *B. pertussis* requires the orchestrated expression of a large number of proteins dedicated to an efficient colonization of the host and defense against host attack. Among these Bvg-activated factors (*vag* genes) there are several mainly secreted or surface associated factors involved in adherence such as the filamentous hemagglutinin (FHA), pertactin (PRN), fimbriae (FIM), and tracheal colonization factor (TcfA),

*Corresponding Author: Roy Gross—Lehrstuhl für Mikrobiologie, Biozentrum, University of Würzburg, Am Hubland, D-97074 Würzburg, Germany.
Email: roy@biozentrum.uni-wuerzburg.de

*Bacterial Signal Transduction: Networks and Drug Targets*, edited by Ryutaro Utsumi.

in serum resistance (BrkA) toxins such as the pertussis toxin (PTX), the adenylate cyclase toxin (CYA), and the dermonecrotic toxin (DNT), a type III secretion system and the *bvgAS* gene locus itself.[4,5] Moreover, the BvgS/BvgA system influences the structure of the lipopolysaccharide and regulates biofilm formation.[6-8]

*B. pertussis* is closely related to other pathogenic *Bordetellae* including *B. parapertussis* also causing whooping cough-like symptoms in infected patients and *B. bronchiseptica* causing respiratory infections of variable consequence for the host ranging from asymptomatic carriage to severe infections in many mammalian species but only occasionally in man.[4] The genomic sequences of these three species show that *B. pertussis* and *B. parapertussis* are independent descendants of *B. bronchiseptica*-like ancestors which during their adaptation to man as the exclusive host have experienced a very significant erosion of their genetic material.[9] All three species encode highly related virulence factors, although there is substantial variation in their capacity to express these factors, as exemplified by PTX which is expressed only by *B. pertussis* despite the presence of the *ptx* operon in the other two species.[10] These species are sometimes referred to as members of the *B. bronchiseptica* complex or as "classical" *Bordetellae*.[11] Somewhat more distantly related is *B. avium* causing respiratory infections in birds.[12] The genome sequence of *B. avium* was established recently and once more shows the presence of many virulence genes orthologous to those of the mammalian pathogens.[13] Several only poorly characterized species have been added to the genus recently with *B. holmesii* being a human pathogen causing severe infections in immunocompromised patients, but also whooping cough-like symptoms in otherwise healthy individuals.[14] *B. hinzii* is a commensal of many birds but can also cause disease in immunocompromised patients.[15] *B. trematum* was repeatedly isolated from wound or ear infections although the etiology of the infections is not known.[16] Finally, *B. petrii* was recently described being the first environmental isolate belonging to the genus *Bordetella*.[11,17] Most interestingly, all of these species encode proteins orthologous to the BvgS/BvgA TCS although in most of the cases no information is available about the factors controlled by the system in these bacteria (see below).[12,18]

## Phenotypic Phases in the Expression of the Virulence Factors of *B. pertussis*

In Lacey's pioneering work the expression of the virulent phenotype of *B. pertussis* was shown to be variable according to changes in the environmental conditions.[19] Various phenotypic phases were described according to morphological and serological criteria. Among the signals identified to reversibly cause switching between the different phases in vitro there are sulphate ions, nicotinic acid and temperature. Only in the relative absence of sulphate ions, nicotinic acid or at body temperature the virulent phenotype (phase I) was found to be expressed, while in the presence of these compounds or at low temperature the bacteria were defined to be in the avirulent phase IV. This phenomenon was termed "phenotypic modulation". Several decades before another regulatory phenomenon had been described which also leads to an avirulent phase IV phenotype, the so-called "phase variation" which occurs at a relatively high incidence and irreversibly locks the bacteria in this avirulent phase.[20] In the early '80s by transposon mutagenesis Weiss and Falkow identified a gene locus (at the time termed *vir*) which is absolutely required for the phase I phenotype and thereby for the expression of the most relevant virulence factors mentioned above.[21] Characterization of the respective gene locus then revealed that it encodes a TCS which was renamed BvgS/BvgA and which subsequently was shown to be responsible for both regulatory phenomena, the "phenotypic modulation" and the "phase variation". The DNA sequence of the *vir* locus revealed a surprising feature of the BvgS/BvgA system: it was one of the first so-called unorthodox TCS, which is characterized by a composite multi domain histidine kinase (BvgS) and the occurrence of a four-step His-Asp-His-Asp phosphorelay (see below) (Fig. 1).[22,23]

In agreement with Lacey's observations, the detailed characterization of the expression pattern of individual BvgS/BvgA-regulated factors revealed that these factors can be classified into different categories regarding their expression kinetics upon environmental changes (Fig. 2).[24] In the laboratory under nonmodulating conditions, i.e., at 37°C and in the absence of nicotinic acid and

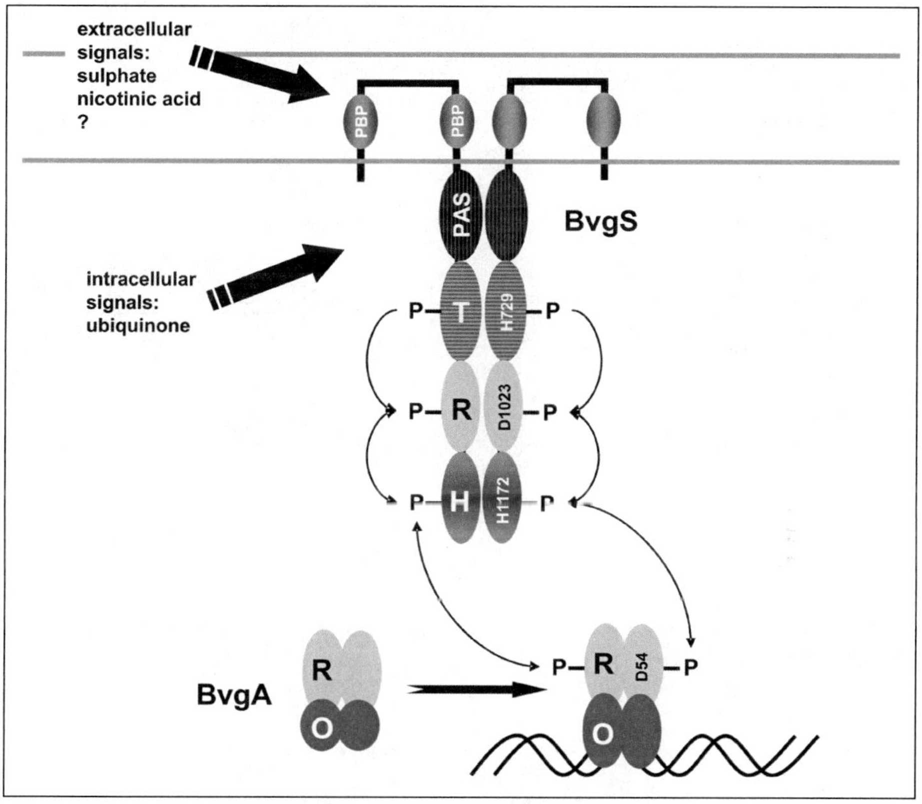


Figure 1. The domain structure of the BvgS and BvgA proteins and the phosphotransfer reactions are shown. The amino acids involved in phosphotransfer reactions are indicated. Abbreviations: PBP, periplasmic binding protein; T, transmitter; R, receiver; H, HPt domain; O, output domain.

sulphate ions, all BvgS/BvgA controlled virulence factors (the so-called virulence activated or *vag* genes) are expressed, while inactivating mutations in the *bvgAS* genes or application of modulating conditions leads to lack of expression of the virulence genes and the induction of expression of another subset of genes (the so-called virulence repressed or *vrg* genes). These two phenotypic stages correspond to Lacey's virulent phase I (vag) and avirulent phase IV (vrg) stages, respectively. However, as an alternative to Lacey's designations the terms Bvg$^+$ and Bvg$^-$ phases, respectively, became common in the relevant literature. The Bvg$^+$ phase *vag* genes could be further classified into two different classes, one class which after a switch from modulating to nonmodulating conditions exhibits a fast kinetics of transcriptional induction (the so-called early or class II genes), while a second subset of genes has a quite slow induction kinetics (the so-called late or class I genes). Interestingly, among the early class II genes there is the autoregulated *bvgAS* locus itself and genes encoding adhesins such as FHA and FIM, while the main factors encoded by late class I genes are PTX and CYA.[24] Most interestingly, recently a novel category of virulence related factors was described for *B. pertussis*. This category was predicted by Lacey's observation of intermediate phases induced by only partially modulating growth conditions. In this so-called intermediate or Bvg$^i$ phase no expression of class I genes (the toxins PTX, CYA) could be observed, while class II genes (adhesins such as FHA) are expressed. By use of a particular BvgS mutant (the BvgS-I1 mutant which has a substitution of threonine by methionine at position 733 within the conserved H-box

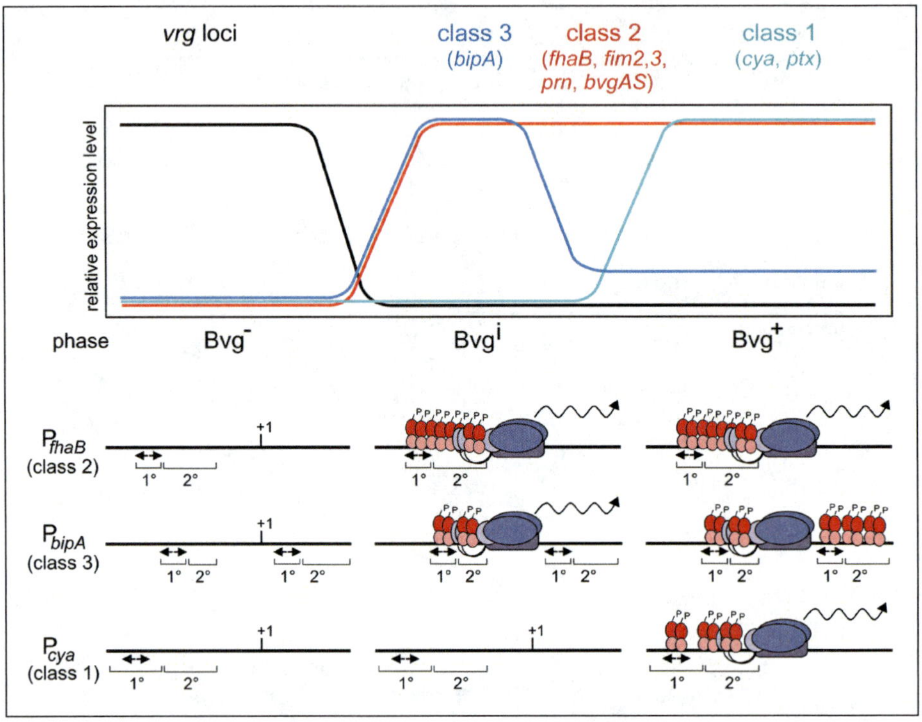


Figure 2. On top of the figure the relative expression level of the different Bvg-controlled classes of genes is shown in the three different phenotypic stages $Bvg^-$, $Bvg^i$ and $Bvg^+$, respectively. On the bottom the structure of the *fhaB*, *bipA* and *cya* promoters is shown. Antidromic arrows indicate BvgA target sites with the parenthesis indicating primary (1°) and secondary (2°) BvgA binding sites. Proposed binding of BvgA~P dimers (in red colour) at the different promoters and their interaction with RNA polymerase (in blue colour) is depicted. +1 indicates the transcriptional start site and rolling arrows represent transcription. The figure is adapted from Jones et al 2005.[58]

of BvgS) which was "frozen" in an intermediate activity state specific "class III" factors expressed only under these semi-modulating conditions could be identified. The representative of these class III or $Bvg^i$ factors is the BipA protein which has sequence similarity to intimin, a virulence factor of several pathogenic Enterobacteriae.[25,26]

In contrast to the regulation of the *vag* genes, very little is known about the negative regulation of the phase IV *vrg* genes. In the mammalian pathogens a gene downstream of the *bvgAS* locus designated *bvgR* was shown by mutation analysis to be involved in the regulation of the *vrg* genes since its inactivation led to constitutive expression of several *vrg* genes.[27] The corresponding 32 kDa BvgR protein is supposed to act as a transcriptional repressor at least at several *vrg* promoters. In agreement with its Bvg-dependent activity, the *bvgR* promoter is activated by the phosphorylated BvgA protein.[28] However, it is interesting to note that the genomes of *B. avium* and *B. petrii* did not reveal the presence of a BvgR ortholog raising the question about the particular importance of this factor and its regulon for the mammalian pathogens.

## Structure Function Relationships in the BvgS/BvgA Phosphorelay System

The 135 kDa BvgS sensor protein was one of the first composite histidine kinases with a complex modular domain architecture described. It consists of a large periplasmic domain followed by a transmembrane, a PAS, a transmitter, a receiver and an HPt domain.[22,23] According to the classification by Grebe and Stock (1999),[29] BvgS belongs to the $HPK_{1b}$ subfamily of histidine kinases. The 23 kDa BvgA protein has the features of a typical response regulator with an N-terminal receiver of the $RC_1$ family and a DNA binding C-terminal output domain with a helix-turn-helix motif that can be classified within the FixJ/NarL family (Fig. 1). The similarity of BvgA and NarL allowed the modelling of its three-dimensional structure on the basis of the NarL crystal structure.[30,31] The output domain contains four alpha-helices (H7 to H10) with the helix-turn-helix motif embedded in the H8-H9 helices. As in the other two-component systems the phosphotransfer from the histidine kinase to the response regulator modulates the activity of the regulatory protein which in its phosphorylated state increases its affinity for DNA binding and allows its specific interaction with the promoters of mainly virulence relevant genes to promote their transcription (see below).

The N-terminal transmembrane domain of BvgS is followed by a very large periplasmic domain which contains two sequence motifs characteristic for periplasmic solute binding proteins (aa 65-525). Directly following the second transmembrane spanning region a PAS domain (aa 591-696) is present.[33] This region of the BvgS protein was formerly termed linker region and mutations within that region affected the activity of the histidine kinase rendering it active constitutively.[34-36] In other signalling proteins it was shown that PAS domains are capable to perceive a variety of different signals including light, redox potential and the presence of small molecules such as FAD.[33] The presence of a large periplasmic portion and a cytoplasmic PAS domain indicates that the BvgS protein may sense extra- and intracellular signals. The transmitter domain has the typical characteristics of such a signalling domain. A subdomain (aa 717-781) containing the conserved histidine known to be phosphorylated (His729) is followed by a second subdomain (aa 837-944) which binds ATP, catalyzes its hydrolysis and transfers the gamma-phosphate to histidine 729. From this residue the phosphoryl group is then transferred to the aspartic acid 1023 of the BvgS receiver domain (aa 976-1095) and finally to His1172 of the HPt domain (aa 1133-1226). From the phosphorylated His1172 the phosphate is then transferred to Asp54 in the BvgA receiver domain.[3,37-41]

Little is known about signals sensed by BvgS which may be relevant in vivo during infection. Interestingly, a link between cysteine metabolism and PTX expression was described recently. It was observed that at early time points of a *B. pertussis* culture a large amount of intracellular sulphate accumulates which is released into the medium at later time points and may down-modulate expression of PTX.[42] It is likely that the periplasmic domain is involved in this signal perception of sulphate, since in vitro the activity of a soluble purified BvgS protein consisting only of the cytoplasmic domains was not diminished in the presence of sulphate.[43] However, with the same purified BvgS protein it could be demonstrated that BvgS activity is strongly inhibited by the presence of a soluble ubiquinone derivative (Q-0) while the reduced form did not have any effect. This indicates that the kinase activity can be reversibly switched on and off depending on the oxidation state of the ubiquinone electron carrier and suggests a relevant link of the BvgS/BvgA signal transduction system with the energy state of the cell.[43]

## BvgA-DNA Interactions

Although not proven experimentally, it is believed that in vivo different levels of BvgA~P (phosphorylated BvgA) control the expression of defined sets of genes in the $Bvg^-$ (IV), $Bvg^i$ (intermediate) and $Bvg^+$ (I) phase. This assumption is based on the in vitro analysis of BvgA binding which revealed that promoters of class 1, 2 and 3 genes are characterized by differences in the affinity and location of BvgA binding sites.[32,44-52] The consensus BvgA binding site consists of the heptanucleotide sequence TTTC(C/T)TA. In the upstream region of the class II *fha* gene this sequence is present as a nearly perfect inverted repeat centred at position –88.5 with respect to the transcriptional start site. Mutational analysis demonstrated that nucleotides at positions 3, 4 and 7 of the binding heptad contribute substantially to sequence specific recognition by BvgA.[53] The heptanucleotide inverted repeat sequence represents the high affinity BvgA binding site in the *fha* promoter which is bound by a BvgA~P dimer. Several lines of experimental evidence indicated that BvgA dimerizes even in the absence of phosphorylation[37,54] and modeling of the sequence of the BvgA output domain on the recently solved crystal structure of the NarL C-terminus bound to its DNA target[55] suggests that BvgA dimers interacting with DNA are formed by the association of the C-terminal α-10 helices.[32] Two additional BvgA~P dimers interact with a secondary low affinity binding region located immediately 3' to the heptanucleotide inverted repeat which overlaps with the –35 region of the *fha* promoter.[32] BvgA binding to this secondary binding region is cooperative and shows only limited DNA binding specificity.[53] In addition, BvgA~P binding to the low affinity binding region is facilitated by cooperative interactions with the C-terminal domain of the α subunit of RNA polymerase (α-CTD) which binds the same stretch of DNA on a different face of the helix.[32] However, binding of the third BvgA~P dimer directly adjacent to the –35 region suggests that the response regulator protein might also make contact with the $\sigma^{70}$ subunit and/or the N-terminal domain of the α subunit of RNA polymerase.[32] The model of three BvgA dimers bound end to end on the same face of the DNA helix with a two-helical turn periodicity is also consistent with deletion analyses showing that 21 bp deletions within the secondary binding region cause hyperactivation of the *fha* promoter.[49]

Like the *fha* promoter the positively autoregulated P1 promoter of the *bvgAS* locus responds quickly to a shift from modulating to nonmodulating conditions. In the upstream region of *bvgAS* a binding region for BvgA~P was identified spanning from position –90 to position –40 with respect to the transcriptional start site.[46] An imperfect inverted repeat of the consensus BvgA binding heptad is present within this region (position –73 to –93) and it was shown that BvgA~P binds to an oligonucleotide containing this inverted repeat sequence.[56]

The *prn* gene shows a kinetics of transcriptional induction upon a shift from modulating to nonmodulating conditions which is delayed as compared to the class 2 *fha* and *bvgAS* genes, but which is clearly faster than the induction kinetics of the class 1 *ptx* and *cya* genes.[48] The upstream region of *prn* contains a high affinity primary BvgA~P binding site ranging from position –52 to –94 comprising an inverted heptanucleotide repeat sequence matching 9 from 14 positions of the consensus BvgA binding site. In addition a downstream low affinity binding region spanning to position +22 was mapped.[48] The location of the primary BvgA binding site which is similar in the *fha* promoter presumably places *prn* among the class 2 genes.

Expression of the class 3 genes is restricted to the $Bvg^i$ phase. Transcription of the best characterized class 3 gene, *bip*A, was shown to be both activated and repressed by BvgA~P.[50-52] Transcriptional activation of *bip*A is achieved by the binding of a BvgA~P dimer to a high affinity BvgA binding site centered at position –66.5 which is nearly identical to the heptanucleotide inverted repeat primary binding site in the *fha* promoter and the interaction of another BvgA~P dimer with the 3' adjacent secondary binding region which overlaps the –35 region of the *bipA* promoter. As observed for the *fha* promoter binding of BvgA~P to the secondary binding region is stabilized by cooperative interactions between BvgA~P dimers and with α-CTD.[50,52] At increased concentrations of BvgA~P a low affinity binding site centred at position +20 is recognized by a BvgA~P dimer which triggers the cooperative binding of two additional BvgA~P dimers to the 3' adjacent secondary region extending to position +75.[50,52] The interaction of BvgA~P with the binding sites located downstream of the transcriptional start site results in repression of *bipA*, however, there is still debate whether BvgA~P interferes with *bipA* expression by competition for binding with RNA polymerase or by inhibiting transcription initiation and/or elongation.[51,52]

Promoters of class 1 genes contain BvgA binding sites of rather low affinity. In vitro binding to the upstream regions of *ptx* and *cya* required significantly higher amounts of BvgA~P than the interaction with the *fha* and *bvgAS* P1 promoters.[46,56] Similarly, in vitro transcription of the toxin genes strictly required BvgA~P, while transcription of *fha*, *bvgAS* and *bipA* was already observed in the presence of unphosphorylated BvgA.[51,57] DNase I footprinting showed that extended regions in the upstream region of *ptx* and *cya* spanning from position −168 to −97 and −137 to −51, respectively, are bound by BvgA~P.[45,46] A more refined analysis revealed that in the *ptx* and *cya* promoters two BvgA~P dimers interact with primary binding sites spanning from position -120 to −165 and −125 to −80, respectively. This relatively low affinity primary binding to sites located further upstream than primary binding sites in the promoters of class 2 and 3 genes is followed by cooperative binding of four or two, respectively, additional dimers to secondary regions extending and overlapping the −35 regions of the promoters.[58]

Taken together, the activation of BvgA-dependent promoters involves the interaction of BvgA~P dimers with primary binding sites showing lower affinity and farther upstream location in class 1 versus class 2 and 3 genes and the subsequent cooperative assembly of additional BvgA dimers down to the −35 promoter region which allows the interaction with RNA polymerase to initiate transcription (Fig. 2).

## Fine Tuning of the Activity of the BvgS/BvgA Phosphorelay

A major question arising from these results is how the BvgS/BvgA system is able to control such a sophisticated transcriptional response which apparently is based on different fine-tuned levels of phosphorylated BvgA. Interestingly, the expression of the *bvgAS* locus is subject to autoregulation by BvgA~P and this may contribute to fine tuning of the system simply by regulating the amount of proteins expressed under different environmental conditions.[59] Recently, an attractive model was proposed based on different activity states of BvgS monomers within a BvgS dimer which in combination with the transcriptional autoregulation of the *bvgAS* operon may further explain the fine-tuned regulation by BvgS/BvgA.[3] According to this model the BvgS sensor protein does not act as a mere molecular switch which is either in the ON or in the OFF state according to the presence or absence of modulating signals. The BvgS protein should rather be able to permit a multitude of activity states which results in fine-tuned differences of the phosphorylation state of BvgA. The model depends on the fact that receiver domains are known to be capable of autophosphorylation with the phosphate deriving either from phosphorylated histidine residues in the transmitter (His729) or in HPt domains (His1172). Moreover, it is known that the BvgS receiver can not only act as a phosphotransferase but also as a phosphatase by the catalysis of the transfer of the phosphoryl group from Asp1023 to water.[60] Assuming that the phosphotransfer from the BvgS HPt to the BvgA receiver domain is reversible, as shown for other TCS such as the ArcB/ArcA system of *E. coli*, then the BvgS receiver domain could also control the phosphorylation state of the BvgA response regulator. In a BvgS dimer such a bifunctional capacity of the receiver domain could result in a further level of control of the phosphorelay, since its phosphotransferase versus phosphatase activities could enable at least three different activity states of the system. In one state (full BvgS activity under permissive conditions) both receiver domains of the BvgS dimer would act as phosphotransferases. In an intermediate activity state of BvgS (under semi-modulating conditions) the receiver of one BvgS protein in the dimer could act as a phosphotransferase, while the other receiver might exert phosphatase activity which would result in a reduced overall phosphorylation state of the BvgA protein. Under fully modulating conditions both of the receivers may be active as phosphatase and the phosphorylation level of BvgA would be negligible. Although not experimentally proven this is an attractive hypothesis and may be of general relevance for the function of complex His-Asp-His-Asp phosphorelay systems which may not only provide additional regulatory circuits to be integrated in the control of the phosphorelay, e.g., by specific phosphatases or anti-histidine kinases as shown for the sporulation control system of *B. subtilis*,[61] but would allow a further level of control by the intrinsic properties of these systems themselves. On the other hand, the view that "normal" TCS are ordinary ON and OFF switches is probably too simplistic and intermediate activity states of such systems may also occur and may, among other possibilities, be achieved by fine tuned signal processing during the initial signal perception step.

## Relevance of BvgS/BvgA Mediated Gene Regulation during the Infection Process

In 1991 Scarlato et al proposed that the different induction kinetics of virulence genes after shift from modulating to nonmodulating conditions observed in vitro may have in vivo relevance, since it reflects the obvious temporal demands of the differentially regulated factors, i.e., adhesins (class II) such as FHA and FIM may play an early role during colonization while toxins (class I) such as PTX and CYA being relevant in later stages to protect the bacteria from host defence mechanisms.[23] There are only a few in vivo studies which taken together do not provide clear cut conclusions, probably due to differences in experimental design and the use of different strains and animal models. However, some indications for the relevance of BvgS/BvgA mediated virulence gene regulation during infection were obtained. For instance, it was shown in a rat model with *B. bronchiseptica* that ectopic expression of a flagellar gene (belonging to the *vrg* genes) during the Bvg$^+$ phase reduced the colonization capacity of the bacteria in the trachea of the animals.[62] More recently, in an intranasal mouse model evidence was provided that *B. pertussis* strains with altered promoter regions of virulence genes interfering with their induction kinetics were impaired in their capacity to colonize the respiratory tract.[63] On the other hand, in a *B. bronchiseptica* rabbit model it was shown that Bvg$^+$-phase locked mutants were able to colonize and it was concluded that the normal BvgS/BvgA-mediated response to modulating signals is not important at least during early colonization.[64] Recently, a possible role for the Bvg$^i$ phase during infection was proposed. It was shown that a *B. pertussis* mutant locked in the Bvg$^i$ phase persisted as well as the wild type strain in the upper respiratory tract of mice, but a deletion of the *bipA* gene (the prototype of a class III factor) caused a reduced capacity to colonize the murine nasal cavity. Moreover, in mixed infections of the Bvg$^i$ phase mutant and the wild type strain, the wild type outcompeted the mutant at late time points. Based on such findings it was suggested that modulation to the Bvg$^i$ phenotype may be relevant in vivo (possibly by slightly different temperature in the nasal cavity versus the trachea) and may contribute to a better transmission of the bacteria to a new host.[65] The role of the BvgR regulon in the virulent phenotype is not well established and only preliminary studies are available. A *B. bronchiseptica bvgR* mutant was shown to colonize the lungs of infected mice as well as the corresponding wild type strain.[66] In a *B. pertussis* respiratory infection model in mice it was concluded that the BvgR-mediated regulation of gene expression influences the success of respiratory infection possibly in late stages of infection.[67] Clearly, further work about the BvgR regulon is required to evaluate its contribution to virulence.

Only recently, in an elegant study, further evidence for an in vivo relevance of Bvg-mediated regulation was obtained. In a mouse model after aerosol challenge with fully modulated *B. pertussis* (phase IV) by a RIVET approach the kinetics of induction of virulence factor expression was monitored.[68] The data clearly show that the in vitro observations about the existence of early and late factors could also be documented in vivo with the adhesins being expressed earlier while CYA was expressed at a later time point. However, in the experimental design the bacteria had to be synchronized in the Bvg$^-$ phase prior to infection, a situation which is not natural, since probably the bacteria transmitted to a new host are in a multitude of expression stages of their virulence factors. It is well conceivable that an efficient infection of the new host in fact requires the presence of a heterogeneous population of infecting agents with a broad expression spectrum of their virulence factors. Therefore, the role of Bvg-mediated fine tuning of virulence gene expression during infection remains undefined.

## Evolutionary Considerations

Recently, TCS orthologous to the *B. pertussis* BvgS/BvgA system have also been identified in the more distantly related members of the genus *Bordetella, B. holmesii, B. avium, B. hinzii, B. trematum* and even in the environmental species *B. petrii*.[12,18] (Gross et al, unpublished results) Interestingly, the gene context surrounding the *bvgAS* operon is different in the "classical" and the "new" *Bordetella* species, respectively. While in the "classical" *Bordetellae* the *bvgAS* system in the upstream region is flanked by the *fha/fim* gene locus and downstream by the *bvgR* gene, in the other pathogenic species, the *bvgAS* locus is flanked by sequences unrelated to those of the classical *Bordetellae*. The structure of the *bvg* locus therefore is in line with the phylogenetic analysis of the different *Bordetella* species which indicates a very close relationship of the mammalian pathogens and a more distant relationship with the other pathogenic *Bordetellae*.

The close relationship between the BvgS/BvgA systems of the "classical" *Bordetellae* is also documented by the fact that the *B. bronchiseptica* system can functionally be interchanged with that of *B. pertussis*.[69] In contrast, the *B. holmesii* BvgS/BvgA system can only partially substitute for the *B. pertussis* system, since only the BvgS proteins are interchangeable, while the *B. holmesii* BvgA protein can not complement a *bvgA* mutant of *B. pertussis*. In fact, the BvgS protein of *B. holmesii* can phosphorylate the *B. pertussis* BvgA protein in vitro and the BvgA protein of *B. holmesii* is not able to bind to BvgA activated promoters of *B. pertussis* such as the *fha* promoter.[18]

Still the relevant signals perceived by the BvgS/BvgA system in vivo are unknown. Temperature could be such a signal since it reflects the difference between an environmental and host associated location of the bacteria. In fact, *B. bronchiseptica* may have the potential to survive outside of a host and the pathogenic *Bordetellae* are probably descendents of environmental relatives such as *B. petrii*.[70,71] Moreover, as mentioned above, there are slight temperature differences within the different colonization niches of *B. pertussis*, e.g., the average temperature within the nasal cavity is slightly lower than in the trachea and may contribute to modulation of virulence gene expression. However, the presence of a PAS domain indicates that BvgS also perceives intracellular signals. In fact, a link of the BvgS/BvgA system to the energy metabolism was recently shown, since the oxidation state of quinones influences BvgS activity. In line with this observation is the fact that the BvgS/BvgA system also controls several genes encoding components of the respiratory electron transport chain including terminal oxidases. For instance, components of the cytochrome$_{bc1}$ complex were found to be expressed mainly in the Bvg$^-$ phase, while components of the cytochrome$_{bd}$ and cytochrome$_{cbb3}$ complexes require an activated BvgS/BvgA system for their expression. The latter terminal oxidases have a high affinity for oxygen and may be particularly relevant under microoxic growth conditions.[71] This connection of the BvgS/BvgA regulon with the respiratory chain is intriguing and may indicate a role of the BvgS/BvgA system in controlling the energy metabolism under different oxygen concentrations, or, initially, in the switch from aerobic to anaerobic growth conditions since *B. petrii*, which appears to be closely related to a common ancestor of the pathogenic *Bordetellae*, has a facultatively anaerobic metabolism. Similar to the structurally related ArcB/ArcA system of *E. coli*, the original function of the BvgS/BvgA system may therefore have been to control the metabolic state of the bacteria according to oxygen availability. In this scenario virulence genes may have been subordinated to BvgS/BvgA regulation, because in the infected host the bacteria encounter microoxic growth conditions sensed by the BvgS/BvgA system.

Based on the genomic sequences of the "classical" *Bordetellae* DNA microarray studies have been performed to define the BvgS/BvgA controlled regulon in these bacteria. Apart from the identification of a series of new *vag* genes with a potential to contribute to virulence also insights in the influence of the BvgS/BvgA system to bacterial metabolism were obtained.[72,73] It was found that the BvgS/BvgA activated gene expression patterns are very similar in *B. bronchiseptica* and *B. pertussis*, while the BvgS/BvgA repressed regulon of the two species showed much difference. In *B. bronchiseptica* a multitude of metabolic, transport, motility and chemotaxis functions are conversely regulated to the virulence genes indicating an important role of this regulon in survival outside of a host. In contrast, the *vrg* regulon of *B. pertussis* seems to be much more limited and it may be on the way of loss of function for the obligate host associated bacteria. An emerging feature from such genome wide studies is that there is very pronounced genome plasticity due to the presence of a large number of mobile genetic elements. This extraordinary genome flexibility may quickly cause different expression profiles of BvgS/BvgA regulated factors in different strains, a feature which may be very important for the evolution of the host-pathogen relationship and the generation of host specificity.[74]

## Acknowledgements

The authors would like to thank the Deutsche Forschungsgemeinschaft (Gr1243/5-2 and Be1543/2-3) for financial support.

## References

1. Beier D, Gross R. Regulation of bacterial virulence by two-component systems. Curr Opin Microbiol 2006; 9:143-152.
2. Cotter PA, DiRita V. Bacterial virulence gene regulation: an evolutionary perspective. Annu Rev Microbiol 2000; 54:519-565.
3. Cotter PA, Jones AM. Phosphorelay control of virulence gene expression in Bordetella. Trends Microbiol 2003; 11:367-373.
4. Mattoo S, Cherry JD. Molecular pathogenesis, epidemiology and clinical manifestation of respiratory infections due to Bordetella pertussis and other Bordetella subspecies. Clin Microbiol Rev 2005; 18:326-382.
5. Bock A, Gross R. The BvgS/BvgA two-component system of Bordetella spp.: a versatile modulator of virulence gene expression. Int J Med Microbiol 2001; 291:119-130.
6. Irie Y, Matto S, Yuk MH. The Bvg virulence control system regulates biofilm production in Bordetella bronchiseptica. J Bacteriol 2004; 186:5692-5698.
7. Mishra M, Parise G, Jackson KD et al. The BvgS/BvgA signal transduction system regulates biofilm development in Bordetella. J Bacteriol 2005; 187:1474-1484.
8. Preston A, Maxin E, Toland E et al. Bordetella bronchiseptica PagP is a Bvg-regulated lipid A palmitoyl transferase that is required for persistent colonization of the mouse respiratory tract. Mol Microbiol 2003; 48:725-736.
9. Parkhill J, Sebaihia M, Preston A et al. Comparative analysis of the genome sequences of Bordetella pertussis, Bordetella parapertussis and Bordetella bronchiseptica. Nat Genet 2003; 35:32-40.
10. Arico B, Rappuoli R. Bordetella parapertussis and Bordetella bronchiseptica contain transcriptionally silent pertussis toxin genes. J Bacteriol 1987; 169:2847-2853.
11. Gerlach G, von Wintzingerode F, Middendorf B et al. Evolutionary trends in the genus Bordetella. Microbes Infect 2001; 3:61-72.
12. Spears PA, Temple LM, Miyamoto DM et al. Unexpected similaritites between Bordetella avium and other pathogenic Bordetellae. Infect Immun 2003; 71:2591-2597.
13. Sebaihia M, Preston A, Maskell DJ et al. Comparison of the genome sequence of the poultry pathogen Bordetella avium with those of B. bronchiseptica, B. pertussis and B. parapertussis reveals extensive diversity in surface structures associated with host interaction. J Bacteriol 2006; 188:6002-6015.
14. Mazengia E, Silva EA, Peppe JA et al. Recovery of Bordetella holmesii from patients with pertussis-like symptoms: use of pulsed-field gel electrophoresis to characterize circulating strains. J Clin Microbiol 2000; 38:2330-2333.
15. Gadea I, Cuenca-Estrella M, Benito N et al. Bordetella hinzii, a "new" opportunistic pathogen to think about. J Infect 2000; 40:298-299.
16. Vandamme P, Heyndrickx M, Vancanneyt M et al. Bordetella trematum sp. nov., isolated from wounds and ear infections in humans and reassessment of Alcaligenes denitrificans Ruger and Tan 1983. Int J Syst Microbiol 1996; 46:849-858.
17. von Wintzingerode F, Schattke A, Siddiqui RA et al. Bordetella petrii sp. nov., isolated from an anaerobic bioreactor and emended description of the genus Bordetella. Int J Syst Evol Microbiol 2001; 51:1257-1265.
18. Gerlach G, Janzen S, Beier D et al. Functional characterization of the BvgS/BvgA two-component system of Bordetella holmesii. Microbiology 2004; 150:3715-3729.
19. Lacey CW. Antigenic modulation of Bordetella pertussis. J Hyg (Lond) 1960; 58:57-93.
20. Leslie PH, Gardner AD. The phases of Haemophilus pertussis. J Hyg (Lond) 1931; 31:423-434.
21 Weiss AA, Falkow S. Genetic analysis of phase change in Bordetella pertussis. Infect Immun 1984; 43:263-269.
22. Arico B, Miller JF, Roy C et al. Sequences required for expression of Bordetella pertussis virulence factors share homology with prokaryotic signal transduction proteins. Proc Natl Acad Sci USA 1989; 86:6671-6675.
23. Stibitz S, Yang MS. Subcellular localization and immunological detection of proteins encoded by the vir locus of Bordetella pertussis. J Bacteriol 1991; 173:4288-4296.
24. Scarlato V, Arico B, Prugnola A et al. Sequential activation and environmental regulation of virulence genes in Bordetella pertussis. EMBO J 1991; 10:3971-3975.
25. Cotter PA, Miller JF. A mutation in the Bordetella bronchiseptica bvgS gene results in reduced virulence and increased resistance to starvation and identifies a new class of Bvg-regulated antigens. Mol Microbiol 1997; 24:671-685.

26. Stockbauer KE, Fuchslocher B, Miller JF et al. Identification and characterization of BipA, a Bordetella Bvg-intermediate phase protein. Mol Microbiol 2001; 39:65-78.
27. Merkel TJ, Stibitz S. Identification of a locus required for the regulation of bvg-repressed genes in Bordetella pertussis. J Bacteriol 1995; 177:2727-2736.
28. Merkel TJ, Boucher PE, Stibitz S et al. Analysis of bvgR expression in Bordetella pertussis. J Bacteriol 2003; 185:6902-6912.
29. Grebe TW, Stock JB. The histidine protein kinase superfamily. Adv Microb Physiol 1999; 41:139-227.
30. Bantscheff M, Perraud AL, Bock A et al. Structure-function relationships in the Bvg and Evg two-component phosphorelay systems. Int J Med Microbiol 2000; 290:317-323.
31. Bock A, Bantscheff M, Perraud AL et al. Rational design and molecular characterization of a chimaeric response regulator protein. J Mol Biol 2001; 310:283-290.
32. Boucher PE, Maris AE, Yang MS et al. The response regulator BvgA and RNA polymerase alpha subunit C-terminal domain bind simultaneously to different faces of the same segment of promoter DNA. Mol Cell 2003; 11:163-173.
33. Taylor BL, Zhulin IB. PAS domains: internal sensors of oxygen, redox potential and light. Microbiol Mol Biol Rev 1999; 63:479-506.
34. Beier D, Deppisch H, Gross R. Conserved sequence motifs in the unorthodox BvgS two-component sensor protein of Bordetella pertussis. Mol Gen Genet 1996; 252:169-176.
35. Manetti R, Arico B, Rappuoli R et al. Mutations in the linker region of BvgS abolish response to environmental signals for the regulation of the virulence factors in Bordetella pertussis. Gene 1994; 150:123-127.
36. Miller JF, Johnson SA, Black WJ et al. Constitutive sensory transduction mutations in the Bordetella pertussis bvgS gene. J Bacteriol 1992; 174:970-979.
37. Beier D, Fuchs TM, Schwarz B et al. In vivo characterization of the unorthodox BvgS two-component sensor protein of Bordetella pertussis. J Mol Biol 1995; 248:596-610.
38. Perraud AL, Kimmel B, Weiss V et al. Specificity of the BvgS/BvgA and EvgAS phosphorelay is mediated by the C-terminal HPt domains of the sensor proteins. Mol Microbiol 1998; 27:875-887.
39. Perraud AL, Weiss V, Gross R. Signalling pathways in two-component phosphorelay systems. Trends Microbiol 1999; 7:115-120.
40. Uhl MA, Miller JF. Autophosphorylation and phosphotransfer in the Bordetella pertussis BvgAS signal transduction cascade. Proc Natl Acad Sci USA 1994; 91:1163-1167.
41. Uhl MA, Miller JF. Integration of multiple domains in a two-component sensor protein: the Bordetella pertussis BvgAS phosphorelay. EMBO J 1996a; 15:1028-1036.
42. Bogdan JA, Nazario-Larrieu J, Sarwar J et al. Bordetella pertussis autoregulates pertussis toxin production through the metabolism of cysteine. Infect Immun 2001; 69:6823-6830.
43. Bock A, Gross R. The unorthodox histidine kinases BvgS and EvgS are responsive to the oxidation status of a quinone electron carrier. Eur J Biochem 2002; 269:3479-3484.
44. Boucher PE, Stibitz S. Synergistic binding of RNA polymerase and BvgA phosphate to the pertussis toxin promoter of Bordetella pertussis. J Bacteriol 1995; 177:6486-6491.
45. Karimova G, Bellalou J, Ullmann A. Phosphorylation-dependent binding of BvgA to the upstream region of the cyaA gene of Bordetella pertussis. Mol Microbiol 1996; 20:489-496.
46. Zu T, Manetti R, Rappuoli R et al. Differential binding of BvgA to two classes of virulence genes of Bordetella pertussis directs promoter selectivity by RNA polymerase. Mol Microbiol 1996; 21:557-565.
47. Boucher PE, Murakami K, Ishihama A et al. Nature of DNA binding and RNA polymerase interaction of the Bordetella pertussis BvgA transcriptional activator at the fha promoter. J Bacteriol 1997; 179:1755-1763.
48. Kinnear SM, Boucher PE, Stibitz S et al. Analysis of BvgA activation of the pertactin gene promoter in Bordetella pertussis. J Bacteriol 1999; 181:5234-5241.
49. Boucher PE, Yang MS, Schmidt DM et al. Genetic and biochemical analyses of BvgA interaction with the secondary binding region of the fha promoter of Bordetella pertussis. J Bacteriol 2001; 183:536-544.
50. Deora R. Differential regulation of the Bordetella bipA gene: distinct roles for different BvgA binding sites. J Bacteriol 2002. 184:6942-6951.
51. Mishra M, Deora R. Mode of action of the Bordetella BvgA protein: transcriptional activation and repression of the Bordetella bronchiseptica bipA promoter. J Bacteriol 2005; 187:6290-6299.
52. Williams CL, Boucher PE, Stibitz S et al. BvgA functions as both an activator and a repressor to control Bvg phase expression of bipA in Bordetella pertussis. Mol Microbiol 2005; 56:175-188.
53. Boucher PE, Yang M-S, Stibitz S. Mutational analysis of the high-affinity BvgA binding site in the fha promoter of Bordetella pertussis. Mol Microbiol 2001b; 40:991-999.
54. Perraud AL, Rippe K, Bantscheff M et al. Dimerization of signalling modules of the EvgAS and BvgAS phosphorelay systems. Biochim Biophys Acta 2000; 1478:341-354.

55. Maris AE, Sawaya, MR, Kaqczor-Grzeskowiak M et al. Dimerization allows DNA target site recognition by the NarL response regulator. Nat Struct Biol 2002; 10:771-778.
56. Karimova G, Ullmann A. Characterization of DNA binding sites for the BvgA protein of Bordetella pertussis. J Bacteriol 1997; 179:3790-3792.
57. Steffen P, Goyard S, Ullmann A. Phosphorylated BvgA is sufficient for transcriptional activation of virulence-regulated genes in Bordetella pertussis. EMBO J 1996; 15:102-109.
58. Jones AM, Boucher PE, Williams CL et al. Role of BvgA phosphorylation and DNA binding affinity in control of BvgA-mediated phenotypic phase transition in Bordetella pertussis. Mol Microbiol 2005; 58:700-713.
59. Scarlato V, Prugnola A, Arico B et al. Positive transcriptional feedback at the bvg locus controls expression of virulence factors in Bordetella pertussis. Proc Natl Acad Sci USA 1990; 87:6753-6757.
60. Uhl MA, Miller JF. Central role of the BvgS receiver as a phosphorylated intermediate in a complex two-component phosphorelay. J Biol Chem 1996b; 271:33176-33180.
61. Errington J. Regulation of endospore formation in Bacillus subtilis. Nat Rev Microbiol 2003; 1:117-126.
62. Akerley BJ, Cotter PA, Miller JF. Ectopic expression of the flagellar regulon alters development of the Bordetella-host interaction. Cell 1995; 80:611-620.
63. Kinnear SM, Marques RR, Carbonetti NH. Differential regulation of Bvg-activated virulence factors plays a role in Bordetella pertussis pathogenicity. Infect Immun 2001; 69:1983-1993.
64. Cotter PA, Miller JF. BvgAS-mediated signal transduction: analysis of phase-locked regulatory mutants of Bordetella bronchiseptica in a rabbit model. Infect Immun 1994; 62:3381-3390.
65. Vergara-Irigaray N, Chavarri-Martinez A, Rodriguez-Cuesta J et al. Evaluation of the role of the Bvg intermediate phase in Bordetella pertussis during experimental respiratory infection. Infect Immun 2005; 73:748-760.
66. Fernandez J, Sisti F, Bottero D et al. Constitutive expression of bvgR-repressed factors is not detrimental to the Bordetella bronchiseptica-host interaction. Res Microbiol 2005; 156:843-850.
67. Merkel TJ, Stibitz S, Keith JM et al. Contribution of regulation by the bvg locus to respiratory infection of mice by Bordetella pertussis. Infect Immun 1998; 66:4367-4373.
68. Veal-Carr WL, Stibitz S. Demonstration of differential virulence gene promoter activation in vivo in Bordetella pertussis using RIVET. Mol Microbiol 2005; 55:788-798.
69. Martinez de Tejada G, Miller JF, Cotter PA. Comparative analysis of the virulence control systems of Bordetella pertussis and Bordetella bronchiseptica. Mol Microbiol 1996; 22:895-908.
70. Porter JF, Parton R, Wardlaw AC. Growth and survival of Bordetella bronchiseptica in natural waters and in buffered saline without added nutrients. Appl Environ Microbiol 1991; 57:1202-1206.
71. von Wintzingerode F, Gerlach G, Schneider B et al. Phylogenetic relationships and virulence evolution in the genus Bordetella. Curr Top Microbiol Immunol 2002; 264:177-199.
72. Cummings CA, Bootsma HJ, Relman DA et al. Species- and strain-specific control of a complex, flexible regulon by Bordetella BvgAS. J Bacteriol 2006; 188:1775-1785.
73. Hot D, Antoine R, Renauld-Mongenie G et al. Differential modulation of Bordetella pertussis virulence genes as evidenced by DNA microarray analysis. Mol Genet Genomics 2003; 269:475-486.
74. Brinig MM, Cummings CA, Sanden GN et al. Significant gene order and expression differences in Bordetella pertussis despite limited gene content variation. J Bacteriol 2006; 188:2375-2382.

# Chapter 11

# Capturing the VirA/VirG TCS of *Agrobacterium tumefaciens*

**Yi-Han Lin, Rong Gao, Andrew N. Binns and David G. Lynn***

## Abstract

Two-component systems (TCS) regulate pathogenic commitment in many interactions and provide an opportunity for unique therapeutic intervention. The VirA/VirG TCS of *Agrobacterium tumefaciens* mediates inter-kingdom gene transfer in the development of host tumors and sets in motion the events that underlie the great success of this multi-host plant pathogen. Significant proof for the feasibility of interventions has now emerged with the discovery of a natural product that effectively "blinds" the pathogen to the host via inhibition of VirA/VirG signal transduction. Moreover, the emerging studies on the mechanism of signal perception have revealed general sites suitable for intervention of TCS signaling. Given the extensive functional homology, it should now be possible to transfer the models discovered for VirA/VirG broadly to other pathogenic interactions.

## Introduction

Biological communities are in constant evolutionary flux and in a very real and practical sense, this adaptive molecular evolution continues to erode the arsenal of effective antibiotics and therapies that underpin our healthcare system. As the human global population grows and more mobile lifestyles emerge, the spread of new pathogens is accelerated. Secondary products synthesized by fungi and bacteria for self-defense provide the essential core of our strategies to fight these threats. However, these current antibiotics target just three functional entities; the bacterial cell wall, the ribosome and DNA gyrase. As this arsenal continues to be weakened by resistance, new targets and strategies are required.

Of the myriad of complex events underpinning pathogenesis, host/pathogen recognition is responsible for mobilizing all subsequent pathogenic programs and therefore provides a critical and obvious target for intervention. Within prokaryotic pathogens, the seductively simple "two-component" signaling modules (TCS) shoulder much of the environmental sensing and host detection responsibility. These histidine auto-kinase sensors and aspartyl phosphate response regulators discussed in this book perceive physical and chemical inputs and mediate outputs that range from gene expression to physical translocation. For successful exploitation in an ever-changing landscape, a more complete and holistic understanding of the strategies and mechanisms of pathogenesis will be necessary. Defining signal perception and transmission does not, in isolation, address what must be the central long-term question—what does the complex set of signal perceiving and integrating activities have to do with a pathogen's success? The temporal and spatial distribution of inducing signals and how this distribution corresponds to tissues and cells within the host that are maximally susceptible to infection must be determined.

*Corresponding Author: David G. Lynn—Center for Fundamental and Applied Molecular Evolution, Departments of Chemistry and Biology, Emory University, Atlanta, GA 30322, USA. Email: dlynn2@emory.edu

*Bacterial Signal Transduction: Networks and Drug Targets*, edited by Ryutaro Utsumi.

*Agrobacterium tumefaciens* initiates tumors on plant hosts and represents a special opportunity to obtain this more holistic information on pathogenesis. The critical pathogenesis genes, including the host detection modules, are localized to the tumor-inducing (Ti) plasmid and are well characterized. The output of pathogenesis—transfer of oncogenic DNA into the eukaryotic host—has been extensively studied due to the technological opportunities for plant genetic engineering. And perhaps most importantly, the multiple molecular signals and key recognition modules regulating pathogenesis have been defined. Nevertheless, these membrane-associated two-component system (TCS) elements have generally proven recalcitrant to structural characterization. We have accordingly developed alternate strategies of defining the global architecture of the domains and these methods have now revealed novel signal perception events and key transmission motions potentially susceptible to therapeutic intervention. Our initial hypothesis that plant tissues, which are constantly exposed to rich, complex and ever-changing physical and biotic challenges, will provide key insights into pathogenic strategies has been rewarding and revealed general elements of TCS signaling. Here we review the emerging understanding of signal perception for the signaling modules in *A. tumefaciens* and show how this understanding can reveal entirely new strategies for specific antibiotic development.

## *Agrobacterium tumefaciens*

The very initial stages of host/pathogen interactions are essential to any developing symbiotic or pathogenic association.[1] A commitment to such intimate associations requires the mobilization of significant resources; minimally the invading organism must detect compatible host organs, tissues and/or cells, respond in a temporally and spatially appropriate manner and stage the avoidance of host defense. This "chemical language" that enables a pathogen's response and evades a host's resistance to colonization must necessarily be complex and fluid. Remarkably, TCS form the basis of many of these environmental sensing strategies in prokaryotes and are widely used by pathogens to control virulence. Accordingly, our initial focus was to understand how this widely distributed class of conserved regulatory systems reviewed in this book and elsewhere [2-4] recognize and integrate signal inputs to ensure both accurate commitment to pathogenesis and host response to colonization.

TCS have, at their core, a 'sensor kinase' that autophosphorylates at a conserved histidine. This $PO_4^{-2}$ is subsequently transferred, directly or indirectly, to a conserved aspartate on a 'response regulator'. Signal input via the sensor kinase generally controls the level of the phosphorylated response regulator, which, in turn, alters the biochemical activity. In pathogenesis, this activity is to induce or repress expression of specific genes through the interaction of the phosphorylated response regulator with specific sites on the regulated promoters. This fairly simple picture may have layers of complexity, depending on whether the sensor exhibits phosphatase activity or involves additional phosphotransfer proteins and exhibit even greater regulation.[5-8] Already, two-component systems of pathogens have been identified and successfully exploited as targets to control pathogenesis.[9-12]

The issue of signal complexity is critical for any biological matrix and questions remain concerning the mechanisms by which a TCS perceives and translates that complexity into useful information for the pathogen. In some cases, multiple two-component systems converge to activate common genes or to individually influence expression of different subsets of genes.[13-15] At present there is a surprising lack of information regarding the signal perception processes and, in particular, the host-derived signals for most systems have not been identified. In fact, the possible role played by multiple signals on any particular two-component system is rarely examined. One exception to this is the NO responsive sensor kinase ResE of *Bacillus subtilus* where genetic evidence suggests multiple signal inputs;[16] one recognized in the cytoplasm and one in the periplasm. Similarly, recent in silico analysis of BvgS (the sensor kinase regulating pathogenesis in *Bordetella pertussis*) indicates that it has two distinct periplasmic protein-binding domains and a cytoplasmic PAS domain, all theoretically capable of responding to separate signals.

*Agrobacterium tumefaciens* is a gram-negative soil bacterium that infects plants, generally at wound sites and initiates the formation of tumors. Because of the ease and relative safety of manipulating this bacterium, it has become an important model system of bacterial pathogenesis.[17-21] Briefly, virulent bacteria contain a large tumor-inducing plasmid (pTi) that carries a variety of virulence (*vir*) genes. Two of the *vir* genes, *virA* and *virG*, comprise the two component system that recognizes signals often associated with plant wound sites, specifically, phenol derivatives, low pH, certain aldose monosaccharides and limited phosphate. VirA is the sensor kinase that processes all these input signals[22-27] and, ultimately, phosphorylates VirG.[28,29] Expression of *virA* and *virG* is autoregulated, while *virG* expression is also enhanced by acidic pH and limited phosphate through their effects on different components of its promoter.[27,30] The phosphorylated version of VirG in turn activates expression of numerous other *vir* genes on the Ti plasmid,[31] most of which are necessary for the optimal transfer of oncogenic DNA from the bacterium into the plant cell (reviewed in ref. 32).

Recognition of host derived signals by VirA is complex. A phenolic signal, e.g., acetosyringone (AS) is essential for *vir* gene induction.[25,26,33,34] The presence of certain sugars (e.g., glucose)[22,24] as well as acidic pH[23,25,27,35] strongly enhances *vir* gene expression if the concentration of phenolic inducer is low, but sugars and low pH are not essential for induction. The chromosomally encoded virulence protein ChvE is necessary for monosaccharides to serve as signals.[22,36] ChvE has significant homology to other periplasmic sugar-binding proteins, in particular the ribose-binding proteins of *E. coli* and *S. typhimurium*.[37] There has been some confusion in the literature about the means by which pH is recognized by VirA, a matter complicated by the fact that the *virG* promoter is responsive to low pH.[27,30] Recent studies however indicate that acidic pH exerts its effect through the periplasmic region and is, at least partly, ChvE-dependent.[23] Although the evidence strongly suggests that ChvE has a physical interaction with the periplasmic domain of VirA,[36] it is not known if this interaction takes place before or after ChvE binds the sugar, but the interaction itself is pH-dependent.

VirA forms membrane bound homodimers,[35,38,39](Fig. 1) and the importance of dimerization for VirA activity is indicated by experiments showing that ATP bound to one subunit is utilized to phosphorylate the other subunit within the dimer.[40-42] In fact, signals resulting from both phenol and sugar perception can be transferred from one monomer to the other.[41,42] Each VirA monomer within a dimer consists of distinct functional domains.[35,39,43] Two transmembrane regions define an N-terminal periplasmic (P) domain. The cytoplasmic portion of VirA consists of three additional regions: the linker (L), the kinase (K) and the receiver (R). The periplasmic domain is involved in sugar signaling through interaction with ChvE;[36] the kinase domain contains the conserved histidine that is the site of autophosphorylation[44] and an ATP-binding site[45] and the receiver domain has some homology with the N-terminus of the VirG response regulator. One of the most intriguing features of the VirA molecule is that several of its domains can function when expressed separately and these can be used to develop models of activation (Fig. 2). This model predicts that the periplasmic domain has a repressive activity on the cytoplasmic domain and is relieved by ChvE/sugar/low pH signaling. The kinase domain, alone, exhibits constitutive activity in stimulating *vir* gene expression that is not regulated by acetosyringone (AS), the phenolic inducer.[23,40,43] Inclusion of the receiver with the kinase fragment (KR) creates a version of the protein that can not be activated,[23] while the linker-kinase-receiver fragment (LKR) requires AS for activity. The contrasting behavior of K, KR and LKR is in accord with the role of the linker in perception of the phenolic inducer and suggests that an inhibitory effect of the receiver is reversed by AS. With LKR and no periplasmic domain, high concentrations of AS are required because the sugar-enhanced phenol response is removed.[23,43] On the other hand, the periplasmic-linker-kinase fragment (PLK) exhibits strong activity in the absence of a phenol inducer, consistent with a model of the receiver as a repressive element, though our recent data (Fang, Lynn, Binns and Wise, submitted) suggests that the receiver can also function as an activator with limiting VirG concentrations. Expression of the receiver in trans to PLK also rescues these regulatory functions.[40,46]

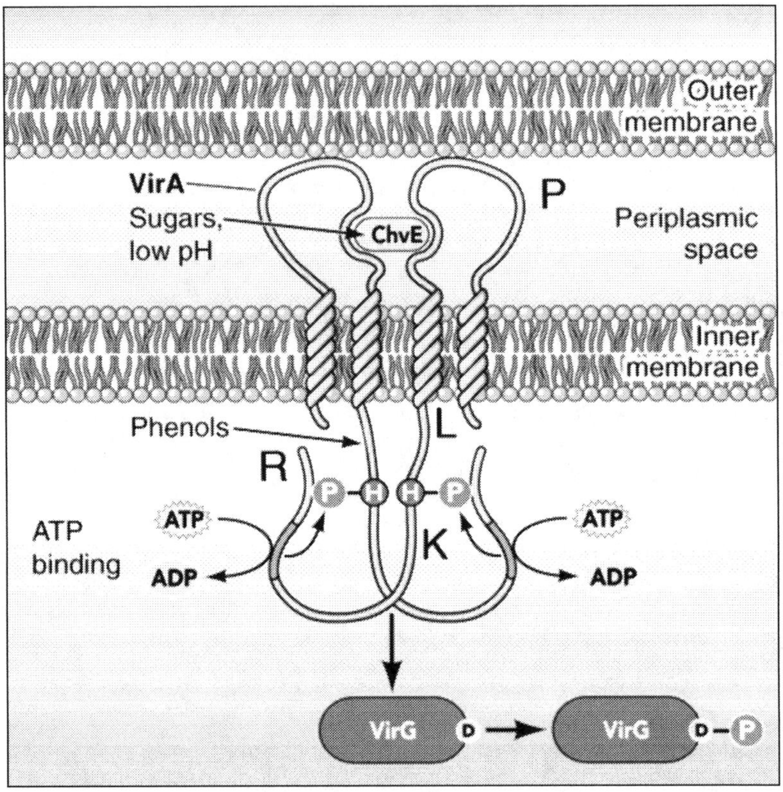


Figure 1. The VirA/VirG TCS in *Agrobacterium*. VirA is located on the inner cell membrane where it exists as a dimer. Phosphorylation occurs between the two subunits, and is followed by phosphor-transfer to the cognate response regulator, VirG (Reproduced with permission from McCullen CA, Binns AN. Annu Rev Cell Dev Biol 2006; 22:101-127).

High-resolution structural data that supports mechanistic models in TCS has been slow to emerge due to limitations in stability and solubility. However, with the accumulating genetic data and information regarding the modularity of TCS functions, we have argued that low-resolution structural data may provide sufficient information to develop therapeutic intervention strategies. Accordingly, we have developed a model whereby helical coils extending from the second transmembrane region into and out of the linker region process and transmit signals to the kinase domain.[47] In vitro signal dependent sensor kinase activities have been developed in relatively few cases[48-50] and only in the case of the related chemotaxis protein, Tar (and Trg), is there strong structural evidence concerning transmembrane signal transduction.[51] Nevertheless, domain swaps with these systems to test functional models of pathogenic TCS can be powerful.

Pathogenic *A. tumefaciens* recognizes and integrates at least four different signals. Such complex and highly regulated recognition system may suggest that a specific location (or time), organ, tissue or cell type is maximally susceptible to the pathogen. Accordingly, these requirements offer many different sites for intervention and the successful strategies may be useful for many different pathogens that have similar but less sophisticated recognition requirements. In *A. tumefaciens*, the requisite wound site certainly provides a 'portal of entry', but other specific processes at the host may be necessary for transformation.[52] For example, the high activity of the phenylpropanoid pathway, low pH and sugars associated with cell wall synthesis are also routinely associated with wound repair and cell division.[53] However, two studies have established that transformation can

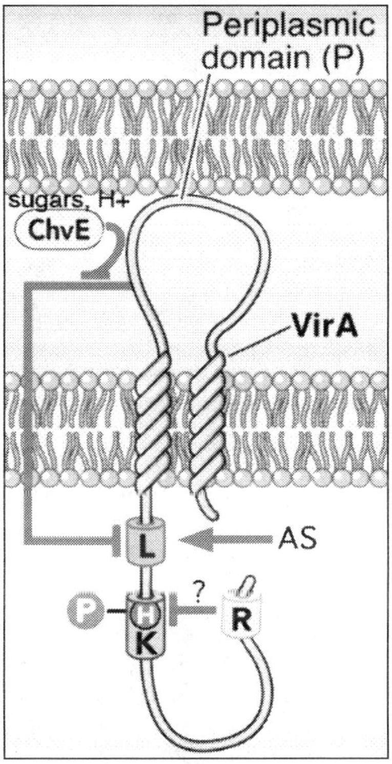


Figure 2. Domain module of VirA, showing only a single monomer for simplicity. AS and ChvE signal inputs into VirA are indicated (Reproduced with permission from McCullen CA, Binns AN. Annu Rev Cell Dev Biol 2006; 22:101-127).

occur in unwounded tobacco seedlings[54,55] and in one case, *vir* gene expression could be observed in the bacteria in the absence of wounding.[54] So, what is the required information? What is the relative timing and frequency of *vir* induction and transformation at wounded vs. unwounded sites? What is the distribution of inducing signals and host defense strategies over both time and space? What is the "signal landscape"[33,56] and how has the pathogen evolved to exploit the information? Certainly *Agrobacterium* offers many opportunities for understanding the complex information exchange that occurs during host invasion and should provide strategies for intervention at the many critical stages of pathogenesis.

## Signal Perception and Transmission

### The Phenol Signal

The most critical and the most studied, but probably the least understood, of the signals required to induce VirA/VirG-mediated gene transfer is the phenol. While acetosyringone (AS) is generally the experimental natural phenol inducer, the diversity of reported inducing phenols is immense.[33] The phenol hydroxyl group is essential for activity, but several auxiliary structural features, including *ortho*-methoxy groups and *para*-acetyl and alkyl substituents, contribute to activity. A simple first-order structure-activity model was proposed to explain the various phenol inducers.[57] This "proton transfer" model involved simple proton transfer from the phenol acid to a VirA base, designated the amine, reversibly switching the kinase activity from "OFF" to "ON" (Fig. 3). This model predicts that any acid with a pKa above the pH of the cytoplasm could access the hydrophobic binding site and protonate a basic amine.

We further evaluated the dose response of *vir* induction and inhibition. These results argue that VirA is directly involved in AS perception, but do not explain the broad phenol specificity. Nor does this assignment explain the inability of the first synthetic inhibitor of TCS induction, ASBr, to physically interact with the VirA.[57]

## The Ratchet Model

In the absence of direct physical or chemical evidence, Lohrke et al provided the best genetic evidence for VirA serving as the phenol receptor by moving the VirA/VirG signaling system into *E. coli*.[58] VirA/VirG alone was sufficient for phenol perception when co-expressed with the RNA polymerase (*rpoA*) gene from *A. tumefaciens*. As outlined in Figure 2, current genetic evidence further identifies the linker domain of VirA as responsible for sensing phenols. Common BLAST methods do not identify functionally homologous sequences, but recent advance in sequence comparison methodology has improved the detection of remote homologous sequences.[59] A search of the VirA linker sequence against the SUPERFAMILY library (www.supfam.org)[60] identified a GAF-like domain as a secondary structure match.[61] Sequence searching using other protein family databases (e.g., Pfam)[62] confirmed the identification, although the sequence homology score was rather low. GAF domains represent a large family of small molecule-binding elements that generally do not display strong sequence conservation.[63] There are now more than 1,400 putative GAF domains identified,[64] each generally coupled with other functional domains and implicating their role in small-molecule-mediated regulation of function.[63]

GAF domains are distantly related to another class of small molecule-binding domains known as PAS.[63,65,66] Both domains appear frequently in histidine kinases (e.g., PAS: FixL, ArcB and BvgS; GAF: LytS (putative regulator of cell autolysis) and ETR1 (plant ethylene receptor)), are usually located immediately prior to the histidine kinase domain and are implicated in small molecule-mediated kinase activation. Like PAS, the sequence conservation among GAF domains is low, which may imply great diversity in their small molecule ligands. GAF domains do have a conserved arrangement of secondary structure elements with three α helices (α2, α3 and α4) separated by three β strands (β1, β2 and β3) between α2 and α3 and another three β strands (β4, β5 and β6) between α3 & α4. A few X-ray crystal structures of GAF domains have now been reported. 1MC0, for PDE2 (cyclic nucleotide phosphodiesterase) in mouse, 1YKD, for *cyaB1* (adenylyl cyclase) in cyanobacterium *Anabaena* as shown in Figure 4 and 1VHM and 1F5M for unknown proteins in *E. coli* and yeast respectively, all contain a central anti-parallel multi-stranded β-sheet (β1-β6), packed on one side with a few α helices (α2, α4 and α1) and on the other with α3 and a β4-β5 loop constituting the potential ligand-binding site.[65,67,68]

The secondary structure of the VirA linker was further predicted with Sam T02 web server (http://www.cse.ucsc.edu/research/compbio/HMM-apps/) to compare with GAF domains of known structure. The structural similarities to GAF domains are well conserved in four different VirA linker sequences examined, except for the absence of the β3 strand and the existence of an additional β strand in the β4-β5 loop. The long flexible β4-β5 loop, usually having some secondary structure features (α-helical in *cyaB1* adenylyl cyclase), has been shown to be involved in ligand binding together with β3-α3 region[67,68] and the β4-β5 loop, β3-α3 and β-sheet of the linker domain may well constitute a potential phenol-binding site. Most importantly, three α-helices (α1,

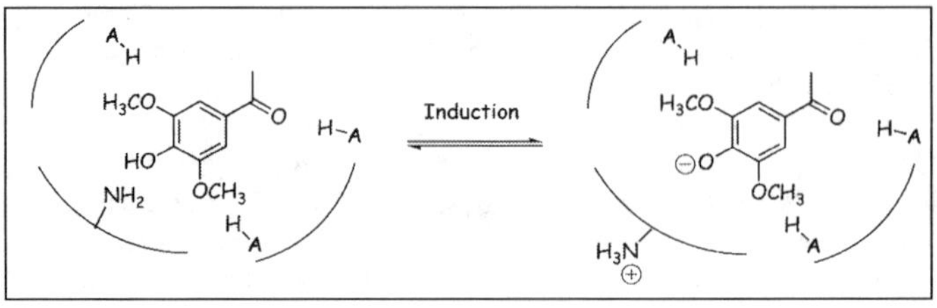


Figure 3. The proton transfer model of AS induction.

α2 and α4) flank the central β element of the VirA linker; α2 and α4 are present in all known GAF structures packing against one side of the β-sheet, while the nonconserved α1 in some GAF structures is in contact with α2/α4 and α1' from another dimer subunit, providing the dimerization interface.[67,68] As developed below, α1 and α4 appear important for signal-mediated kinase activation in VirA. Since the predictions at Figure 5 have α1 connecting to the pH/sugar-sensing periplasmic domain via TM2 and α4 as a continuous helix through the H-box of the kinase domain, the interaction and the relative motion among these helices could be critical for integrating the pH/sugar and phenol perception to regulate the kinase activity.

Dimerization appears key for histidine kinase function[38] and this association is maintained in VirA (Fig. 1). As with other histidine kinases, VirA autophosphorylation is an inter-subunit reaction, whereby a phosphate group is transferred from the ATP-binding region on one dimer subunit to a conserved histidine on the second subunit.[40,41,69] While dimerization alone seems insufficient for signaling in histidine kinases,[38,70-72] the signal-mediated displacement of one subunit relative to the other has been suggested to be critical for the activation of histidine kinases.[73-75] Previous analyses by PHD-sec algorithms identified helical coiled-coil-like heptad patterns,[47] now assigned α1 and α4 by the GAF nomenclature. Initial tests of the functional role of the helical interface in signal transduction were explored by fusing the leucine zipper (LZ) of GCN4 in register at the beginning of helix α1. The large folding free energy of the leucine zipper with adapter sequences (0, 3, or 4 residues inserted between LZ and the site of fusion) was predicted to bias subunit position and enforce alignment of three different interfaces. Indeed, the resulting LZ-(0/3/4)-linker-kinase-reciever (LZ-LKR) constructs displayed phenol independent activities, consistent with LZ restricting the flexibility of the interfaces to activate/suppress the phenol response. These initial studies led to the "ratchet" model for signal perception/transmission where a signal-induced twisting of the dimer subunits propagated through the kinase domain to mediate phosphorylation.[47]

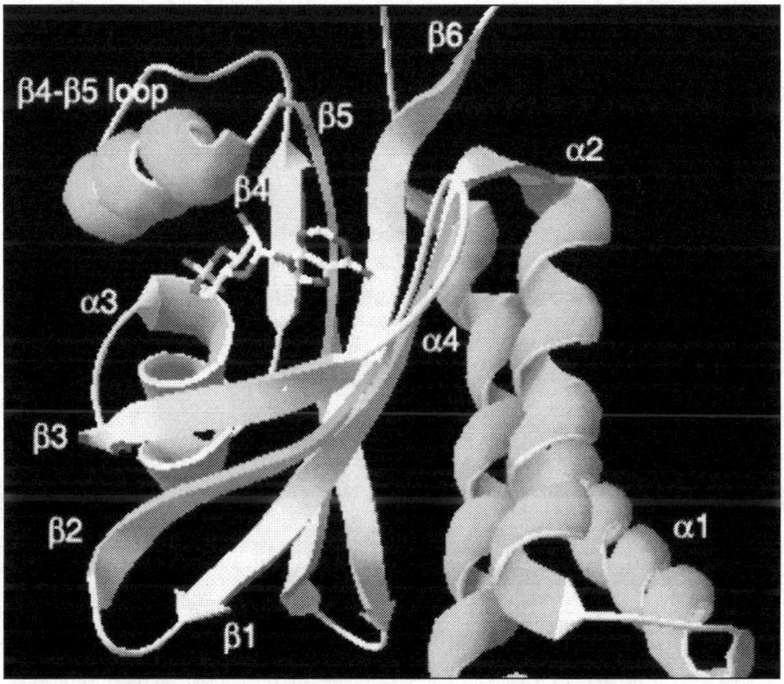


Figure 4. Structure of GAF domain from *cyaB1*.[65] PDB ID: 1YKD; ligand, cAMP.

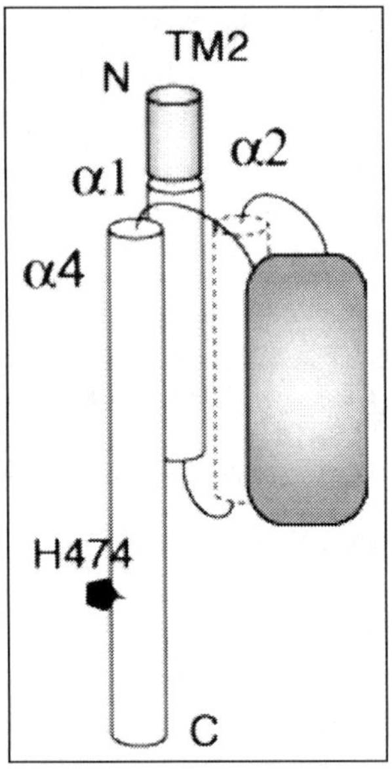


Figure 5. Schematic representation of the VirA linker domain. In this model, α1 transmits signal perception from the periplasmic domain through α4 and into the conserved histidine at 474. This model for the monomer has allowed a more detailed α1/α4 four-helix bundle model to be developed, as shown in Figure 7.

**Signal Integration**

The global motions of VirA revealed by the LZ fusions provided the first insight into signal perception by VirA/VirG and opened several further lines of analysis. For example, G471E and G665D represent two known substitutions located within the kinase domain of VirA that alter signal perception.[76] G471E is located near the conserved histidine (H474, site of auto-phosphorylation) and creates a VirA allele that is activated by sugar alone, but is further inducible by phenols. G665D lies near the predicted ATP-binding site and unlike $VirA_{G471E}$, $VirA_{G665D}$ shows low constitutive *vir* gene expression in the absence of host signals, but whose activity is also further induced by sugar and/or phenol signals. Remarkably G471E and G665D altered the activity of the LZ-LKR constructs very differently.[47] The LZ-(0/3)-$LKR_{G471E}$ allele turned the activity off completely while LZ-(3)-$LKR_{G665D}$ was ON and LZ-(4)-$LKR_{G665D}$ was OFF. These initial results suggested the very real possibility of mapping the interacting faces across the helices using these LZ fusions.

As the developing model predicted direct helix/helix interactions between α1 and α4, leucine zipper fusions in α4, creating the LZ-(0/3/4)-KR constructs, were also explored.[61] As with the LZ-(0/3/4)-$LKR_{G665D}$ constructs, the expression of a $P_{virB}$-lacZ reporter was dependent on the adapter sequence inserted between LZ and α4. LZ-(4)-$KR_{G665D}$ and LZ-(0)-$KR_{G665D}$ were ON in *vir* expression while the activity of LZ-(3)-$KR_{G665D}$ was repressed. Removal of the receiver domain (LZ-$K_{G665D}$ series) resulted in much higher activity for LZ-(0)-$K_{G665D}$ and LZ-(4)-$K_{G665D}$, consistent with the receiver functioning as a repressive element, while LZ-(3)-$K_{G665D}$ was again inactive. Importantly, the LZ-α4 chimeras without an intact receiver displayed the same activity dependence on the α4 interface as the LZ-(0/3/4)-$KR_{G665D}$ constructs, suggesting a ratchet model of rotational motion that unites α1 and α4 in regulating activation of the kinase domain. In other words, the two helices appear to be functionally coupled; a phase shift of α1 propagates to α4 and regulates phosphorylation of the kinase. Moreover, the insertions appeared to be specific to α1 and α4 as equivalent insertions in α2 had no effect on the activity.

The emerging GAF model of the linker domain positions α1 and α4 as parallel helices in a four-helix bundle within the VirA dimer, much like the parallel four-helical coiled-coil present in another widely distributed signaling domain in histidine kinases, the HAMP domain.[77] Indeed, four-helix bundle structures are common in bacterial signaling and the H-box-containing the subdomain of the histidine kinase itself likely forms a four-helix bundle. The chemotaxis receptors appear to transmit periplasmic signals through a piston-like sliding motion of one helix in a four-helix bundle formed by transmembrane helices.[51,73] Based on the similarity between the periplasmic domains of VirA and the chemoreceptors Tar and Trg, a transmembrane signaling model with the same helical sliding movement along the TM2 axis was imagined for pH/sugar transduction. Since α1 connects immediately to TM2, such a helical sliding motion could move α1 either toward or away from the inner membrane. A reasonable hypothesis might be that the α1/α4/α1'/α4' four-helix bundle coordinates the phenol-induced rotating motion with the pH/sugar-induced sliding motion to give an integrated response. Since phenol is absolutely required for virulence induction, while pH/sugar synergistically enhance phenol induction, the rotating motion appears indispensable for kinase activation, while a sliding motion, caused by sugar and pH signaling, may significantly lower the energy barrier of helix rotation.

Closer analysis of α1 and α4 revealed a feature that may integrate the proposed sugar-induced helical sliding with the rotation we propose is induced by phenol. α1 and α4 contain a high density of negatively charged residues clustered at the N-terminus of both helices and most of these acidic residues are conserved among different VirA sequences (Fig. 6). Particularly for α1, 5 out of 14 residues are D or E. A similar high density of negatively charged residues have been reported in the adaptation subdomain of chemoreceptor Tar, where the negative charge may destabilize the local four-helical structure and enhance the conformational dynamics for signaling.[78] The presence of so many D and E residues in α1 and α4 would invariably place some negatively charged residues in the otherwise hydrophobic packing interface of the four-helix bundle. During the rotation of helices, two negatively charged residues could collide such that the electrostatic repulsion would restrict rotation (Fig. 7, lower scheme). However, a sliding motion of one helix relative to the other would separate the negatively charged residues a few Å along the helix axis, reducing the electrostatic repulsion and allowing for rotation (Fig. 7). Extending the initial hypothesis would suggest that the periplasmic domain of VirA anchors the α1 helix in an unfavorable position for rotation and limits the capacity of AS to induce kinase activity, while pH/sugar perception relieves this inhibition by sliding the α1 helix to displace negatively charged residues in the four-helical interface, easing phenol-induced rotation.

To explore the functional roles of these negatively charged residues in α1/α4, conservative substitutions of the D and E residues to N or Q were constructed.[61] If the electrostatic repulsion among the negatively charged residues does play a role in limiting the rotation, the charge-neutral substitutions should lower the energy barrier for rotation and show a higher activity than wild type. Indeed both E295Q and E299Q showed significantly higher activity when tested in the absence of inducing sugar. Moreover, in the presence of inducing sugar, the increase in phenol responsiveness was eliminated, consistent with the positioning of these residues being responsible for the repressive nature of the periplasmic domain. These data are further consistent with the general working model for the motions of VirA and support the general approach in that it allows direct in vivo mapping of the interfaces of the helices before and after activation.

```
     *  **   *     *
α1   DYEELIKEIGVCFE
     *    *
α4   EIQLLELATACLCH...
```

Figure 6. The acidic residues as indicated by * at the N-terminus of both α1 and α4.

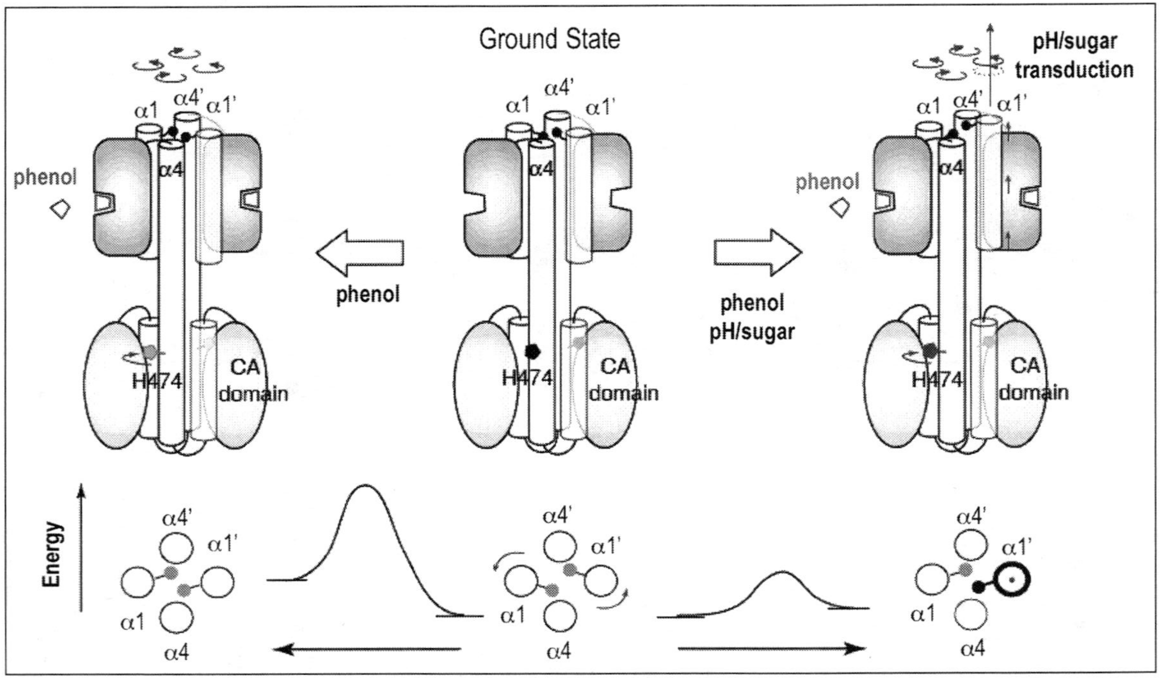


Figure 7. Signal integration model for VirA signaling. A central four-helix bundle formed by α1 and α4 is critical for both periplasmic and cytoplasmic signaling. α4 is directly connecting to the histidine containing helix of the kinase domain (CA) and the rotational motion modulates the phosphorylation of the histidine residue (pentagon). Phenol perception by the small molecule-binding GAF structure of the linker domain is postulated to initiate the rotation within the four-helix bundle but the rotation energy barrier is high due to the restriction by the periplasmic domain possibly via the positioning of clustered charged residues (see Fig. 6). The periplasmic sensing of pH/sugar induces a sliding of signaling helices to lower the energy barrier for rotation to enhance the phenol response.[61]

Histidine kinase hybrids with Tar have also been used previously to probe transmembrane signaling mechanisms. The successful construction of an aspartate-responsive Tar-EnvZ hybrid, Taz1, has been used to support a piston mechanism operating in EnvZ signaling.[79,80] On the other hand, Tar-ArcB hybrids do not respond to aspartate, consistent the absence of the transmembrane piston signaling in a protein that does not have a periplasmic sensing domain; rather a cytoplasmic PAS domain appears to be used for sensing redox signals.[74] Tar-VirA hybrid proteins have previously been constructed to investigate VirA signaling, but none of the Tar-VirA hybrids showed responsiveness to aspartate.[35] However, we felt the insensitivity to aspartate may result from incorrect fusion points in these chimeras, therefore Tar-VirA hybrids were re-investigated and based on the TM2 sequence analogy of VirA and Tar, Tar(aa1-214) was fused at the N-terminus of the cytoplasmic portion of VirA. Since the helical interface appears critical for the phenol response, multiple starting points of VirA (aa281-287), specifically those previously selected in the LZ fusions, dictated the interface. Fusions at 281, 282 and 283 showed a gradual increase of kinase activity, consistent with the rotation mechanism for signaling. Adding residues shifted the registry. Alleles carrying a single amino acid insertion at 283 gave the 282-fusion phenotype, further suggesting that the difference in activities is specifically the result of the points of fusion. Moreover, fusions at 285, 286, 287, with four residues less than the corresponding 281, 282 and 283 (approximately one turn of a helix), displayed a similar pattern of increasing activity.

The TarA fusions required phenols (AS), were insensitive to sugar and low pH and showed a decrease of activity in response to aspartate; the $IC_{50}$ between 0.5~1 mM was comparable to Taz1, the Tar-EnvZ fusion.[79,80] Therefore, the TarA fusions respond specifically to the natural ligand of Tar, aspartate and appear to integrate phenol and aspartate sensing. Since the transmembrane signaling mechanism of Tar has been suggested to operate through the helical sliding into the cytoplasmic domain, these data further implicate helical sliding in VirA signaling. Moreover, the direction of helical sliding may be different because aspartate decreases the activity of the TarA fusions, but induces the activity of Tar. Like the sugar response of wild type VirA, aspartate also altered AS sensitivity, decreasing the $ED_{50}$ of AS responses of 283 and 282 fusions. Therefore, the ability of TarA fusions to integrate aspartate and phenol and VirA to integrate pH/sugar and phenol, may well rely on the VirA linker domain to ratchet the interfaces of $\alpha 1$ and $\alpha 4$ and modulate the energy barrier of helical rotation through transmembrane helical sliding.

In addition to the piston mechanism, significant evidence suggests that helical rotation is essential for histidine kinase signaling in Sln1 and ArcB.[74,81] VirA, Sln1 and ArcB all contain a cytoplasmic "linker" domain (~120aa) between the kinase and TM2. Two helical segments invariably flank this cytoplasmic linker, corresponding to $\alpha 1$ and $\alpha 4$ in VirA. Helical-interface-dependent activities have been observed for the second helix in Sln1 (AMP, the 2nd helix of HAMP) and the first helix in ArcB (aa73-115).[74,81] Therefore, the helical rotation within the four-helix bundle may be a common signaling mechanism for histidine kinases with a domain organization of two helices flanking the cytoplasmic linker. Such a mechanism may have evolved for cytoplasmic sensing tasks, as seen in VirA and ArcB, even though the function of the linker from Sln1 is not clear. Additionally, a variety of sensor kinases have a periplasmic domain postulated to influence their linker domain and transmembrane sliding movement has been hypothesized as critical to this activity (e.g., NarX, ResE, DifA).[16,82] NarX/DifA chimeras have yielded nitrate activation of Dif regulon members.[82] Interestingly, NarX from *E. coli* also has a central linker module and we have found it to be GAF-like by the SUPERFAMILY web server (www.supfam.org).[60]

Two additional critical discoveries not yet published from this approach are worth mentioning. First, the amino acids just prior to $\alpha 1$, residues 288-293, regulate phenol structural specificity. Substitutions in this region narrow the active phenols to those missing the *para*-acyl substituent. This discovery is important both as it helps pin-point the region involved in phenol perception and also provides insight into the ability of the molecule to detect such a broad range of structures. As we have argued that this ability to detect a broad structural class is important for multi-host pathogenesis,[1] alleles that narrow the structural range of the signal will be important in testing this assertion. Second, the C-terminal receiver domain of VirA is unusual among the TCS and is not shown in Figure 7. The discovery of "psuedo-receiver" domains, similar in sequence to the receiver domains of typical response regulators but without the conserved aspartate,[83-85] suggests that these elements may have functions that go beyond phosphate transfer. Indeed, now physical and genetic evidence has suggested that the receiver of VirA associates with the kinase (Fig. 2), the linker, as well as VirG to direct phosphate transfers. The exact role the receiver plays in the motions of the linker is currently being investigated.

## Antibiotic Development

The mechanistic model for signal perception and transmission in *Agrobacterium* pathogenesis has developed to the point where drug design and evolutionary approaches could be employed to identify lead inhibitory compounds. The process of harnessing the Ti-encoded DNA transfer machinery as a natural vector for the incorporation of foreign genes into higher eukaryotes revealed several hosts resistant to *Agrobacterium* pathogenesis. The observation that maize and in fact most cereal crops, were resistant to transformation enabled the discovery of 2-hydroxy-4,7-dimethoxy-benzoxazin-3-one (HDMBOA) (Fig. 8, 1).[86] HDMBOA, which constitutes >98% of the organic exudate of young maize seedling, targets the VirA/VirG two-component signal transduction system with anti-virulence activity at concentrations orders of magnitude below the compounds general toxicity threshold. Reaction-diffusion models have now shown that HDMBOA defines a steady-state zone of inhibition circumscribing the young maize seedling, creating an unusual innate immunity where the sensing machinery for virulence is disrupted and no direct inhibition of growth or viability are contributing to the development of resistance.

Further studies have now revealed that the exuded natural product HDMBOA is not the active inhibitor of *vir* gene expression. Rather an intermediate produced along HDMBOA's two decomposition pathways, specifically pathway **a** to intermediate 2, creates the actual inhibitor (Fig. 9).[87] These results led to the proposal that HDMBOA functions as a "pro-drug" for regulating *Agrobacterium*'s virulence. The relatively stability in the rhizosphere (~pH 5.5) allows for accumulation of HDMBOA around the maize root, but as the compound accesses the more neutral (~pH 7.0) bacterial cytoplasm, the imidoquinone intermediate 2 forms rapidly and subsequently inhibits the VirA kinase to represses *vir* gene expression.

While the instability of HDMBOA appears to be critical to its function in the soil, the use of this strategy to develop lead compounds may well require stable analogs. Based on the inhibition mechanism of HDMBOA, key structural elements of the inducer were developed by structure-activity analyses (SAR) and are shown in the box above. These elements were then synthetically incorporated into a novel benzofuranone skeleton and the first stable synthetic inhibitor, HDI was created.[88] Unlike HDMBOA that functions on the kinase domain of VirA, HDI appears to function only when the receiver domain is present (Maresh and Lynn, unpublished). While binding small molecule (e.g., benzoquinone) on the psudo-receiver domain of CikA to regulate

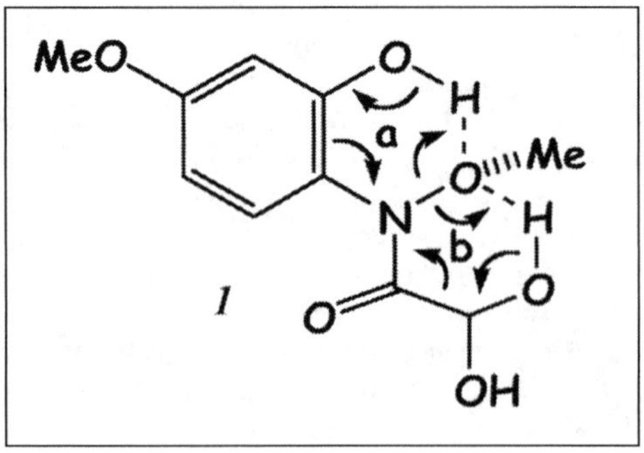


*1*

Figure 8. The ring-opened acetal structure of HDMBOA. The two decomposition pathways are indicated as a and b.

kinase activity has been discussed recently,[89] the distinction between the sites of action of these two inhibitors, HDMBOA and HDI, remains to be determined. Just as HDMBOA served as a valuable lead for the discovery of HDI, HDI sets the stage for the discovery of an entirely new class of potential inhibitors. We anticipate that the detailed mechanism of HDI's inhibition will facilitate the rational design of additional chemistry that can be extended to TCS more generally.

## Perspective

The rate at which entire genomes are sequenced continues to increase and most of the predicted proteins remain of unknown structure and function. In this climate, structural information is rate limiting in our efforts to understand molecular biology. TCS, in many ways, are prototypical of this limitation. Several components of TCS are integral membrane proteins and most of the other elements function at interfaces or with cognate partners with poor in vitro solubility. These limitations compromise both structural and functional analyses. While these environmental sensing modules are abundant in prokaryotes, where many serve essential roles in pathogenesis, most of the activating ligands are unknown. This book focuses on TCS as potential targets for the development of new antibiotics and alternate strategies of structural and functional analyses of these proteins will be necessary for these efforts to be successful.

The plant pathogen *Agrobacterium tumefaciens* can be studied safely and effectively and has served as a model pathogen in many contexts. The bacterium's unique capacity to mediate lateral gene transfer across kingdoms has motivated intense investigations for plant biotechnology and the VirA/VirG system continues to provide very useful mechanistic information about signal perception and transmission and to inform our understanding of TCS in pathogenesis. These in vivo strategies have successfully created a model for the complex motions necessary to transmit and integrate multiple input signals from diverse cellular compartments. Further, the identification of natural inhibitors has revealed a unique strategy for regulating pathogen commitment and enabled the development of a novel lead compound for the discovery of entirely new and unexpected potential antibiotics.

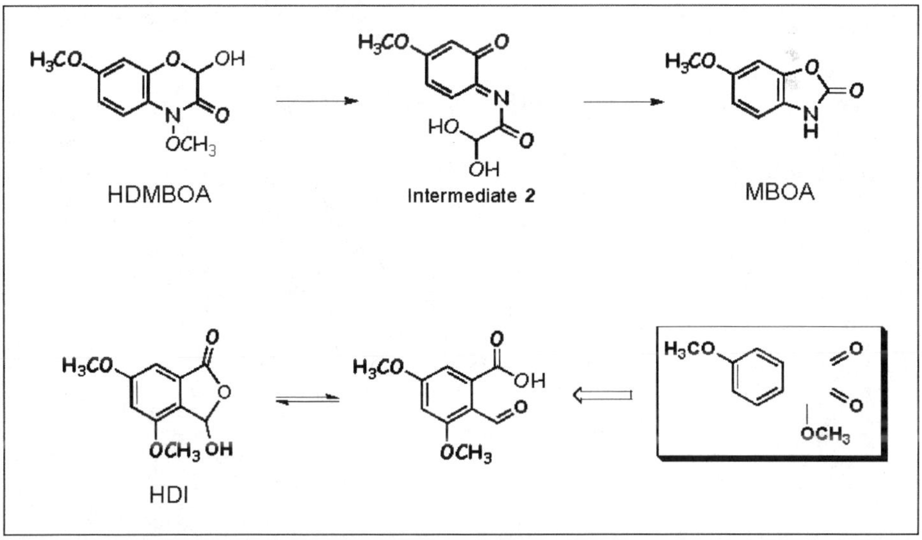


Figure 9. The upper scheme shows the decomposition pathway a in Figure 8 of HDMBOA gives intermediate 2. Intermediate 2 can further decompose into MBOA. Lower scheme shows the retro-synthetic analysis of the critical functional groups necessary for the construction of the stable inhibitor, HDI.

Most importantly however, this foundation established in *Agrobacterium* will uniquely stage our ability to ask why pathogens evolve systems that require the presence of several host-derived signals simultaneously to activate virulence gene expression. Using *Agrobacterium* as an in vivo detector carrying different constructs to respond to different signals, the plant host can be analyzed to determine which signals are actually available in the host tissue and when they emerge. Through this development of signal specific versions of VirA, it will be possible to probe the chemical, temporal and spatial dimensions of the host "signal landscape". Moreover, such studies will reveal which specific features of the landscape are critical for successful pathogenesis. Building further on this information, a "synthetic biology" engineering different detectors that respond to desired signals, can be envisioned. Such detectors, expressed within an organism able to survive in either soil or plant interstitial spaces, could report on biological threats, chemical toxins, or developmental alterations within a cellular matrix. We can now imagine a platform of fundamental and applied questions being asked and technology being developed directly from a basic understanding of the VirA/VirG system in *Agrobacterium tumefaciens*.

## References


1. Palmer AG, Gao R, Maresh J et al. Chemical biology of multi-host/pathogen interactions: chemical perception and metabolic complementation. Annu Rev Phytopathol 2004; 42:439-464.
2. Stock JB, Stock AM, Mottonen JM. Signal transduction in bacteria. Nature 1990; 344:395-400.
3. West AH, Stock AM. Histidine kinases and response regulator proteins in two-component signaling systems. Trends Biochem Sci 2001; 26:369-376.
4. Wolanin P, Thomason P, Stock J. Histidine protein kinases: key signal transducers outside the animal kingdom. Genome Biol 2002; 3:1-8.
5. Appleby JL, Parkinson JS, Bourret RB. Signal transduction via the multi-step phosphorelay: not necessarily a road less traveled. Cell 1996; 86:845-848.
6. Cavicchioli R, Schroder I, Constanti M et al. The NarX and NarQ sensor-transmitter proteins of Escherichia coli each require two conserved histidines for nitrate-dependent signal transduction to NarL. J Bacteriol 1995; 177:2416-2424.
7. Diep DB, Johnsborg O, Risøen PA et al. Evidence for dual functionality of the operon plnABCD in the regulation of bacteriocin production in Lactobacillus plantarum. Mol Microbiol 2001; 41:633-644.
8. Kreikemeyer B, Boyle MDP, Buttaro BA et al. Group A streptococcal growth phase-associated virulence factor regulation by a novel operon (Fas) with homologies to two-component-type regulators requires a small RNA molecule. Mol Microbiol 2001; 39:392-406.
9. Stephenson K, Hoch JA. Two-component and phosphorelay signal-transduction systems as therapeutic targets. Curr Opin Pharmacol 2002; 2:507-512.
10. Stephenson K, Hoch JA. Developing inhibitors to selectively target two-component and phosphorelay signal transduction systems of pathogenic microorganisms. Curr Med Chem 2004; 11:765-773.
11. Barrett JF, Goldschmidt RM, Lawrence LE et al. Antibacterial agents that inhibit two-component signal transduction systems. Proc Natl Acad Sci USA 1998; 95:5317-5322.
12. Lyon GJ, Mayville P, Muir TW et al. Rational design of a global inhibitor of the virulence response in Staphylococcus aureus, based in part on localization of the site of inhibition to the receptor-histidine kinase, AgrC. Proc Natl Acad Sci USA 2000; 97:13330-13335.
13. Bronner S, Monteil H, Prevost G. Regulation of virulence determinants in Staphylococcus aureus: complexity and applications. FEMS Microbiol Rev 2004; 28:183-200.
14. Cotter PA, Jones AM. Phosphorelay control of virulence gene expression in Bordetella. Trends Microbiol 2003; 11:367-373.
15. Kato A, Groisman EA. Connecting two-component regulatory systems by a protein that protects a response regulator from dephosphorylation by its cognate sensor. Genes Dev 2004; 18:2302-2313.
16. Baruah A, Lindsey B, Zhu Y et al. Mutational analysis of the signal-sensing domain of ResE histidine kinase from Bacillus subtilis. J Bacteriol 2004; 186:1694-1704.
17. Cascales E, Christie PJ. The versatile bacterial type IV secretion systems. Nature Rev Microbiol 2003; 1:137-149.
18. Christie PJ. Type IV secretion: the Agrobacterium VirB/D4 and related conjugation systems. Biochim Biophys Acta 2004; 1694:219-234.
19. Guttman DS. Plants as models for the study of human pathogenesis. Biotechnol Adv 2004; 22:363-382.
20. Li J, Wolf SG, Elbaum M et al. Exploring cargo transport mechanics in the type IV secretion systems. Trends Microbiol 2005; 13:295-298.

21. Winans SC. Two-way chemical signaling in Agrobacterium-plant interactions. Microbiol Mol Biol Rev 1992; 56:12-31.
22. Cangelosi GA, Ankenbauer RG, Nester EW. Sugars induce the Agrobacterium virulence genes through a periplasmic binding protein and a transmembrane signal protein. Proc Natl Acad Sci USA 1990; 87:6708-6712.
23. Gao R, Lynn DG. Environmental pH sensing: resolving the VirA/VirG two-component system inputs for Agrobacterium pathogenesis. J Bacteriol 2005; 187:2182-2189.
24. Shimoda N, Toyoda-Yamamoto A, Nagamine J et al. Control of expression of Agrobacterium vir genes by synergistic actions of phenolic signal molecules and monosaccharides. Proc Natl Acad Sci USA 1990; 87:6684-6688.
25. Stachel SE, Messens E, Van Montagu M et al. Identification of the signal molecules produced by wounded plant cells that activate T-DNA transfer in Agrobacterium tumefaciens Nature 1985; 318:624-629.
26. Stachel SE, Nester EW, Zambryski PC. A plant cell factor induces Agrobacterium tumefaciens vir gene expression. Proc Natl Acad Sci USA 1986; 83:379-383.
27. Winans SC. Transcriptional induction of an Agrobacterium regulatory gene at tandem promoters by plant-released phenolic compounds, phosphate starvation and acidic growth media. J Bacteriol 1990; 172:2433-2438.
28. Jin SG, Prusti RK, Roitsch T et al. Phosphorylation of the VirG protein of Agrobacterium tumefaciens by the autophosphorylated VirA protein: essential role in biological activity of VirG. J Bacteriol 1990; 172:4945-4950.
29. Mukhopadhyay A, Gao R, Lynn DG. Integrating input from multiple signals: the VirA/VirG two-component system of Agrobacterium tumefaciens. Chembiochem 2004; 5:1535-1542.
30. Mantis NJ, Winans SC. The Agrobacterium tumefaciens vir gene transcriptional activator virG is transcriptionally induced by acid pH and other stress stimuli. J Bacteriol 1992; 174:1189-1196.
31. Jin SG, Roitsch T, Christie PJ et al. The regulatory VirG protein specifically binds to a cis-acting regulatory sequence involved in transcriptional activation of Agrobacterium tumefaciens virulence genes. J Bacteriol 1990; 172:531-537.
32. Binns AN, Costantino P. The Agrobacterium oncogenes. In: Spaink HP, Kondorosi A, Hooykaas PJJ, eds. The Rhizobiaceae: Molecular Biology of Model Plant-Associated Bacteria. Dordrecht: Kluwer Academic Publishers, 1998:251-266.
33. Duban ME, Lee KH, Lynn DG. Strategies in pathogenesis—mechanistic specificity in the detection of generic signals. Mol Microbiol 1993; 7:637-645.
34. Melchers LS, Regensburgtuink AJG, Schilperoort RA et al. Specify of signal molecules in the activation of Agrobacterium virulence gene-expression. Mol Microbiol 1989; 3:969-977.
35. Melchers LS, Regensburg-Tuïnk TJ, Bourret RB et al. Membrane topology and functional analysis of the sensory protein VirA of Agrobacterium tumefaciens. EMBO J 1989; 8:1919-1925.
36. Shimoda N, Toyoda-Yamamoto A, Aoki S et al. Genetic evidence for an interaction between the VirA sensor protein and the ChvE sugar-binding protein of Agrobacterium. J Biol Chem 1993; 268:26552-26558.
37. Tam R, Saier MH Jr. Structural, functional and evolutionary relationships among extracellular solute-binding receptors of bacteria. Microbiol Rev 1993; 57:320-346.
38. Pan SQ, Charles T, Jin S et al. Preformed dimeric state of the sensor protein VirA is involved in plant-Agrobacterium signal transduction. Proc Natl Acad Sci USA 1993; 90:9939-9943.
39. Winans SC, Kerstetter RA, Ward JE et al. A protein required for transcriptional regulation of Agrobacterium virulence genes spans the cytoplasmic membrane. J Bacteriol 1989; 171:1616-1622.
40. Brencic A, Xia Q, Winans SC. VirA of Agrobacterium tumefaciens is an intradimer transphosphorylase and can actively block vir gene expression in the absence of phenolic signals. Mol Microbiol 2004; 52:1349-1362.
41. Toyoda-Yamamoto A, Shimoda N, Machida Y. Genetic analysis of the signal-sensing region of the histidine protein kinase VirA of Agrobacterium tumefaciens. Mol Gen Genet 2000; 263:939-947.
42. Wise AA, Voinov L, Binns AN. Intersubunit complementation of sugar signal transduction in VirA heterodimers and posttranslational regulation of VirA activity in Agrobacterium tumefaciens. J Bacteriol 2005; 187:213-223.
43. Chang C, Winans S. Functional roles assigned to the periplasmic, linker and receiver domains of the Agrobacterium tumefaciens VirA protein. J Bacteriol 1992; 174:7033-7039.
44. Jin S, Roitsch T, Ankenbauer RG et al. The VirA protein of Agrobacterium tumefaciens is autophosphorylated and is essential for vir gene regulation. J Bacteriol 1990; 172:525-530.
45. Parkinson JS, Kofoid EC. Communication modules in bacterial signaling proteins. Ann Rev Genet 1992; 26:71-112.
46. Chang CH, Zhu J, Winans SC. Pleiotropic phenotypes caused by genetic ablation of the receiver module of the Agrobacterium tumefaciens VirA protein. J Bacteriol 1996; 178:4710-4716.

47. Wang Y, Gao R, Lynn DG. Ratcheting up vir gene expression in Agrobacterium tumefaciens: coiled coils in histidine kinase signal transduction. Chembiochem 2002; 3:311-317.
48. Bader MW, Sanowar S, Daley ME et al. Recognition of antimicrobial peptides by a bacterial sensor kinase. Cell 2005; 122:461-472.
49. Dunham CM, Dioum EM, Tuckerman JR et al. A distal arginine in oxygen-sensing heme-PAS domains is essential to ligand binding, signal transduction and structure. Biochemistry 2003; 42:7701-7708.
50. Malpica R, Franco B, Rodriguez C et al. Identification of a quinone-sensitive redox switch in the ArcB sensor kinase. Proc Natl Acad Sci USA 2004; 101:13318-13323.
51. Peach ML, Hazelbauer GL, Lybrand TP. Modeling the transmembrane domain of bacterial chemoreceptors. Protein Sci 2002; 11:912-923.
52. Braun AC. Conditioning of the host cell as a factor in the transformation process in crown gall. Growth 1952; 16:65-74.
53. Baron C, Zambryski PC. The plant response in pathogenesis, symbiosis and wounding: variations on a common theme? Annu Rev Genet 1995; 29:107-129.
54. Brencic A, Angert ER, Winans SC. Unwounded plants elicit Agrobacterium vir gene induction and T-DNA transfer: transformed plant cells produce opines yet are tumour free. Mol Microbiol 2005; 57:1522-1531.
55. Escudero J, Hohn B. Transfer and integration of T-DNA without cell injury in the host plant. Plant Cell 1997; 9:2135-2142.
56. McCullen CA, Binns AN. Agrobacterium tumefaciens and plant cell interactions and activities required for interkingdom macromolecular transfer. Annu Rev Cell Dev Biol 2006; 22:101-127.
57. Hess KM, Dudley MW, Lynn DG et al. Mechanism of phenolic activation of Agrobacterium virulence genes: development of a specific inhibitor of bacterial sensor/response systems. Proc Natl Acad Sci USA 1991; 88:7854-7858.
58. Lohrke SM, Yang H, Jin S. Reconstitution of acetosyringone-mediated Agrobacterium tumefaciens virulence gene expression in the heterologous host Escherichia coli. J Bacteriol 2001; 183:3704-3711.
59. Karplus K, Barrett C, Hughey R. Hidden Markov models for detecting remote protein homologies. Bioinformatics 1998; 14:846-856.
60. Gough J, Karplus K, Hughey R et al. Assignment of homology to genome sequences using a library of hidden Markov models that represent all proteins of known structure. J Mol Biol 2001; 313:903-919.
61. Gao R, Lynn DG. Integration of rotation and piston motions in coiled-coil signal transduction. J Bacteriol 2007; 189:6048-6056.
62. Bateman A, Coin L, Durbin R et al. The Pfam protein families' database. Nucleic Acids Res 2004; 32: D138-141.
63. Anantharaman V, Koonin EV, Aravind L. Regulatory potential, phyletic distribution and evolution of ancient, intracellular small-molecule-binding domains. J Mol Biol 2001; 307:1271-1292.
64. Zoraghi R, Corbin JD, Francis SH. Properties and functions of GAF domains in cyclic nucleotide phosphodiesterases and other proteins. Mol Pharmacol 2004; 65:267-278.
65. Ho YJ, Burden LM, Hurley JH. Structure of the GAF domain, a ubiquitous signaling motif and a new class of cyclic GMP receptor. EMBO J 2000; 19:5288-5299.
66. Taylor BL, Zhulin IB. PAS domains: internal sensors of oxygen, redox potential and light. Microbiol Mol Biol Rev 1999; 63:479-506.
67. Martinez SE, Bruder S, Schultz A et al. Crystal structure of the tandem GAF domains from a cyanobacterial adenylyl cyclase: modes of ligand binding and dimerization. Proc Natl Acad Sci USA 2005; 102:3082-3087.
68. Martinez SE, Wu AY, Glavas NA et al. The two GAF domains in phosphodiesterase 2A have distinct roles in dimerization and in cGMP binding. Proc Natl Acad Sci USA 2002; 99:13260-13265.
69. Ninfa EG, Atkinson MR, Kamberov ES et al. Mechanism of autophosphorylation of Escherichia coli nitrogen regulator II (NRII or NtrB): trans-phosphorylation between subunits. J Bacteriol 1993; 175:7024-7032.
70. Cai S, Inouye M. Spontaneous subunit exchange and biochemical evidence for trans-autophosphorylation in a dimer of Escherichia coli histidine kinase (EnvZ). J Mol Biol 2003; 329:495-503.
71. Falke JJ, Koshland DE Jr. Global flexibility in a sensory receptor: a site-directed cross-linking approach. Science 1987; 237:1596-1600.
72. Hidaka Y, Park H, Inouye M. Demonstration of dimer formation of the cytoplasmic domain of a transmembrane osmosensor protein, EnvZ, of Escherichia coli using Ni-histidine tag affinity chromatography. FEBS Lett 1997; 400:238-242.
73. Falke JJ, Hazelbauer GL. Transmembrane signaling in bacterial chemoreceptors. Trends Biochem Sci 2001; 26:257-265.
74. Kwon O, Georgellis D, Lin ECC. Rotational on-off switching of a hybrid membrane sensor kinase Tar-ArcB in Escherichia coli. J Biol Chem 2003; 278:13192-13195.

75. Murphy OJ, Kovacs FA, Sicard EL et al. Site-directed solid-state NMR measurement of a ligand-induced conformational change in the serine bacterial chemoreceptor. Biochemistry 2001; 40:1358-1366.
76. McLean BG, Greene EA, Zambryski PC. Mutants of Agrobacterium VirA that activate vir gene expression in the absence of the inducer acetosyringone. J Biol Chem 1994; 269:2645-2651.
77. Aravind L, Ponting CP. The cytoplasmic helical linker domain of receptor histidine kinase and methyl-accepting proteins is common to many prokaryotic signaling proteins. FEMS Microbiol Lett 1999; 176:111-116.
78. Starrett DJ, Falke JJ. Adaptation mechanism of the aspartate receptor: electrostatics of the adaptation subdomain play a key role in modulating kinase activity. Biochemistry 2005; 44:1550-1560.
79. Jin T, Inouye M. Ligand binding to the receptor domain regulates the ratio of kinase to phosphatase activities of the signaling domain of the hybrid Escherichia coli transmembrane receptor, Taz1. J Mol Biol 1993; 232:484-492.
80. Utsumi R, Brissette RE, Rampersaud A et al. Activation of bacterial porin gene expression by a chimeric signal transducer in response to aspartate. Science 1989; 245:1246-1249.
81. Tao W, Malone CL, Ault AD et al. A cytoplasmic coiled-coil domain is required for histidine kinase activity of the yeast osmosensor, SLN1. Mol Microbiol 2002; 43:459-473.
82. Xu Q, Black WP, Ward SM et al. Nitrate-dependent activation of the Dif signaling pathway of Myxococcus xanthus mediated by a NarX-DifA interspecies chimera. J Bacteriol 2005; 187:6410-6418.
83. O' Hara BP, Norman RA, Wan PTC et al. Crystal structure and induction mechanism of AmiC-AmiR: a ligand-regulated transcription antitermination complex. EMBO J 1999; 18:5175–5186.
84. Williams SB, Vakonakis I, Golden SS et al. Structure and function from the circadian clock protein KaiA of Synechococcus elongatus: A potential clock input mechanism. Proc Natl Acad Sci USA 2002; 99:15357-15362.
85. Zhang X, Dong G, Golden SS. The pseudo-receiver domain of CikA regulates the cyanobacterial circadian input pathway. Mol Microbiol 2006; 60:658-668.
86. Zhang J, Boone L, Kocz R et al. At the maize/Agrobacterium interface: natural factors limiting host transformation. Chem Biol 2000; 7:611-621.
87. Maresh J, Zhang J, Lynn DG. The innate immunity of maize and the dynamic chemical strategies regulating two-component signal transduction in Agrobacterium tumefaciens. ACS Chem Biol 2006; 1:165-175.
88. Maresh J, Zhang J, Tzeng Y et al. Rational design of inhibitors of VirA-VirG two-component signal transduction. Bioorg Med Chem Lett 2007; 17:3281-3286.
89. Gao T, Zhang X, Ivleva NB et al. NMR structure of the pseudo-receiver domain of CikA. Protein Sci 2007; 16:465-475.

# CHAPTER 12

# Quorum Sensing and Biofilm Formation by *Streptococcus mutans*

**Dilani Senadheera and Dennis G. Cvitkovitch***

## Abstract

*Streptococcus mutans* is the primary causative agent involved in dental caries in humans. Among important virulence factors of this pathogen, its ability to form and sustain a polysaccharide-encased biofilm (commonly called dental plaque) is vital not only to its survival and persistence in the oral cavity, but also for its pathogenicity as well. This chapter focuses on the *S. mutans'* biofilm phenotype and how this mode of growth is regulated by its density-dependent quorum sensing (QS) system primarily comprised of the Competence Stimulating Peptide (CSP) and the ComD/ComE two-component signal transduction system. In addition to biofilm formation, the CSP-mediated QS system in *S. mutans* also affects its acidogenicity, aciduricity, genetic transformation and bacteriocin production. Interestingly, it has also been discovered that these properties are optimally expressed in cells derived from a biofilm as opposed to a free-floating planktonic mode of growth. Hence, strategies targeting *S. mutans'* QS system to attenuate biofilm formation and/or virulence are currently being used to develop therapeutic or preventative measures against dental caries. Recently, it was discovered that the addition of CSP in large concentrations (relative to amounts used for normal competence development) resulted in growth arrest and eventual cell death, thus paving way for CSP-mediated targeted killing of *S. mutans*. In addition to the QS system, effects of other two-component signal transduction systems on the biofilm phenotype of *S. mutans* are also discussed.

## Introduction

In nature, bacteria live in diverse habitats that are often subjected to various environmental fluxes. The ability to detect and adapt their cell physiology or behavior to signals reflecting such changes can ultimately determine the survival and persistence of these organisms. The two-component signal transduction paradigm, illustrated by the chapters in this book, facilitates this adaptive response via a sophisticated sensor and responder. More specifically, a particular stimulus or a combination of stimuli can autophosphorylate a membrane-located histidine kinase sensor, which can then convey this message into the cell by transferring the phosphate group to a responder protein (response regulator).[1,2] The adaptive phenotype is elicited when the resulting phosphorylation triggers a conformational change in the response regulator prompting its binding to specific DNA promoter regions, thereby rendering transcriptional regulation of genes under their control. In recent years, the availability of complete bacterial genomes has facilitated the discovery of novel two-component signal transduction systems, which in combination with effective genetic tools has vastly enhanced our understanding of how bacteria sense and respond

*Corresponding Author: Dennis G. Cvitkovitch—Oral Microbiology, Faculty of Dentistry, University of Toronto, 124 Edward St., Toronto, Ontario M5G 1G6, Canada.
Email: d.cvitkovitch@dentistry.utoronto.ca

*Bacterial Signal Transduction: Networks and Drug Targets*, edited by Ryutaro Utsumi.

to various signals. From a clinical standpoint, these systems provide us with potential targets that can be manipulated to combat pathogenic infections.

The quorum sensing (QS) phenomenon, which enables bacteria to alter their gene expression when a critical density of the cell population is reached, is one of the most fascinating behaviors observed utilizing signal transduction.[3-8] Ubiquitous in both Gram-negative and Gram-positive bacteria, QS is induced by chemical messengers (autoinducer molecules or pheromones) that are released into the environment and increased in concentration proportionate to a growing population. Once the density reaches a particular threshold, these signal molecules can trigger an expression cascade, which can alter the cell populations' physiology or behavior as a collective unit instead of an individual entity. Various physiological activities controlled by this cell-cell communication circuitry include antibiotic production, competence development, sporulation, biofilm differentiation, conjugation and bioluminescence in bacteria.[6,9-17] Such highly coordinated group behavior can have profound implications on the survival as well as the pathogenicity of a bacterial population. In addition to mounting stress responses and coping mechanisms under high cell density, it may be advantageous for these organisms to synchronize the release of toxins and other antigenic virulence factors with cell density and thus overwhelm the host immune system with the sudden unanimous release of virulence factors. Accumulating evidence of infectious models conducted by using animals harboring mutant bacterial strains defective in one or more components of their respective QS systems have shown reduced or attenuated virulence.[6,18]

In this ensuing chapter, we will discuss how *Streptococcus mutans*—a primary etiologic agent responsible for human dental caries as well as endocarditis, utilizes its QS signaling system to regulate biofilm formation. Biofilms are surface-attached microbial communities that are protected by a self-generated organic polymer matrix.[19-24] By resorting to a biofilm lifestyle, *S. mutans* can modulate various physiological and physical properties that are beneficial to its growth, survival and persistence in the oral cavity. For example, the biofilm matrix can act as a diffusion barrier and limit the penetration of antimicrobials to the innermost cells. Also, biofilm cells have a transcriptional profile that is markedly distinct from their planktonic counterpart, likely giving rise to their inherently resilient or resistant phenotype that is widely documented in the scientific literature.[20,25-30] On a different note, it has been shown that *S. mutans* cells grown as a biofilm have a dramatic increase in genetic competence and are highly transformable by exogenous DNA.[38] They also exhibit an increased ability to tolerate acid challenges.[13,31] Much to the fascination of researchers in this field, it was recently discovered that the addition of the biologically active synthetic autoinducer QS peptide pheromone in excess of concentration required for the induction of genetic transformation caused growth arrest and eventually cell death, followed by disruption of the biofilm.[32] This finding was an exciting discovery for the scientific community investigating the possibility of targeting *S. mutans* to control its pathogenicity.

## Virulence Properties of *S. mutans*

*Streptococcus mutans* is a well known bacterium in the fields of oral microbiology and clinical dentistry as the etiologic agent of dental caries in humans. *S. mutans* was first described as small, chained coccobacilli by J. K. Clark in 1924, although its relationship with caries was clearly distinguished only in the 1960s following a series of experiments by Fitzgerald and Keyes, who demonstrated its capability to induce dental caries in rats and hamsters.[33] However, a link between *S. mutans* and human dental caries was only established after a series of longitudinal studies that showed a statistically significant increase in *S. mutans*' counts in teeth that were destined to develop carious lesions in humans.[34-38] After recognizing the association of *S. mutans* with human dental caries, decades of research has focused not only on identifying its virulence traits, but also on understanding how these factors are regulated in the plaque biofilm to cause disease. A growing evidence of literature has highlighted three virulence traits of *S. mutans* as being vital to the initiation and progression of caries. These include: (1) the ability to metabolize dietary carbohydrates and produce lactic acid (acidogenicity), (2) ability to grow and survive in a low pH environment (aciduricity) and (3) ability to utilize dietary sugars to produce glucan polymers and form the plaque biofilm.[39-42] Of the first two factors, acidogenicity or the release of acid into its growth environment results in corroding the calcium-phosphate matrix of the tooth structure, whereas high aciduricity of *S. mutans* enables it to survive and further utilize sugars for metabolic activities under low pH that is usually detrimental to a large number of plaque organisms. The third attribute, which is the ability to form a biofilm offers *S. mutans*' protection from transient environmental changes and mechanical forces, thereby facilitating its growth and survival in the oral cavity. In fact, *S. mutans* is so attuned to the biofilm lifestyle that it cannot be found in the mouths of people without teeth or dentures for it to adhere to. The carbohydrate polymer matrix in the biofilm also serves as an excellent food reserve from which *S. mutans* and other bacteria can derive nutrients by digesting these organic polymers. Interestingly, in *S. mutans*, aciduricity and biofilm formation are controlled by a peptide-induced QS system,[13,43] which we will discuss in the following section.

## Quorum Sensing System in *S. mutans*

### *Identification of the Pneumococcal Quorum Sensing System*

The discovery of the QS network in *S. mutans* was largely dependent on information derived from its closely related species, *Streptococcus pneumoniae*, which comprises the best-characterized cell-cell communication circuitry among members of the genus *Streptococcus*. In *S. pneumoniae*, QS was believed to function mainly to acquire and incorporate foreign DNA from the environment by induction of a physiological state known as genetic competence. In this bacterium, the link between cell population density and transformation was first discovered in the 1960s by Pakula and Walczak (1963) and Tomasz and Hotchkiss (1964).[44,45] Upon noticing that the induction of genetic competence took place only at a particular cell population density, it was also discovered that supplementing cell-free supernatant derived by filtering a competent culture was capable of inducing competence in otherwise noncompetent cells. It was later determined that the competent state was induced by a protein-like "activator" compound present in the supernatant of the competent cultures.[44,46] Later this peptide was identified as a 17-residue, cationic peptide and was referred to as the competence stimulating peptide (CSP) due to its role in genetic competence development in *S. pneumoniae*.[47,48] By using the reverse translated amino acid sequence of the CSP, the *comC* precursor peptide gene (*com* for competence) and its contiguous genes were identified.[47,49-51] Interestingly, the *comC* was part of the *comCDE* tricistronic operon, in which the latter genes encoded a membrane-bound histidine kinase sensor protein (ComD) and its cognate respond regulator (ComE).[52,53]

### *Competence Stimulating Peptide*

The QS circuit in *S. mutans* was discovered by searching its genome using the *comCDE* sequence of *S. pneumoniae*. Despite the high homology shown by the orthologous genes in *S. mutans*, the orientation of these genes was distinct from that in *S. pneumoniae*: the *comC* was located 59 bp proximal and in the reversed orientation of the complementary strand harboring *comDE*.[31] In addition to *S. pneumoniae*, the organization of these genes in *S. mutans* also differed from those in other streptococci including *Streptococcus mitis*, *Streptococcus oralis*, *Streptococcus gordonii* and *Streptococcus sanguinis*.[54] In these bacteria, the *comC*, *comD* and *comE* genes are organized as an operon starting with *comC* and followed by *comD* and *comE* genes in the 5′ to 3′ direction.

The deduced primary translational product of *comC*, which was derived by using its consensus nucleotide sequences from various *S. mutans* strains, was identified as a 46-amino acid cationic peptide, whose C-terminus harbored the biologically active secreted CSP, 21-aa in length.[31] As in the case of pneumococci, posttranslational processing was required for the maturation of the propeptide by cleaving the Gly-Gly site located at -1 and -2 positions (relative to the cleavage site) in the N-terminus of the precursor CSP. Interestingly, the biological activity of the mature synthetic CSP was retained when added to growing cultures as judged by the increased frequency in transformation.[31] Hence, this finding provided researchers with a novel tool to use synthetic, exogenously supplied CSP to simulate population growth in *S. mutans*' cultures in the laboratory, as well as further investigate the physiological properties controlled by CSP-induced QS in

*S. mutans*. The ability to use synthetic CSPs for QS experiments was demonstrated prior to this study by Håvarstein et al who showed that distinct species of streptococci from the anginosus group commonly encoded and responded to identical CSPs, thereby belonging to the same 'pherotype'.[55] In contrast, it was discovered that CSPs from other groups were most often species-specific and in some cases strain-specific.

In a recent study, Syvitski et al[56] investigated the structure-function relationship of the *S. mutans* signaling peptide based on three-dimensional structures of CSP derived from the UA159 wild-type strain and a C-terminally truncated peptide (TPC3) from the JH1005 strain, which was defective in genetic competence development. By synthesizing a series of peptides that harbored aa-substitutions or aa-deletions and by using them in competitive inhibition assays, these researchers showed that the C-terminal structural motif comprising of polar-hydrophobic charged residues was crucial for QS activation, whereas the core alpha-helical structure present in the mature peptide was important for receptor binding. It was demonstrated that the lack of 3 or more C-terminal peptides resulted in noncompetent cells and that these mutant peptides, at a higher concentration, were able bind and competitively inhibit and eventually abolish the activation of the CSP-induced QS pathway in *S. mutans*.

### *ComDE Signaling Pathway*

In *S. mutans*, the QS pathway encompasses at least two genetic loci—*comCDE* and the *comAB*.[31] While *comC* encodes the precursor CSP, the *comDE* genes encode a two-component signal transduction system comprising of a membrane-bound histidine kinase (ComD) and its cognate response regulator (ComE), respectively. The *comA* and *comB* genes encode the secretion apparatus necessary for processing and export of the signaling molecule. More specifically, the ComA consists of an ATP-binding cassette transporter that utilizes ComB as an accessory protein for processing of the precursor CSP. Based on the QS mechanism of *S. pneumoniae*, a model for QS in *S. mutans* has been proposed[31] (Fig. 1). It is believed that when mature CSP reaches a threshold concentration, it is detected by the ComD receptor, which undergoes autophosphoryaltion at a conserved histidine residue. Consequently, the phosphorelay of ComD to its responder protein, results in its activation, thus transducing the density-dependent message into a cellular response by altering the transcription of *comAB, comC* and *comDE*, as well as that of an alternative sigma-factor (designated *comX*) that can regulate the so-called "late genes".[57] Recently, it has been shown that in the absence or in the presence of low CSP concentrations, unphosphorylated ComE can repress the transcription of *comC*, whereas under high CSP concentration phosphorylated ComE can increase the comC transcription via a de-repression mechanism.[58] The ComE-dependent activation of the late-phase is one of four temporally distinct transcription profiles observed in response to CSP in *S. pneumoniae*.[59] These include the early, late, delayed gene induction and gene repression profiles that are yet to be identified in *S. mutans*. Of these transcriptional profiles, the early genes encode proteins necessary for CSP production, export and recognition and include the *comAB, comCDE* and *comX* genes whose activation is likely dependent on the phosphorylated state of ComE. In the pneumococcal competence development model, the ComE-binding site consensus sequence includes aCAtTTca/gG-$N_{12}$-ACAt/gTTgAG.[60] Interestingly, activated ComX together with RNA polymerase (which are stabilized by ComW) can recognize and bind to a conserved promoter sequence of downstream target genes under their control. In *S. pneumoniae*, this consensus sequence (TACGAATA) is known as the com-box or cin-box and are known to control genes important for DNA processing, uptake and recombination.[61-63] Interestingly, CSP-induced ComX activation also promotes the cell-lysis and release of DNA from a sub-fraction of the bacterial population.[64,65] This observation is not surprising since, by regulating the availability of donor DNA in concert with genetic competence, these bacteria ensure that the energy and resources utilized by cells for genetic transformation is worthwhile. A similar phenomenon of DNA release via QS was observed in *S. mutans*,[32] which will be discussed in the following sections along with other phenotypes controlled by the CSP in this bacterium.

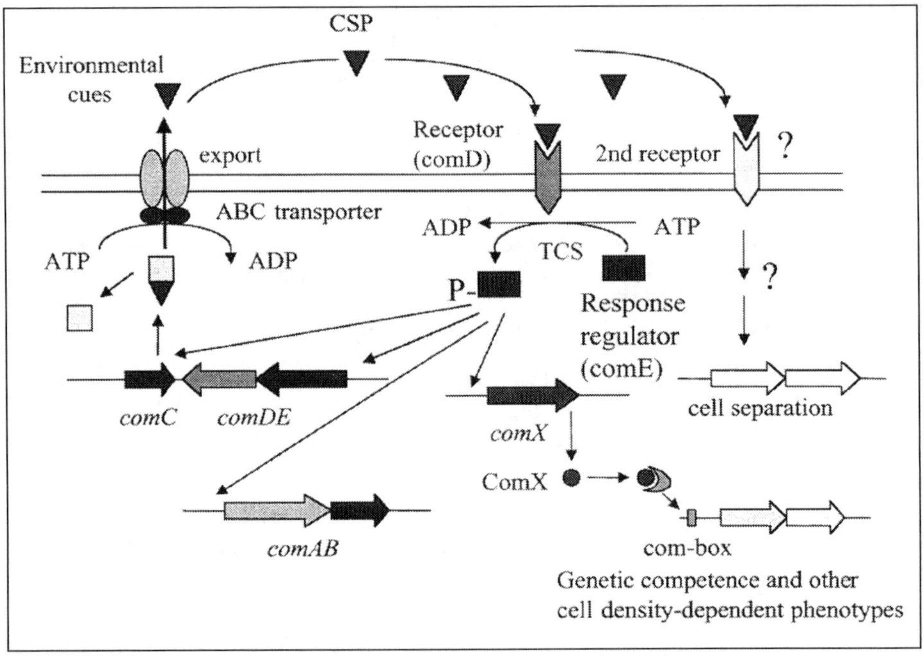


Figure 1. Schematic representation of the putative molecular mechanism of the QS system in *S. mutans*. From Li YH et al. J Bacteriol 2002:2699-708;[43] with permission of the American Society for Microbiology.

## Quorum Sensing and Biofilm Formation in *S. mutans*

Activation of the QS system in *S. mutans* is known to regulate several phenotypes: biofilm formation, competence induction, acid tolerance response (ATR) and bacteriocin production.[13,31,43,66] Interestingly, evidence suggests that transformability and ATR function optimally when *S. mutans* is grown as a biofilm in contrast to free-living planktonic cells.[13,31] In this section, we will give prominence to discussing the CSP-mediated biofilm phenotype and how the QS pathway can be manipulated to control the formation of dental biofilms.

In Gram-positive bacteria, the first report to establish a link between QS and biofilm formation came from an investigation by Loo et al involving *S. gordonii*.[14] This bacterium is a pioneer organism in the oral microflora that initiates the formation of dental plaque. In an attempt to identify various genetic factors involved in biofilm formation in the *S. gordonii* Challis strain, these researchers used Tn*916* transposon mutagenesis to screen for mutants defective in biofilm formation using an in vitro biofilm assay. It was revealed that one of the mutants, which had a transposon inserted in the *comD* signal peptide histidine kinase sensor gene formed defective biofilms that lacked the three-dimensional structure present in wild-type biofilms, thus linking QS to biofilm formation.

When the genetic components of the *S. mutans*' QS system were first identified, it was observed that the genetic transformability was 10- to 600-fold higher in cells derived from biofilms, compared with those obtained from planktonic cultures.[31] Hence, it was hypothesized that in addition to competence development, this system might also be involved in biofilm formation. To test this assumption, mutants deficient in the *comC*, *comD*, *comE* and *comX*, as well as the entire *comCDE* operon were constructed and assayed for their ability to initiate biofilm formation.[43] Interestingly, compared with the wild-type, all mutants formed biofilms that lacked the architectural integrity of the wild-type biofilm, whereas the *comD*-, *comE*- and *comX*-deficient mutants formed biofilms with reduced biomass. Notably, only biofilms formed by the *comC* mutant and not those formed by the *comD, comE* or *comX* mutants were restored to the wild-type architecture by complementing with synthetic CSP or a plasmid containing the wild-type *comC* (Fig. 2). This discrepancy was likely caused by the involvement of multiple signal transduction pathways in detecting and/or responding to CSP. However, the identity and the mode of regulation of a second putative CSP-responsive system are yet to be established. Moreover, expression analysis of the *comCDE* genes in *S. mutans* and *S. gordonii* showed that they were upregulated in the biofilm mode of growth relative to their planktonic counterpart further supporting the QS-biofilm link.[14,43]

Recently, in an attempt to understand the molecular basis of biofilm regulation by the CSP-dependent QS system, the expression of several biofilm-associated genes were tested using 18-h biofilms derived from *S. mutans* wild-type UA159 and its isogenic mutant derivatives deficient in the *comC, comD* or *comE* genes.[67] Expression analysis showed that the ComD/ComE signal transduction system had a positive regulatory effect on the expression of these genes, which include glucosyltransferase B/C/D (*gtfB/C/D*), fructosyltransferase (*ftf*) and glucan-binding protein B (*gbpB*). Of these, the glucosyltransferases utlilze the glucan moiety of the sucrose molecule to produce insoluble (GtfB/C) and soluble (GtfD) glucan polymers that promote plaque formation. In a different study, it was discovered that *gtfB/C* expression were significantly upregulated in *S. mutans* UA159 when supplemented with CSP relative to the no-CSP control in a ComE-dependent manner.[68] Interestingly, since previous studies have linked these genes with cariogenicity in rat models, their regulation by the QS system in *S. mutans* is indeed insightful and could potentially lead to the design of a new drug target that may affect QS-mediated biofilm formation.

In a different study, Petersen et al investigated the specificity of the CSP effect on biofilm formation utilizing *S. mutans* and *Streptococcus intermedius*.[69] They observed that each species only responded to its native CSP and showed no biofilm response when exposed to the nonnative peptide. For example *S. mutans* wild-type LT-11 responded to its native peptide by enhancing biofilm formation but was unaffected by the signal peptide produced by *S. intermedius*. This kind of specificity displayed by QS peptides can be used to our advantage to specifically target biofilm formation by potentially pathogenic bacteria. In fact, such a technique would be less invasive and possibly have less influence in disrupting the ecological imbalance of the oral microflora by killing particular opportunistic pathogens using antibacterial drugs.

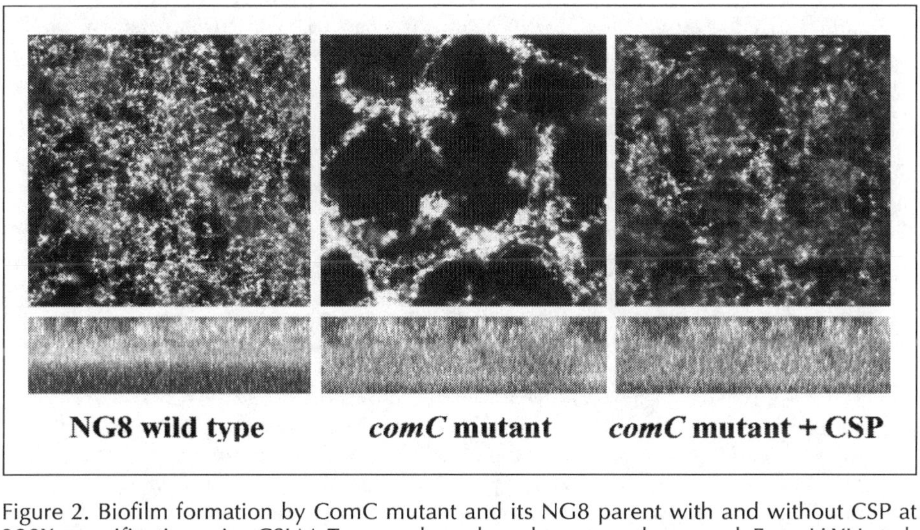


Figure 2. Biofilm formation by ComC mutant and its NG8 parent with and without CSP at 300X magnification using CSLM. Top panel: *xy* plane, lower panel: *xz* panel. From Li YH et al. J Bacteriol 2002:2699-708;[43] with permission of the American Society for Microbiology.

## Density-Dependent Production of Bacteriocins: Implications on Survival in Plaque

Recent discoveries pertaining to the QS system in *S. mutans* have demonstrated a link to bacteriocin or bacteriocin-like gene expression in this organism. Bacteriocins (mutacins) are ribosomally synthesized proteins, which have antimicrobial activity against closely related bacteria.[70] In *S. mutans*, two types of mutacins have been identified: (1) lantibiotics represented by mutacin I, II and III (2) nonlantibiotics represented by mutacin IV.[71-74] While the former type can kill most of the Gram-positive bacteria, the latter has a narrower spectrum killing mostly streptococci belonging to the sanguinis and mitis types. As a resident in the oral biofilm, the activity of bacteriocins can help *S. mutans* compete for limited nutrients available in its ecological niche. The high genetic diversity observed among different *S. mutans* strain might be a result of the QS-dependent bacteriocin production that is coupled with horizontal gene transfer. Moreover, by synchronizing bacteriocin production with population density, *S. mutans* can not only reduce its competition for food, but also ensure that a high percentage of heterologous DNA is present during genetic competence development. In addition to producing bacteriocins, *S. mutans* was recently shown to produce bacteriocin-immunity proteins (encoded by *bip* and *smbG*), which modulate the sensitivity to antimicrobials in *S. mutans*.[75] In the presence of low antibiotic concentrations, genes encoding these bacteriocin-immunity proteins showed an increased expression, whereas their optimal expression was seen when cells were grown as a biofilm as opposed to planktonic cells.

## Quorum Sensing-Dependent Growth Arrest and Cell Death

Based on multiple phenotypes controlled by the *S. mutans* QS system, it seems rational that manipulating QS can possibly reduce its effects on various virulence attributes controlled by this system. Recently, it was demonstrated that the addition of exogenous CSP in excess of the levels necessary for competence induction resulted in the inhibition of *S. mutans* growth.[32] Not surprisingly, this effect was ComDE-dependent and when CSP concentration was further increased, the cells underwent growth arrest and cell death. Mutational analyses suggested that cell death was mediated by ComD/ComE in a ComX-independent manner. More recently, the effect of CSP (0 to 10 μg/ml) on *S. mutans* biofilm architecture was investigated using Scanning Electron microscopy.[68] Supplementing *S. mutans* biofilms with high dose of synthetic CSP (5 or 10 μg/ml) resulted in a high fraction of ruptured and swollen cells that also showed a different cell shape and surface texture. Hence, these observations reveal that the effects of CSP on biofilm formation are multifactorial. It is yet to be determined whether it is possible to selectively control *S. mutans* in a multi-species biofilm by varying the CSP concentration. Moreover, using CSP to control *S. mutans* to modulate virulence obviously warrants further in vivo testing.

## Effect of Other Signal Transduction Systems on *S. mutans* Biofilm Formation

In addition to the ComD/ComE system, other two component signal transduction systems in *S. mutans* have been associated with biofilm formation (Fig. 3). Of these, the VicR/VicK system has been shown to control the expression of *gtfB/C/D*, *ftf* and *gbpB* genes.[76] The *vicK* and *vicR* genes encode a putative histidine kinase and putative response regulator, respectively. While the *vicR* serves an essential function in *S. mutans*, mutational analyses using a *vicK*-deletion strain revealed that it affected biofilm architecture, cell growth and sucrose-dependent adhesion in *S. mutans*. Interestingly, although this mutant produced an excess of extracellular polymer, specific pathogen-free rats harboring the *vicK* mutant strain showed no difference in dental caries relative to those that harbored the parent strain. However, more smooth-surface plaque and significantly diminished mutant CFUs were observed relative to the control group. Since it is known that the cariogenicity in these animals are affected by the microbial composition, the diet and the nature of the polysaccharide matrix, it is not surprising that the excess plaque extent observed for the *vicK* mutant did not necessarily result in hypercariogenicity. However, it is noteworthy that without

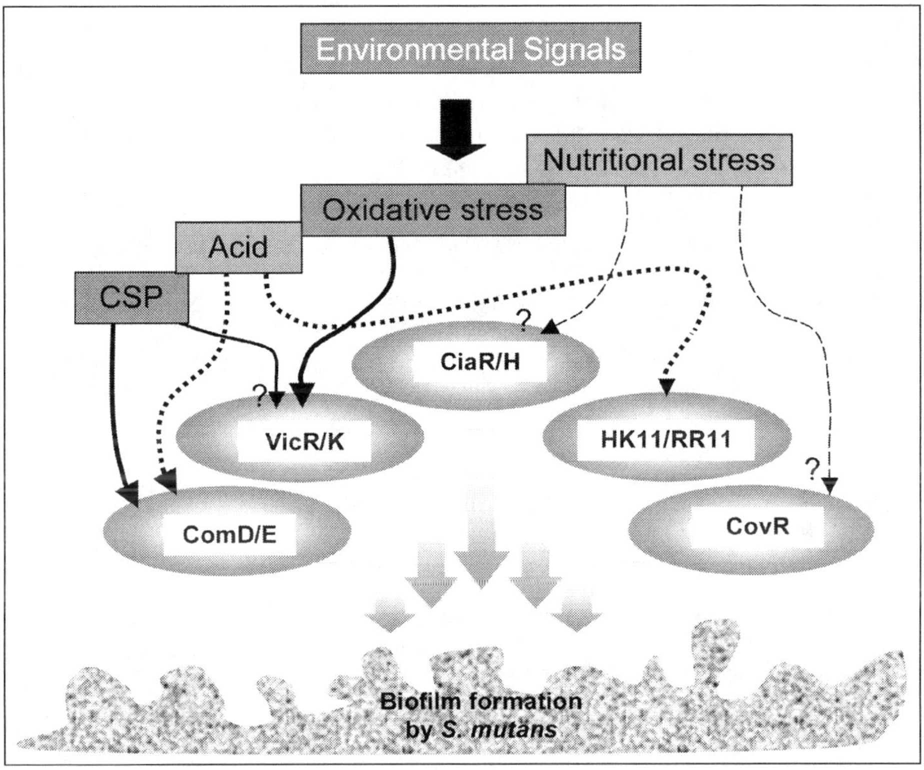


Figure 3. QS-mediated phenotypes in *S. mutans* and two-component signal transduction systems that modulate the biofilm phenotype. From Li YH et al. J Bacteriol 2002:2699-708;[43] with permission of the American Society for Microbiology.

*vicK* the excess plaque that was generated by the mutant strain was not cariogenic, thus making it interesting to identify the nature and biochemical properties of this polymer.

Studies pertaining to two other signal transduction systems, HK11/RR11[77] and CiaR/CiaH,[78] as well as an orphan response regulator CovR[79,80] have also shown to influence the biofilm phenotype in *S. mutans*. In each signal transduction system, the *hk11/rr11* and the *ciaH/ciaR* genes encode a histidine kinase sensor protein and its cognate response regulator, respectively. The hk11/rr11 system was investigated by Li et al who showed that deletion of *hk11* or *rr11* caused defects in biofilm formation with 50% to 70% reduction in biofilm biomass and resistance to acidic pH.[77] Mutant biofilms observed by Scanning electron microscopy revealed biofilms having a sponge-like architecture with large open areas throughout the biofilm. In a different study, inactivation of either of the *ciaR/ciaH* genes resulted in reduced biofilm biomass, whereas the absence of *ciaH* altered sucrose-dependent biofilm formation.[78] Interestingly, this histidine kinase mutant also showed reduced mutacin production, deficiency in transformability, as well as diminished ability to tolerate stress. Involvement of the CovR mutant (also called GcrR) in biofilm formation was first discovered by Idone et al.[80] These researchers showed that a mutant deficient in *covR* was defective in sucrose-dependent adhesion and was hypocariogenic in a germ-free rat model relative to its parent strain. Consequently, it was discovered by Biswas et al that the CovR negatively regulated the expression of the *gtfB* and *gtfC* genes by directly binding to the promoter region.[79] While these signal transduction systems possibly cross-communicate to regulate biofilm formation in *S. mutans* more studies are warranted to investigate if and how they function in response to cell density and other environmental signals.

## Future Perspectives

The prospect of targeting QS systems to control biofilm-mediated infections is receiving a considerable amount of attention. Chemical agents that mimic QS molecules are constantly being evaluated for their ability to 'confuse' the bacteria and hopefully attenuate their ability to cause disease. This is indeed a paradigm shift in the way infections are treated and managed. Current strategies of antimicrobial action are focused on molecules with broad-spectrum indiscriminate activity. The treatment of patients with these compounds often results in massive alteration of the host's microflora often with health affecting consequences (such as ulcerative pseudomembranous colitis). By specifically targeting QS systems of pathogenic bacteria we may hopefully be able to surgically remove the pathogens while leaving the normal flora intact with hopefully little or no detrimental side effects to the host. As we learn more about the subtle differences between bacteria and the way they sense their environments these approaches will very likely become commonplace. The day that a diagnostic lab identifies a pathogen and prescribes a highly specific treatment that has the highest effectiveness with the least likelihood of causing secondary problems is within reach.

## References

1. Stock AM, Robinson VL, Goudreau PN. Two-component signal transduction. Annu Rev Biochem 2000:183-215.
2. West AH, Stock AM. Histidine kinases and response regulator proteins in two-component signaling systems. Trends Biochem Sci 2000:369-376.
3. Bassler BL. How bacteria talk to each other: regulation of gene expression by quorum sensing. Curr Opin Microbiol 1999:582-587.
4. Bassler BL. Small talk. Cell-to-cell communication in bacteria. Cell 2002:421-424.
5. Henke JM, Bassler BL. Bacterial social engagements. Trends Cell Biol 2004:648-656.
6. Kleerebezem M, Quadri LE. Peptide pheromone-dependent regulation of antimicrobial peptide production in Gram-positive bacteria: a case of multicellular behavior. Peptides 2001:1579-1596.
7. Kleerebezem M, Quadri LE, Kuipers OP et al. Quorum sensing by peptide pheromones and two-component signal-transduction systems in Gram-positive bacteria. Mol Microbiol 1997:895-904.
8. Suntharalingam P, Cvitkovitch DG. Quorum sensing in streptococcal biofilm formation. Trends Microbiol 2005:3-6.
9. Alloing G, Martin B, Granadel C et al. Development of competence in Streptococcus pneumoniae: pheromone autoinduction and control of quorum sensing by the oligopeptide permease. Mol Microbiol 1998:75-83.
10. De Kievit TR, Iglewski BH. Quorum sensing, gene expression and Pseudomonas biofilms. Methods Enzymol 1999:117-128.
11. Kleerebezem M. Quorum sensing control of lantibiotic production; nisin and subtilin autoregulate their own biosynthesis. Peptides 2004:1405-1414.
12. Lazazzera BA. Quorum sensing and starvation: signals for entry into stationary phase. Curr Opin Microbiol 2000:177-182.
13. Li YH, Hanna MN, Svensäter G et al. Cell density modulates acid adaptation in Streptococcus mutans: implications for survival in biofilms. J Bacteriol 2001:6875-6884.
14. Loo CY, Corliss DA, Ganeshkumar N. Streptococcus gordonii biofilm formation: identification of genes that code for biofilm phenotypes. J Bacteriol 2000:1374-1382.
15. Fuqua WC, Winans SC, Greenberg EP. Quorum sensing in bacteria: the LuxR-LuxI family of cell density-responsive transcriptional regulators. J Bacteriol 1994:269-275.
16. Bassler BL, Greenberg EP, Stevens AM. Cross-species induction of luminescence in the quorum-sensing bacterium Vibrio harveyi. J Bacteriol 1997:40,43-45.
17. Dunlap PV. Quorum regulation of luminescence in Vibrio fischeri. J Mol Microbiol Biotechnol 1999:5-12.
18. Finch R, Pritchard D, Bycroft B et al. Quorum sensing: a novel target for anti-infective therapy. J Antimicrob Chemother 1998:569-571.
19. Costerton JW, Stewart PS, Greenberg EP. Bacterial biofilms: a common cause of persistent infections. Science 1999:1318-1322.
20. Fux CA, Costerton JW, Stewart PS et al. Survival strategies of infectious biofilms. Trends Microbiol 2005:34-40.
21. Stoodley P, Sauer K, Davies DG et al. Biofilms as complex differentiated communities. Annu Rev Microbiol 2002:187-209.
22. Sutherland I. Biofilm exopolysaccharides: a strong and sticky framework. Microbiology 2001:3-9.
23. Kolenbrander PE. Oral microbial communities: biofilms, interactions and genetic systems. Annu Rev Microbiol 2000:413-437.

24. O'Toole G, Kaplan HB, Kolter R. Biofilm formation as microbial development. Annu Rev Microbiol 2000:49-79.
25. Fux CA, Stoodley P, Hall-Stoodley L et al. Bacterial biofilms: a diagnostic and therapeutic challenge. Expert Rev Anti Infect Ther 2003:667-683.
26. Costerton W, Veeh R, Shirtliff M et al. The application of biofilm science to the study and control of chronic bacterial infections. J Clin Invest 2003:1466-1477.
27. Stewart PS, Costerton JW. Antibiotic resistance of bacteria in biofilms. Lancet 2001:135-138.
28. Gilmore KS, Srinivas P, Akins DR et al. Growth, development and gene expression in a persistent Streptococcus gordonii biofilm. Infect Immun 2003:4759-4766.
29. Keren I, Shah D, Spoering A et al. Specialized persister cells and the mechanism of multidrug tolerance in Escherichia coli. J Bacteriol 2004:8172-8180.
30. Donlan RM. Role of biofilms in antimicrobial resistance. Asaio J 2000:S47-52.
31. Li YH, Lau PC, Lee JH, Ellen RP et al. Natural genetic transformation of Streptococcus mutans growing in biofilms. J Bacteriol 2001:897-908.
32. Qi F, Kreth J, Lévesque CM, Kay O et al. Peptide pheromone induced cell death of Streptococcus mutans. FEMS Microbiol Lett 2005:321-326.
33. Tanzer JM. Dental caries is a transmissible infectious disease: the Keyes and Fitzgerald revolution. J Dent Res 1995:1536-1542.
34. Loesche WJ, Straffon LH. Longitudinal investigation of the role of Streptococcus mutans in human fissure decay. Infect Immun 1979:498-507.
35. Masuda N, Tsutsumi N, Sobue S et al. Longitudinal survey of the distribution of various serotypes of Streptococcus mutans in infants. J Clin Microbiol 1979:497-502.
36. Walter RG. A longitudinal study of caries development in initially caries-free naval recruits. J Dent Res 1982:1405-1407.
37. Alaluusua S, Renkonen OV. Streptococcus mutans establishment and dental caries experience in children from 2 to 4 years old. Scand J Dent Res 1983:453-457.
38. Kingman A, Little W, Gomez I et al. Salivary levels of Streptococcus mutans and lactobacilli and dental caries experiences in a US adolescent population. Community Dent Oral Epidemiol 1988:98-103.
39. Curtiss R III. Genetic analysis of Streptococcus mutans virulence. Curr Top Microbiol Immunol 1985:253-277.
40. Kuramitsu HK. Virulence factors of mutans streptococci: role of molecular genetics. Crit Rev Oral Biol Med 1993:159-176.
41. Kuramitsu HK. Virulence properties of oral bacteria: impact of molecular biology. Curr Issues Mol Biol 2001:35-36.
42. Tanzer JM, Freedman ML, Fitzgerald RJ et al. Diminished virulence of glucan synthesis-defective mutants of Streptococcus mutans. Infect Immun 1974:197-203.
43. Li YH, Tang N, Aspiras MB et al. A quorum-sensing signaling system essential for genetic competence in Streptococcus mutans is involved in biofilm formation. J Bacteriol 2002:2699-2708.
44. Pakula R, Walczak W. On the nature of competence of transformable streptococci. J Gen Microbiol 1963:125-133.
45. Tomasz A, Hotchkiss RD. Regulation of the transformability of pneumococcal cultures by macromolecular cell products. Proc Natl Acad Sci USA 1964:480-487.
46. Tomasz A. Control of the competent state in Pneumococcus by a hormone-like cell product: an example for a new type of regulatory mechanism in bacteria. Nature 1965:155-159.
47. Havarstein LS, Coomaraswamy G, Morrison DA. An unmodified heptadecapeptide pheromone induces competence for genetic transformation in Streptococcus pneumoniae. Proc Natl Acad Sci USA 1995:11140-11144.
48. Havarstein LS, Diep DB, Nes IF. A family of bacteriocin ABC transporters carry out proteolytic processing of their substrates concomitant with export. Mol Microbiol 1995:229-240.
49. Cheng Q, Campbell EA, Naughton AM et al. The com locus controls genetic transformation in Streptococcus pneumoniae. Mol Microbiol 1997:683-692.
50. Håvarstein LS, Gaustad P, Nes IF et al. Identification of the streptococcal competence-pheromone receptor. Mol Microbiol 1996:863-869.
51. Pestova EV, Havarstein LS, Morrison DA. Regulation of competence for genetic transformation in Streptococcus pneumoniae by an auto-induced peptide pheromone and a two-component regulatory system. Mol Microbiol 1996:853-862.
52. Claverys JP, Havarstein LS. Extracellular-peptide control of competence for genetic transformation in Streptococcus pneumoniae. Front Biosci 2002:d1798-1814.
53. Havarstein LS, Morrison DA. Quorum sensing and peptide pheromones in streptococcal competence for genetic transformation in Cell-cell signaling in Bacteria. G.M.a.W. Dunny SC, Ed, ASM Press, Washington, DC 1999:9-26.

54. Cvitkovitch DG. Genetic competence and transformation in oral streptococci. Crit Rev Oral Biol Med 2001:217-243.
55. Havarstein LS, Hakenbeck R, Gaustad P. Natural competence in the genus Streptococcus: evidence that streptococci can change pherotype by interspecies recombinational exchanges. J Bacteriol 1997:6589-6594.
56. Syvitski RT, Tian XL, Sampara K et al. Structure-Activity Analysis of Quorum-Sensing Signaling Peptides from Streptococcus mutans. J Bacteriol 2006.
57. Aspiras MB, Ellen RP, Cvitkovitch DG. ComX activity of Streptococcus mutans growing in biofilms. FEMS Microbiol Lett 2004:167-174.
58. Kreth J, Hung DC, Merritt J et al. The response regulator ComE in Streptococcus mutans functions both as a transcription activator of mutacin production and repressor of CSP biosynthesis. Microbiology 2007, In press.
59. Peterson SN, Sung CK, Cline R et al. Identification of competence pheromone responsive genes in Streptococcus pneumoniae by use of DNA microarrays. Mol Microbiol 2004:1051-1070.
60. Ween O, Gaustad P, Havarstein LS. Identification of DNA binding sites for ComE, a key regulator of natural competence in Streptococcus pneumoniae. Mol Microbiol 1999:817-827.
61. Lee MS, Morrison DA. Identification of a new regulator in Streptococcus pneumoniae linking quorum sensing to competence for genetic transformation. J Bacteriol 1999:5004-5016.
62. Luo P, Li H, Morrison DA. ComX is a unique link between multiple quorum sensing outputs and competence in Streptococcus pneumoniae. Mol Microbiol 2003:623-633.
63. Luo P, Morrison DA. Transient association of an alternative sigma factor, ComX, with RNA polymerase during the period of competence for genetic transformation in Streptococcus pneumoniae. J Bacteriol 2003:349-358.
64. Steinmoen H, Knutsen E, Havarstein LS. Induction of natural competence in Streptococcus pneumoniae triggers lysis and DNA release from a subfraction of the cell population. Proc Natl Acad Sci USA 2002:7681-7686.
65. Steinmoen H, Teigen A, Havarstein LS. Competence-induced cells of Streptococcus pneumoniae lyse competence-deficient cells of the same strain during cocultivation. J Bacteriol 2003:7176-7183.
66. Kreth J, Merritt J, Zhu L et al. Cell density- and ComE-dependent expression of a group of mutacin and mutacin-like genes in Streptococcus mutans. FEMS Microbiol Lett 2006:11-17.
67. Senadheera D, Lee AWC, Hung DC et al. Streptococcus mutans quorum-sensing system controls gtfB/C/D, ftf and gbpB expression. General meeting hosted by the American Society for microbiology. Toronto, Canada 2007.
68. Senadheera D, Mair R, Cvitkovitch DG. Streptococcus mutans competence stimulating peptide affects gtfB/C and vicR/K expression and biofilm structure. Biofilms conference hosted by the American Society for Microbiology. Quebec, Canada 2007.
69. Petersen FC, Pecharki D, Scheie AA. Biofilm mode of growth of Streptococcus intermedius favored by a competence-stimulating signaling peptide. J Bacteriol 2004:6327-6331.
70. Sablon E, Contreras B, Vandamme E. Antimicrobial peptides of lactic acid bacteria: mode of action, genetics and biosynthesis. Adv Biochem Eng Biotechnol 2000:21-60.
71. Novak J, Caufield PW, Miller EJ. Isolation and biochemical characterization of a novel lantibiotic mutacin from Streptococcus mutans. J Bacteriol 1994:4316-4320.
72. Qi F, Chen P, Caufield PW. Purification of mutacin III from group III Streptococcus mutans UA787 and genetic analyses of mutacin III biosynthesis genes. Appl Environ Microbiol 1999:3880-3887.
73. Qi F, Chen P, Caufield PW. Purification and biochemical characterization of mutacin I from the group I strain of Streptococcus mutans, CH43 and genetic analysis of mutacin I biosynthesis genes. Appl Environ Microbiol 2000:3221-3229.
74. Qi F, Chen P, Caufield PW. The group I strain of Streptococcus mutans, UA140, produces both the lantibiotic mutacin I and a nonlantibiotic bacteriocin, mutacin IV. Appl Environ Microbiol 2001:15-21.
75. Matsumoto-Nakano M, Kuramitsu HK. Role of bacteriocin immunity proteins in the antimicrobial sensitivity of Streptococcus mutans. J Bacteriol 2006:8095-8102.
76. Senadheera MD et al. A VicRK signal transduction system in Streptococcus mutans affects gtfBCD, gbpB and ftf expression, biofilm formation and genetic competence development. J Bacteriol 2005:4064-76.
77. Li YH, Lau PC, Tang N et al. Novel two-component regulatory system involved in biofilm formation and acid resistance in Streptococcus mutans. J Bacteriol 2002:6333-6342.
78. Qi F, Merritt J, Lux R et al. Inactivation of the ciaH Gene in Streptococcus mutans diminishes mutacin production and competence development, alters sucrose-dependent biofilm formation and reduces stress tolerance. Infect Immun 2004:895-899.
79. Biswas S, Biswas I. Regulation of the glucosyltransferase (gtfBC) operon by CovR in Streptococcus mutans. J Bacteriol 2006:988-998.
80. Idone V, Brendtro S, Gillespie R et al. Effect of an orphan response regulator on Streptococcus mutans sucrose-dependent adherence and cariogenesis. Infect Immun 2003:4351-4360.

# Chapter 13

# The Roles of Two-Component Systems in Virulence of Pathogenic *Escherichia coli* and *Shigella* spp.

**Toru Tobe***

## Abstract

Two-component systems (TCSs) are well conserved among *E. coli* strains, including pathogenic *E. coli* and also closely related *Shigella* spp. Although 25% of the genome of pathogenic *E. coli* is strain-specific, only small number of strain-specific TCSs is found. Regulation of virulence genes in response to environmental stimuli is partly dependent on TCSs commonly present in nonpathogenic *E. coli* strains. Some virulence genes are directly regulated by response regulator of TCS but some are affected at posttranscriptional steps of production or assembly of macromolecule by TCS-induced products. In the process of acquiring virulence traits, regulatory systems for virulence genes expression seem to be built by integrating *E. coli* backbone TCSs with the virulence regulatory network via transcription regulatory gene.

## Introduction

Although most of *Escherichia coli* strains are harmless to humans and animals and even some are inhabitant of the human intestinal tract, certain population of *E. coli* strains cause infectious diseases to humans and animals. Pathogenic *E. coli* strains possess specific virulence factors that enable them to cause diarrhoea or extraintestinal infections, such as urinary tract infection or meningitis.[1] Diarrheagenic *E. coli* faces intensive competition with other enteric bacteria to occupy particular niches in intestine. Although they are not resident of intestine, once entered to intestinal tract they successfully colonize and multiply using their strategies different from members of normal flora. On the other hand, uropathogenic *E. coli* (UPEC), which causes urinary tract infection, colonizes from feces or perineal region and ascend the urinary tract to the bladder, which environment is completely different from intestine.[2] Thus, pathogenic strains have the ability to rapidly adapt to environmental changes and survive against host defense systems including barrier function of intestinal microflora. Specifics of bacterial virulence factors used and the effect on the host vary from pathogen to pathogen, but the functions of factors are classified to several categories: adhesins are required for colonization on epithelial cells, secreted toxins for disruption of host function and effectors translocated to host cells by specific secretion system for modulation of host physiology.

Six classes of diarrheagenic *E. coli* that cause intestinal diseases are recognized[3]: enteropathogenic *E. coli* (EPEC), enterohaemorrhagic *E. coli* (EHEC), enterotoxigenic *E. coli* (ETEC), enteroinvasive *E. coli* (EIEC) and enteroaggregative *E. coli* (EAggEC), diffusely adherent *E. coli* (DAEC). Although pathogenic strategies used by pathogens are different, they use a limited array of

*Toru Tobe—Division of Applied Bacteriology, Graduate School of Medicine, Osaka University, 2-2 Yamadaoka, Suita, Osaka 565-0871, Japan. Email: torutobe@bact.med.osaka-u.ac.jp

*Bacterial Signal Transduction: Networks and Drug Targets*, edited by Ryutaro Utsumi.

macromolecular systems. Virulence factors used for adherence or modification of host function by each pathogen are different from each other, but underlying platforms for presenting these factors are similar. For example, most of EPEC strains produce type IV pili known as the bundle forming pili (BFP), to make bacterial aggregates and to form microcolony on surface of the epithelial cells, while, ETEC strains from human produce CFA (colonization factor antigen)/ I fimbriae or CFA/III type IV pili. Toxins produced by ETEC are ST (heat-stable) and LT (heat-labile), which is very similar to cholera toxin, while, EHEC produces shiga toxins, which is very similar to toxin produced by *Shigella* spp. Although types of diarrhea induced are different, both EPEC and EHEC induce the identical histopathological effect on epithelial cells of intestine, which is called attaching and effacing (A/E) lesion. The A/E lesion is induced as a consequence of involvement of various virulence factors, such as type III secreted proteins and specific adhesin, intimin. EIEC and *Shigella* spp. possess the same virulence plasmid encoding many virulence factors involved in invasion of host cells and intracellular spreading. Variety of virulence factors for invasion is translocated to host cells through type III secretion system.

Expression of virulence factors is altered in response to changes in environmental conditions. Pathogens have various habitats during their life cycle and need to sense proximity and location within the host. This made them to express virulence traits appropriately. Also, virulence factors, especially on surface exposed macromolecules, are silent until reaching to target site to avoid recognition by and attack from host defense mechanisms. In addition, pathogens encounter various stress conditions in host to be overcome for survival. Stress responses that enable pathogen to sense and adapt to changing environmental conditions are also important for successful infection. Some environmental factors commonly affect the expression in many pathogens but some are specific for pathogen, reflecting infectious strategies. Most common factor must be temperature.[4] Expression of virulence genes is repressed at low temperature, such as 27°C, but is activated when grown at 37°C for human pathogenic *E. coli* and *Shigella* spp.[5-7] Temperature must be one of environmental factors to sense host since expression of virulence genes in human pathogens is maximum at 37°C whereas 42°C for rabbit enteropathogenic *E. coli.*[8] Osmolarity and pH also affect the expression in many pathogens.[9-11] Expression of *Shigella* virulence genes necessary for invasion is enhanced at high osmolarity condition and activated at alkaline pH conditions.[12] Other environmental factors are described for specific virulence genes of each pathogen. In EPEC, expression of BFP is repressed by ammonium ion while activated by calcium.[13] In EHEC, the level of LEE genes expression is affected by broth medium; grown in tissue culture medium, such as Dulbecco's modified Eagles Medium (DMEM), increase the expression as compared LB.[14] This is partly due to the presence of sodium bicarbonate in DMEM, since addition of the same concentration of sodium bicarbonate enhanced the expression in EHEC grown in LB. Type III secretion in EPEC and EHEC is also affected by variety of environmental factors.[6] Secretion of EspB, one of type III secreted protein, is maximal at 37°C, pH7, physiological osmolarity and enhanced by sodium bicarbonate and calcium. The expression of Shiga-toxin has shown to be affected by concentration of iron in the medium.[15]

## Genome Structure of Pathogenic *E. coli* and *Shigella* spp.

The genome sequences of several *E. coli* strains and *Shigella* strains were determined. Comparison of genome sequences revealed high levels of diversity resulting from gross chromosomal changes. Sizes of chromosome of pathogenic *E. coli* strains are larger than that of nonpathogenic *E. coli* K-12 strain.[16-18] Genome sequence common in *E. coli* strains, which determines characteristics of *E. coli* physiology, is predicted as about 4 Mb including about 3000 genes by comparing *E. coli* strains. Strain-specific sequences are found as insert of DNA fragment ranging from several bp to 100 kbp to *E. coli* backbone genome. The inserted sequences containing cluster of virulence genes is called pathogenicity island. LEE (locus for enterocyte effacement) of EPEC and EHEC contains over 30 genes encoding type III secretion machinery, secreted proteins and adhesin, which are required for induction of A/E lesion. LEE is found at *selC* or *pheV* tRNA gene.[19] The PAI (pathogenicity island) of UPEC containing genes for Pap pilus biosynthesis and haemolysin is inserted at *pheV* site in a certain UPEC strain. Although possessing the same set of virulence genes, combination of virulence genes in PAIs and location of PAIs are variable among strains.[18] In addition, there are variations in virulence genes even among strains in the same pathotype.[20] These genome island often have a different G + C contents and codon usage from the *E. coli* backbone genome. The accumulation of these inserted sequences occurred via numerous, independent horizontal transfer events at many discrete chromosomal sites. Indeed, of these genome islands most of large islands are prophages and prophage-like elements in EHEC O157:H7 sequenced strain.[16] On the other hand, most important virulence determinants of EIEC and *Shigella* spp. are on the large plasmid. The plasmid carries genes for type III secretion system and around 20 genes for effector, type III-dependent secreted protein.[21] Genome of *Shigella flexneri* is slightly smaller than that of *E. coli* K-12 and relatively little number of genome islands as compared to those of EHEC and UPEC.[22,23] Henceforce, pathogenicity of *E. coli* and *Shigella* spp. are obtained by extensive rearrangement of genome structure via horizontal gene transfer.

## Two-Component Systems (TCSs) in Virulence Expression

Two-component systems are used to sense and respond to extracytoplasmic conditions. The systems are generally composed of a sensor kinase and a response regulator. *E. coli* K-12 strain possesses 30 genes for sensor kinase and 34 genes for response regulator to adapt to changes in variety of chemical or physical conditions. Although extensive diversity of genome structure, repertoire of genes for two-component systems are not much altered (Table 1). For example, all of genes for sensor kinase and response regulator except AtoS/AtoC system are present in EHEC O157:H7 Sakai strain. In addition to these genes this strain possesses one gene for sensor kinese and two genes for response regulator. Only one of genes for response regulator is missing from UPEC CFT073 strain and three sets of genes for two-component systems are found. Size of *Shigella flexneri* genome is slightly smaller than that of *E. coli* K-12, but 18% of genome is different from *E. coli* K-12. *S. flexneeri* loses several of two-component regulatory systems as compared to *E. coli* K-12 strain. In pathogenic *E. coli* and UPEC, it remains unknown about the role of strain-specific two-component regulatory systems. Whereas, involvement of TCSs, which are common in *E. coli* K-12 strains, in regulation of virulence gene expression is reported in pathogenic *E. coli* and *Shigella* spp (Table 2). Response regulator is usually a transcription regulator and controls restricted number of target genes. Some of virulence genes are regulated directly by response regulator at transcriptional step, while expressions of other genes are affected indirectly at posttranscriptional steps by the products of TCS-regulated genes. Linking additional genes to *E. coli* backbone regulatory systems facilitate the ability of *E. coli* to develop new strategy of multiplication. Even there is not any effect on virulence gene expression, TCS seems to be involved in pathogenicity. Since success in bacterial infection of host requires multiple abilities, disability in one of the steps, such as adaptation to new environment or stress produced by host or other microflora, affect the infectivity of pathogen.

### *CpxA/CpxR System*

The Cpx two-component signal transduction pathway responds specifically to stress caused by disturbances in the cell envelope and activates genes encoding periplasmic protein folding and degrading factors. The Cpx system is activated by stresses that cause CpxP to dissociate from CpxA at periplasmic site, activating CpxA kinase activity at cytoplasmic site and phosphorylate CpxR. Phosphorylated CpxR regulates expression of many genes by binding to upstream sequence. Members of Cpx regulon include *dsbA* (encodes periplasmic disulphide isomerase), *ppiA* and *ppiD* (periplasmic peptidyl propyl isomerases) and *degP* (periplasmic protease and chaperone). The connection of Cpx system to the regulation of expression of surface structures that are necessary for virulence have been reported in UPEC, EPEC and *Shigella* spp.[24] Pap or P pili of UPEC are the typical chaperone-usher class of pili, which are assembled via a complex process mostly at periplasm. The pilus subunits are initially synthesized in the cytoplasm and transported through the general secretory system to the periplasm. The P pili specific chaperone interacts with the subunits at the periplasm to assist folding, prevent from premature aggregation by temporally

*Table 1. Distribution of TCS in pathogenic* E. coli *and* Shigellae

| TCS | | Strain | | | |
|---|---|---|---|---|---|
| **Sensor Kinase** | **Response Regulator** | ***E. coli* K-12 MG1655** | **EHEC O157 Sakai** | **UPEC CFT037** | ***S. flexneri* 2a 2457T** |
| OmpR family | | | | | |
| PhoR | PhoB | Present | Present | Present | Present |
| PhoQ | PhoP | Present | Present | Present | Present |
| EnvZ | OmpR | Present | Present | Present | Present |
| RstB | RstA | Present | Present | Present | Present |
| CpxA | CpxR | Present | Present | Present | Present |
| CreC | CreB | Present | Present | Present | Present |
| BaeS | BaeR | Present | Present | Present | Present |
| CusS | CudR | Present | Present | Present | Present |
| QseC | QseB | Present | Present | Present | Present |
| KdpD | KdpE | Present | Present | Present | absent |
| TorS | TorR | Present | Present | Present | absent (TorS)<br>Present (TorR) |
| ArcB | ArcA | Present | Present | Present | Present |
| BasS | BasR | Present | Present | Present | Present |
| YedV | YedW | Present | Present | Present | Present |
| NarL family | | | | | |
| NarX | NarL | Present | Present | Present | Present |
| NarQ | NarP | Present | Present | Present | absent (NarQ)<br>Present (NarP) |
| UhpB | UhpA | Present | Present | Present | Present |
| RcsC | RcsB | Present | Present | Present | Present |
| BarA | UvrY | Present | Present | Present | Present |
| EvgS | EvgA | Present | Present | Present | Present |
| NtrC family | | | | | |
| GlnL | GlnG | Present | Present | Present | Present |
| HydH | HydG | Present | Present | Present | Present |
| AtoS | AtoC | Present | absent | Present | absent |
| YfhK | YfhA | Present | Present | Present | Present |
| Chemotaxis family | | | | | |
| CheW | CheY | Present | Present | Present | Present |
| CheA | CheB | Present | Present | Present | Present |
| CitB family | | | | | |
| CitA | CitB | Present | Present | Present | absent |
| DcuS | DcuB | Present | Present | Present | Present |
| LytT family | | | | | |
| LytS | LytR | Present | Present | Present | Present |
| YpdA | YpdB | Present | Present | Present | Present |
| | HyfR | Present | Present | absent | absent |

*continued on next page*

***Table 1. Continued***

| TCS | | Strain | | | |
|---|---|---|---|---|---|
| **Sensor Kinase** | **Response Regulator** | ***E. coli* K-12 MG1655** | **EHEC O157 Sakai** | **UPEC CFT037** | ***S.flexneri* 2a 2457T** |
| | YgaA | Present | Present | Present | Absent |
| | YifB | Present | Present | Present | Absent |
| | BglJ | Present | Present | Present | Absent |
| Pathogen-specific | | | | | |
| | ECs5067 | Absent | Present | Absent | Absent |
| ECs0417 | ECs0418 | Absent | Present | Absent | Absent |
| c3564 | c3565 | Absent | Absent | Present | Absent |
| c4545 | c4546 | Absent | Absent | Present | Absent |
| c5041 | c5040 | Absent | Absent | Present | Absent |

The data from TTDB was modified. (http://spock.genes.nig.ac.jp/%7 Ettdb -tf/two/).

providing peptide to interactive groove of the subunit and promote assembly into the growing pilus at an outer membrane usher. In the process of assembly, misfolded subunits, failed to interact with chaperone, leave the pathway to form nonproductive interactions and are degraded subsequently. Expression of P pilus in *E. coli* K-12 strain activates Cpx system, which in turn induces periplasmic factors, such as DegP and DsbA, required for P pilus assembly.[25] In the absence of Cpx system, only short pili are assembled, whereas the activation of Cpx system by introducing constitutive active CpxR mutants enhanced production of pilus in *E. coli* K-12. In addition, phosphorylated CpxR binds to the pap promoter and affect the phase variation of pilus expression. Thus, it is presumed that Cpx system enables proper production of pili by activating periplasmic assembly factors and maintaining the ON state of phase variation at urinary tract.

Type IV bundle-forming pili (BFP) of EPEC is poorly produced from exogenous promoter in *E. coli* K-12 strain unless the Cpx system is constitutively activated.[26] In EPEC, elimination of Cpx system by mutation of *cpxR* results in diminished BFP production, consequently reduction in pili-mediated autoaggregation. Furthermore, the *cpxR* mutation has a significant effect on the adherence to epithelial cells, which is the first step in EPEC pathogenesis. Process of type IV pili assembly is completely different from that of P pili. Periplasmic assembly intermediates have not been identified for BFP. Premature pilin subunits (prebundlin) are initially synthesized in cytoplasm and translocated to cytoplasmic membrane through general secretion pathway. Then, leader sequence of the subunit is cleaved by BfpP leader peptidase and disulfide bond at C-terminus, which is essential for assembly, is formed by DsbA. Possibly, amount of DsbA in *E. coli* K-12 is not enough for maturation of bundlin, whereas, basal level of Cpx-regulated factors are high enough for assembly process in EPEC. Alternatively, other periplasmic factors may be required or BFP assembly factors are affected. Since expression of bundlin in *E. coli* K-12 activates Cpx system, Cpx system sense the production of bundlin and supports the efficient assemble of BFP in EPEC.

Curli fibers are produced by nonpathogenic and pathogenic *E. coli* and *Salmonella enterica*. Curli are involved in adherence to surfaces of mammalian host tissue and also to inorganic materials, which leads to formation of biofilm. Hence, Curli are thought to play a role in virulence. Cpx system is implicated in the regulation of curli biogenesis.[27] CpxR binds to the *csgD* promoter region, resulting in repression of *csg* gene expression. Cpx system is activated in response to high osmolarity shift and CpxR represses *csgD* expression in high salt medium. However, repression by CpxR is restricted to high salt conditions. Response to other high osmolarity conditions, created by high sucrose, is mediated by H-NS. Although regulation of *csg* genes transcription is mediated

*Table 2. TCS in virulence expression*

| TCS | Regulated Step of Expression | Mode | Regulated Gene | Pathogen | Virulence Determinants/Role in Virulence | Reference |
|---|---|---|---|---|---|---|
| CpxA/CpxR | Assembly* | Positive | NA | EPEC | Type IV bundle-forming pili | 26 |
| CpxA/CpxR | Assembly and transcription | Positive | NA | UPEC | P pili | 25 |
| CpxA/CpxR | Transcription | Negative | *csgD* | *E. coli* | Cruli fimbriea | 27 |
| CpxA/CpxR | Transcription | Positive | *virF* | *S. sonnei* | TTSS and effectors | 12 |
| CpxA/CpxR | posttranscription | Positive | *invE* | *S. sonnei* | TTSS and effectors | 29 |
| BarA/UvrY | Unknown | Positive | Unknown | UPEC | Carbon metabolism, Growth in urinary tract | 35 |
| BarA/UvrY | Unknown pathogenic *E. coli* | Positive | Unknown | avian | Survival in host | 46 |
| EnvZ/OmpR | Transcription | Positive | *csgD* | *E. coli* | Cruli fimbriea | 27 |
| EnvZ/OmpR | Transcription | Positive | *ompC* | *S. flexneri* | Invasion | 37 |
| EvgS/EvgA | Unknown | Negative | Unknown | EPEC, EHEC | TTSS, effectors, adhesin | 47 |
| RcsC/RcsD/RcsB | Transcription* | Positive | *grvA* | EHEC | TTSS, effectors, adhesin | 40 |
| RcsC/RcsD/RcsB | Transcription* | Negative | *pchA*, pchB | EHEC | TTSS, effectors, adhesin | 40 |
| RcsC/RcsD/RcsB | Transcription* | Negative | *csgD* | *E. coli* | Cruli fimbriea | 48 |
| QseC/QseB | Transcription | Positive | *flhDC* | EHEC | Flagella | 42 |
| PhoQ/PhoP | Unknown | Positive | unknown | *S. flexneri* | Resistance to killing by PMN | 43 |

NA: not applied; *: uncertain

by several factors, at least the Cpx system regulates curli expression at transcriptional level in response to high salt conditions.

Adherence to surface of host tissue or abiotic material is essential for infection or biofilm formation. Bacteria response to contact to the surface and adjust their cell structure and physiology during adhension. This response is partly mediated by Cpx system.[28] It has been shown that Cpx system is activated upon attachment of *E. coli* to abiotic hydrophobic surface. This activation requires NlpE, an outer membrane lipoprotein, which is known as an inducer of Cpx system. Although underlying mechanism remained unclear, activation of Cpx system is important to establish novel bacteria-surface interaction for efficient adherence.

Type III secretion system (TTSS) of *Shigella* spp. is necessary for invasion of host cells. Expression of the TTSS genes is affected by pH and Cpx system has been found to activate regulatory gene for TTSS genes.[12] Phosphorylated CpxR activates transcription of *virF*, which encodes transcriptional activator for *invE* (*virB*) encoding another regulator for TTSS genes, by binding to promoter region of the *virF* gene. Cpx system also affect the production of InvE protein.[29] Consequently, Cpx system is involved in virulence expression in *Shigella* spp. by enhancing TTSS expression at both transcriptional and posttranscriptional processes.

### *BarA/UvrY System*

The BarA/UvrY two-component signal transduction pathway regulates the expression of CsrA/CsrB system, which is a global regulatory system controlling carbon metabolism, flagella biosynthesis and biofilm formation.[30] UvrY positively controls transcription of noncoding RNA genes, *csrB*. Binding of *csrB* RNA to CsrA protein prevent its activity that block access of ribosome to certain mRNA. The homologues of BarA/UvrY system in other bacteria are involved in virulence of the bacteria. For example, BarA/SirA in *Salmonella enterica* affects the expression of virulence genes in SPI-1 pathogenicity island.[31] Deletion of ExpS/ExpA in *Erwinia carotovora*, VarS/VarA in *Vibrio cholerae* and GacS/GacA in *Pseudomonus* spp. diminishes their virulence.[32-34] Involvement of BarA/UvrY system in UPEC was demonstrated in monkey cystitis competition model.[35] It has been shown that the *uvrY* mutant is outcompeted in the monkey bladder when equal number of the mutant and wild type of UPEC are inoculated. Outcompetition of the *uvrY* mutant in the medium containing urine suggested that BarA/UvrY system affects bacterial ability to efficiently switch between carbon sources. Thus, virulence of UPEC is dependent on BarA/UvrY system because the system may support efficient growth in urinary tract.

### *EnvZ/OmpR System*

The EnvZ/OmpR signal transduction system regulates expression of porin genes, *ompF* and *ompC*, in response to changes in osmolarity. In addition to these porin genes, OmpR regulates expression of many genes in *E. coli*. Disruption of EnvZ/OmpR system in *Shigella flexneri* reduces its virulence and decreases expression of virulence genes.[36] But the effect of EnvZ/OmpR system seems to be indirect. Virulence of the mutant is restored by multicopy of *ompC*, which is barely expressed in the mutant. These results suggest that the main contribution of EnvZ/OmpR system to virulence of *Shigella* spp. is through the unknown role of OmpC.[37]

Constitutively active OmpR mutation stimulates biofilm formation by *E. coli*. OmpR binds promoter of *csgD*, which encodes transcriptional regulator for *csgAB* operon encoding curli and enhances transcription. Thus, activated OmpR promotes biofilm formation by enhancing the initial adhesion of bacteria to abiotic surface.[38,39] The regulation of biofilm formation and expression of curli genes in wild type *E. coli* is complicated by involvement of other regulators. The formation of biofilm by wild type *E. coli* is inhibited by high osmolarity condition, which activates EnvZ/OmpR system.[39] The *csgD* gene is also regulated by CpxR, a response regulator of CpxA/CpxR system. Since Cpx system is also activated in response to high osmolarity stress conditions and CpxR represses *csgD* expression by binding to the promoter, it is likely that expression of *csgD* gene is strongly repressed by CpxR antagonizing positive effect of OmpR at the *csgD* promoter at least in high salt conditions. Hence, EnvZ/OmpR system can positively regulate expression of *csg* genes and biofilm formation in *E. coli* but the expression is dependent on combination with other regulators that are activated by different or the same stimuli.

## *RcsC/RcsD/RcsB System*

The RcsC/RcsD/RcsB signal transduction system regulates genes encoding enzymes involved in synthesis of extrapolysaccharide, such as *cps* genes for colanic acid polysaccharide synthesis and *ugd* gene for UDP-D-glucose dehydrogenase. The expression of virulence genes in EHEC is affected by the RcsC/RcsD/RcsB system.[40] Overexpression of *rcsB*, encoding response regulator, or *rcsD*, encoding phosphotransfer protein, drastically increase the level of expression of genes in LEE pathogenicity island. Consequently, virulence of EHEC is enhanced by the increase of ability of adherence to epithelial cells and type III secretion of effectors. Further, activation of RcsC/RcsD/RcsB system by overproduction of periplasmic protein RcsF or introduction of constitutively active RcsC enhances the expression of LEE genes. This activation is dependent on master regulator gene, *ler*, in LEE and upstream regulator gene *grvA*, which is located outside of LEE. The *grvA* gene is on the EHEC strain-specific chromosomal segment and encodes transcriptional regulator for *LEE1* operon, which includes *ler* gene (see Fig.1). Therefore, activation of RcsC/RcsD/RcsB system activates about 40 LEE genes through cascade of positive regulation: activation of *grvA* expression leads increase of *LEE1* operon transcription, which increases production of Ler. Ler then activates other LEE genes transcription. Involvement of RcsC/RcsD/RcsB system in regulation of LEE genes is complicated because LEE genes are regulated by several other regulators and by various environmental factors. This activation is achieved only when bacteria is grown in the Dulbecco's Modified Eagle medium (DMEM) but not in LB. Moreover, deletion of *rcsB* is resulted in increased expression of LEE genes when bacteria are grown in DMEM. These results indicate that activation of other regulatory system(s) for LEE genes by growth in DMEM is primarily required for further regulation by RcsB. And basal level of RcsB activity represses LEE genes expression, while further activation of RcsB enhances the expression. Repression of LEE genes by the RcsB at basal level is dependent on other regulator genes, *pchA* and *pchB*, for LEE genes.

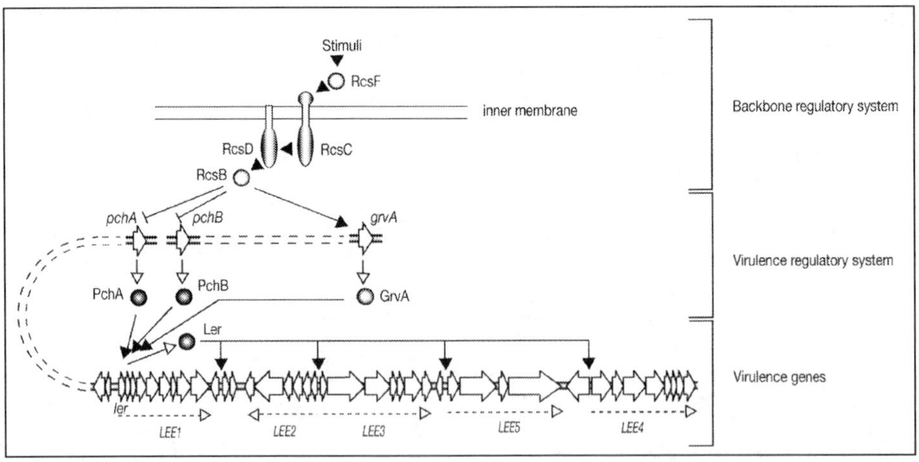


Figure 1. Integration of virulence regulatory system with backbone TCS. Dual regulatory cascades mediates response of LEE genes to Rcs-activating stimuli. RcsC/RcsD/RcsB system affects LEE genes expression in positive and negative manner. RcsB at normal level affects negatively on LEE genes through repressing *pchA* and *pchB* genes expression. Activated RcsB affects positively on LEE genes through activation of *grvA* gene expression. Environment-response of virulence genes in LEE is mediated by regulatory genes at other loci, which transfer the signal from TCS. Thus, regulatory systems for horizontally acquired virulence genes are integrated into backbone regulatory systems via virulence regulators.

Amino acid sequences of PchA and PchB share 98% similarity and both can activate transcription of *LEE1* operon, which includes *ler* gene. Enhancement of LEE gene expression in DMEM-grown EHEC is at least partly caused by the increase of *pch* genes expression, which is activated by DMEM. Negative effect of RcsB on *pch* transcription was demonstrated. Transcription of *pch* genes was enhanced in *rcsB* deletion mutant and reduced by overproduction of RcsB. Accordingly, LEE genes are both positively and negatively regulated by RcsC/RcsD/RcsB system through two different regulatory cascade (Fig. 1). Basal activity of RcsC/RcsD/RcsB system negatively affect LEE genes expression by repressing *pch* regulator genes, while, when RcsC/RcsD/RcsB system is activated by environmental stimuli, activated RcsB enhances the LEE genes expression through activation of another positive regulator gene *grvA*.

### *Other TCS in Virulence Expression*

QseC (YgiY)/QseB (YgiX) system is a homolog of PreA-PreB system of *Salmonella enterica*, which modulates expression of another TCS PmrA/PmrB.[41] PmrA/PmrB system regulates the modification of lipopolysaccharide leading to resistance against polymyxim B in *Salmonella enterica*. The QseC (YgiY)/QseB (YgiX) system in *E. coli* regulates production of flagella and ability of motility.[42] The system senses autoinducer, which is produced by *E. coli* and used as signal molecule for cell cell communication or quorum sensing. Expression of virulence in EHEC is also affected by the same autoinducer, but the QseC (YgiY)/QseB (YgiX) system is not involved in the regulation of virulence expression.

PhoQ/PhoP system is known to regulate transcription of *Salmonella* virulence genes. Although phoP mutant of *Shigella flexneri* has the same ability of cell invasion, intercellular spreading, induction of apoptosis in macrophages and resistance to extreme acid pH, the mutant *S. flexneri* are cleared more rapidly than wild type from infected animals. Although precise mechanisms remain to be unknown, the PhoQ/PhoP system is responsible for resistance to killing by polymorphonuclear leucocytes (PMN) and antimicrobial peptide.[43]

## Conclusion

Pathogenic *E. coli* or *Shigella* spp. encounters variety of environmental conditions during its life cycle. Even in gastointestine, where is niches of diarrheagenic pathogens, environmental conditions changes dramatically from extremely acid to neutral pH, rich to very restricted nutrient conditions, normal to high osmolarity, As nonpathogenic *E. coli* is a member of intestinal microflora, pathogenic *E. coli* and *Shigella* have to adapt to the same environmental changes by using the same response and adaptation systems. In addition, because pathogenic strains are newcomer in intestine, they have to increase their population and colonize on intestinal mucosa by overcoming already existing microflora. Thus, variety of genes are required for pathogenicity by involving in survival or multiplication as well as for virulence to host. By these reasons, TCSs, which are required for growth in environments of infection sites, could be described as essential for pathogenicity. This case is exemplified by BarA/UvrY system in UPEC, which affect the growth of bacteria in urinary tract. More specifically, it has been reported that production of virulence determinant are directly or indirectly affected by activation of some TCSs. Virulence associated genes are strictly regulated to express at target site of infection. Sensing environmental conditions and transferring signals to regulatory systems for virulence genes are important for successful infection. Response of virulence genes expression to changes in environmental conditions is achieved through various ways as shown for house-keeping (*E. coli* backbone) genes. For example, thermoregulation of *Shigella flexneri* virulence genes is mediated by sensing changes in DNA superhelicity at regulatory gene *virF*.[44] Activation of virulence genes expression in EHEC at conditions inducing growth arrest is mediated by increase of ppGpp, which is a signal molecule for stringent response.[45] TCS is also involved in regulation of virulence genes to respond to environmental factors. At least several systems are involved in regulation of virulence genes expression directly, such as the CpxA/CpxR system in transcription of Pap-pilus genes in UPEC and in virulence genes in *Shigella sonnei*, OmpR in curli genes in *E. coli*. In these cases transcription of target virulence genes are affected by response regulator, which binds promoter region of the gene. In other cases, TCS-regulated gene products affect production of virulence factors or assembly of macromolecules. Since virulence genes in *E. coli* are acquired by horizontal transfer of external DNA and usually possess different codon usage from *E. coli* backbone genes, virulence factors may requires stress-induced functions for efficient production.

Genome structure of pathogenic *E. coli* suggested that most of virulence-associated genes are located at strain-specific chromosomal segments or on plasmids. Although genes for macromolecules, such as pilus or type III secretion machinery, are located in the same segment and tightly regulated by a specific regulator, which is also encoded by the same segment, regulatory system for responding to environmental stimuli is dependent upon the *E. coli* backbone regulatory system. Integration of acquired virulence genes to the *E. coli* backbone response regulatory systems must be an important process to establish a successful pathogenic strain. It has been shown that placing the virulence regulatory gene into the *E. coli* backbone response regulatory network renders the virulence regulatory system to respond to signals processed through *E. coli* backbone regulatory system.

## References


1. Kaper JB, Nataro JP, Mobley HL. Pathogenic Escherichia coli. Nat Rev Microbiol 2004; 2:123-140.
2. Emody L, Kerenyi M, Nagy G. Virulence factors of uropathogenic Escherichia coli. Int J Antimicrob Agents 2003; 22:29-33.
3. Nataro JP, Kaper JB. Diarrheagenic Escherichia coli. Clin Microbiol Rev 1998; 11:142-201.
4. Konkel ME, Tilly K. Temperature-regulated expression of bacterial virulence genes. Microbes Infect 2000; 2:157-166.
5. Umanski T, Rosenshine I, Friedberg D. Thermoregulated expression of virulence genes in enteropathogenic Escherichia coli. Microbiology 2002; 148:2735-2744.
6. Ebel F, Deibel C, Kresse AU et al. Temperature- and medium-dependent secretion of proteins by Shiga toxin-producing Escherichia coli. Infect Immun 1996; 64:4472-4479.
7. Maurelli AT, Sansonetti PJ. Identification of a chromosomal gene controlling temperature-regulated expression of Shigella virulence. Proc Natl Acad Sci USA 1988; 85:2820-2824.
8. Abe A, Kenny B, Stein M et al. Characterization of two virulence proteins secreted by rabbit enteropathogenic Escherichia coli, EspA and EspB, whose maximal expression is sensitive to host body temperature. Infect Immun 1997; 65:3547-3555.
9. Schwan WR, Lee JL, Lenard FA et al. Osmolarity and pH growth conditions regulate fim gene transcription and type 1 pilus expression in uropathogenic Escherichia coli. Infect Immun 2002; 70:1391-1402.
10. Olsen A, Arnqvist A, Hammar M et al. Environmental regulation of curli production in Escherichia coli. Infect Agents Dis 1993; 2:272-274.
11. Geerlings SE, Brouwer EC, Gaastra W et al. Effect of glucose and pH on uropathogenic and nonuropathogenic Escherichia coli: studies with urine from diabetic and nondiabetic individuals. J Med Microbiol 1999; 48:535-539.
12. Nakayama S, Watanabe H. Involvement of cpxA, a sensor of a two-component regulatory system, in the pH-dependent regulation of expression of Shigella sonnei virF gene. J Bacteriol 1995; 177:5062-5069.
13. Puente JL, Bieber D, Ramer SW et al. The bundle-forming pili of enteropathogenic Escherichia coli: transcriptional regulation by environmental signals. Mol Microbiol 1996; 20:87-100.
14. Abe H, Tatsuno I, Tobe T et al. Bicarbonate ion stimulates the expression of locus of enterocyte effacement-encoded genes in enterohemorrhagic Escherichia coli O157:H7. Infect Immun 2002; 70:3500-3509.
15. Calderwood SB, Mekalanos JJ. Iron regulation of Shiga-like toxin expression in Escherichia coli is mediated by the fur locus. J Bacteriol 1987; 169:4759-4764.
16. Hayashi T, Makino K, Ohnishi M et al. Complete genome sequence of enterohemorrhagic Escherichia coli O157:H7 and genomic comparison with a laboratory strain K-12. DNA Res 2001; 8:11-22.
17. Perna NT, Plunkett G, 3rd, Burland V et al. Genome sequence of enterohaemorrhagic Escherichia coli O157:H7. Nature 2001; 409:529-533.
18. Welch RA, Burland V, Plunkett G 3rd et al. Extensive mosaic structure revealed by the complete genome sequence of uropathogenic Escherichia coli. Proc Natl Acad Sci USA 2002; 99:17020-17024.
19. Dougan G, Haque A, Pickard D et al. The Escherichia coli gene pool. Curr Opin Microbiol 2001; 4:90-94.
20. Ogura T, Kurokawa K, Ooka T et al. Complexity of the genome diversity in enterohemorrhagic Escherichia coli O157 revealed by the combinational use of the O157 Sakai oligoDNA microarray and the whole genome PCR scanning. DNA Res 2006; 13:3-14.

21. Venkatesan MM, Goldberg MB, Rose DJ et al. Complete DNA sequence and analysis of the large virulence plasmid of Shigella flexneri. Infect Immun 2001; 69:3271-3285.
22. Wei J, Goldberg MB, Burland V et al. Complete genome sequence and comparative genomics of Shigella flexneri serotype 2a strain 2457T. Infect Immun 2003; 71:2775-2786.
23. Jin Q, Yuan Z, Xu J et al. Genome sequence of Shigella flexneri 2a: insights into pathogenicity through comparison with genomes of Escherichia coli K12 and O157. Nucleic Acids Res 2002; 30:4432-4441.
24. Rowley G, Spector M, Kormanec J et al. Pushing the envelop: extracytoplasmic stress responses in bacterial pathogens. Nat Rev Microbiol 2006; 4:383-394.
25. Hung DL, Raivio TL, Jones CH et al. Cpx signaling pathway monitors biogenesis and affects assembly and expression of P pili. EMBO J 2001; 20:1508-1518.
26. Nevesinjac AZ, Raivio TL. The Cpx envelope stress response affects expression of the type IV bundle-forming pili of enteropathogenic Escherichia coli. J Bacteriol 2005; 187:672-686.
27. Jubelin G, Vianney A, Beloin C et al. CpxR/OmpR interplay regulates curli gene expression in response to osmolarity in Escherichia coli. J Bacteriol 2005; 187:2038-2049.
28. Otto K, Silhavy TJ. Surface sensing and adhesion of Escherichia coli controlled by the Cpx-signaling pathway. Proc Natl Acad Sci USA 2002; 99:2287-2292.
29. Mitobe J, Arakawa E, Watanabe H. A sensor of the two-component system CpxA affects expression of the type III secretion system through posttranscriptional processing of InvE. J Bacteriol 2005; 187:107-113.
30. Suzuki K, Wang X, Weilbacher T et al. Regulatory circuitry of the CsrA/CsrB and BarA/UvrY systems of Escherichia coli. J Bacteriol 2002; 184:5130-5140.
31. Fortune DR, Suyemoto M, Altier C. Identification of CsrC and characterization of its role in epithelial cell invasion in Salmonella enterica serovar Typhimurium. Infect Immun 2006; 74:331-339.
32. Eriksson AR, Andersson RA, Pirhonen M et al. Two-component regulators involved in the global control of virulence in Erwinia carotovora subsp. carotovora. Mol Plant Microbe Interact 1998; 11:743-752.
33. Lenz DH, Miller MB, Zhu J et al. CsrA and three redundant small RNAs regulate quorum sensing in Vibrio cholerae. Mol Microbiol 2005; 58:1186-1202.
34. Kitten T, Kinscherf TG, McEvoy JL et al. A newly identified regulator is required for virulence and toxin production in Pseudomonas syringae. Mol Microbiol 1998; 28:917-929.
35. Tomenius H, Pernestig AK, Jonas K et al. The Escherichia coli BarA-UvrY two-component system is a virulence determinant in the urinary tract. BMC Microbiol 2006; 6:27.
36. Bernardini ML, Fontaine A, Sansonetti PJ. The two-component regulatory system ompR-envZ controls the virulence of Shigella flexneri. J Bacteriol 1990; 172:6274-6281.
37. Bernardini ML, Sanna MG, Fontaine A et al. OmpC is involved in invasion of epithelial cells by Shigella flexneri. Infect Immun 1993; 61:3625-3635.
38. Prigent-Combaret C, Vidal O, Dorel C et al. Abiotic surface sensing and biofilm-dependent regulation of gene expression in Escherichia coli. J Bacteriol 1999; 181:5993-6002.
39. Prigent-Combaret C, Brombacher E, Vidal O et al. Complex regulatory network controls initial adhesion and biofilm formation in Escherichia coli via regulation of the csgD gene. J Bacteriol 2001; 183:7213-7223.
40. Tobe T, Ando H, Ishikawa H et al. Dual regulatory pathways integrating the RcsC-RcsD-RcsB signalling system control enterohaemorrhagic Escherichia coli pathogenicity. Mol Microbiol 2005; 58:320-333.
41. Merighi M, Carroll-Portillo A, Septer AN et al. Role of Salmonella enterica serovar Typhimurium two-component system PreA/PreB in modulating PmrA-regulated gene transcription. J Bacteriol 2006; 188:141-149.
42. Sperandio V, Torres AG, Kaper JB. Quorum sensing Escherichia coli regulators B and C (QseBC): a novel two-component regulatory system involved in the regulation of flagella and motility by quorum sensing in E. coli Mol Microbiol 2002; 43:809-821.
43. Moss JE, Fisher PE, Vick B et al. The regulatory protein PhoP controls susceptibility to the host inflammatory response in Shigella flexneri. Cell Microbiol 2000; 2:443-452.
44. Tobe T, Yoshikawa M, Sasakawa C. Thermoregulation of virB transcription in Shigella flexneri by sensing of changes in local DNA superhelicity. J Bacteriol 1995; 177:1094-1097.
45. Nakanishi N, Abe H, Ogura Y et al. ppGpp with DksA controls gene expression in the locus of enterocyte effacement (LEE) pathogenicity island of enterohaemorrhagic Escherichia coli through activation of two virulence regulatory genes. Mol Microbiol 2006; 61:194-205.
46. Herren CD, Mitra A, Palaniyandi SK et al. The BarA-UvrY Two-Component System Regulates Virulence in Avian Pathogenic Escherichia coli O78:K80:H9. Infect Immun 2006; 74:4900-4909.
47. Nadler C, Shifrin Y, Nov S et al. Characterization of enteropathogenic Escherichia coli mutants that fail to disrupt host cell spreading and attachment to substratum. Infect Immun 2006; 74:839-849.
48. Vianney A, Jubelin G, Renault S et al. Escherichia coli tol and rcs genes participate in the complex network affecting curli synthesis. Microbiology 2005; 151:2487-2497.

# Chapter 14

# Vancomycin Resistance VanS/VanR Two-Component Systems

**Hee-Jeon Hong, Matthew I. Hutchings and Mark J. Buttner***

## Abstract

Vancomycin is a member of the glycopeptide class of antibiotics. Vancomycin resistance (*van*) gene clusters are found in human pathogens such as *Enterococcus faecalis*, *Enterococcus faecium* and *Staphylococcus aureus*, glycopeptide-producing actinomycetes such as *Amycolotopsis orientalis*, *Actinoplanes teichomyceticus* and *Streptomyces toyocaensis* and the nonglycopeptide producing actinomycete *Streptomyces coelicolor*. Expression of the *van* genes is activated by the VanS/VanR two-component system in response to extracellular glycopeptide antibiotic. Two major types of inducible vancomycin resistance are found in pathogenic bacteria; VanA strains are resistant to vancomycin itself and also to the lipidated glycopeptide teicoplanin, while VanB strains are resistant to vancomycin but sensitive to teicoplanin. Here we discuss the enzymes the *van* genes encode, the range of different VanS/VanR two-component systems, the biochemistry of VanS/VanR, the nature of the effector ligand(s) recognised by VanS and the evolution of the *van* cluster.

## Introduction

Vancomycin is clinically important for treating enterococcal infections arising after abdominal surgery and is vital as the only widely effective treatment for infections caused by methicillin-resistant *Staphylococcus aureus* (MRSA), a major killer in hospital-acquired infections. Vancomycin and other glycopeptide antibiotics inhibit cell wall biosynthesis by binding to the D-alanyl-D-alanine (D-Ala-D-Ala) terminus of lipid-attached peptidoglycan precursors on the outside of the cytoplasmic membrane (Fig. 1A).[1,2] This interaction blocks formation of mature peptidoglycan, principally by denying transpeptidase access to its substrate, thereby preventing formation of the peptide crosslinks between polysaccharide strands that give the cell wall its rigidity.

The first clinical isolates of vancomycin-resistant strains of pathogenic *Enterococcus faecalis* and *Enterococcus faecium* (VRE) appeared in the late 1980s and were shown to reprogramme cell wall biosynthesis such that the 'stem' pentapeptide of peptidoglycan precursors terminated in D-alanyl-D-lactate (D-Ala-D-Lac), rather than in D-Ala-D-Ala (Fig. 1B).[3-6] The affinity of vancomycin for precursors terminating in D-Ala-D-Lac is ~1000-fold lower than for precursors terminating in D-Ala-D-Ala, rendering the modified bacteria resistant.[3] Because vancomycin is the front-line therapy for treating problematic infections caused by MRSA, the spread of vancomycin resistance through bacterial populations is an acute public health issue, highlighted by the recent emergence of vancomycin-resistant, methicillin-resistant *Staphylococcus aureus* (VRSA) in hospitals.[7-10]

*Corresponding Author: Mark J. Buttner—Department Molecular Microbiology, John Innes Centre, Norwich Research Park, Colney, Norwich NR4 7UH, UK.
Email: mark.buttner@bbsrc.ac.uk

*Bacterial Signal Transduction: Networks and Drug Targets*, edited by Ryutaro Utsumi.

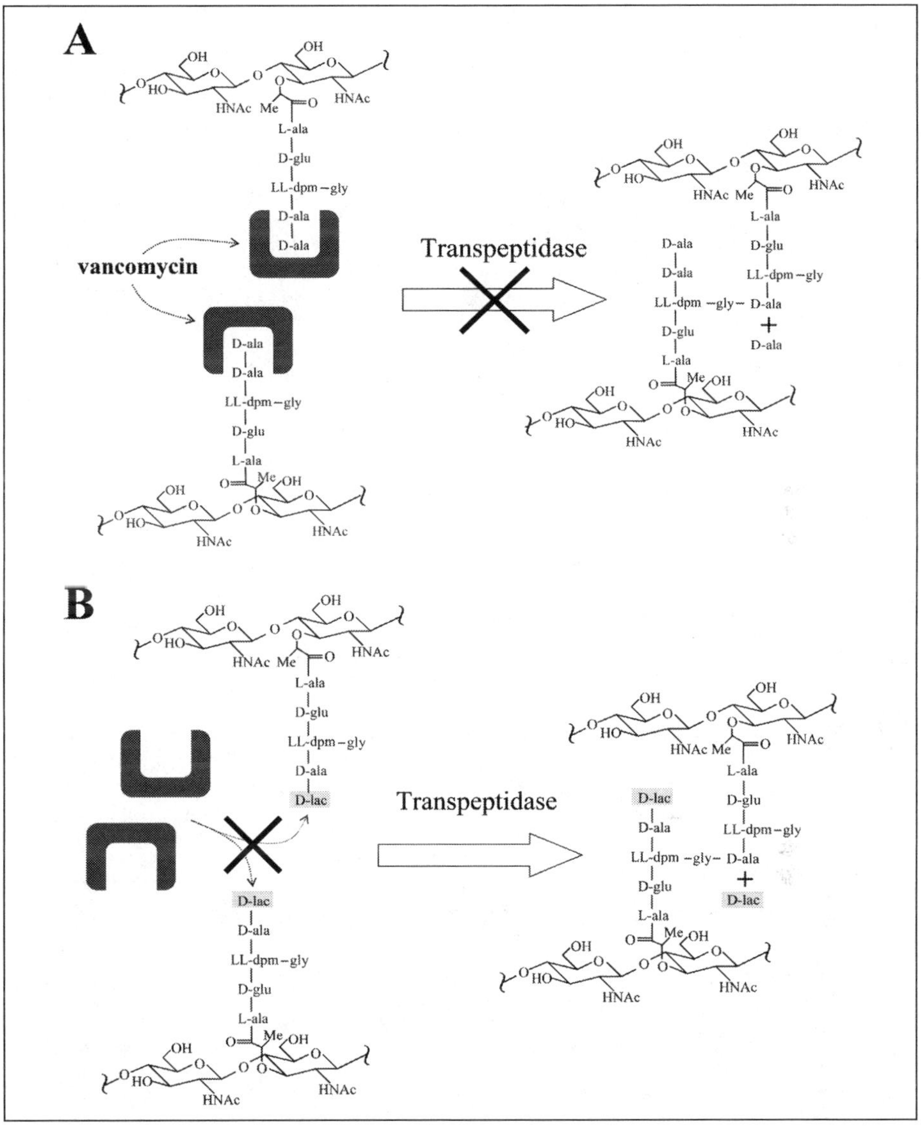


Figure 1. Transpeptidase and the mode of action of vancomycin. A) Transpeptidase recognises the sequence D-alanyl-D-alanine (D-Ala-D-Ala) at the end of the pentapeptide chain, cleaves off the terminal alanine and joins the remainder to the branch of a stem peptide from an adjacent polysaccharide chain. Vancomycin and other glycopeptide antibiotics inhibit cell wall biosynthesis by binding to the D-Ala-D-Ala terminus of lipid-attached peptidoglycan precursors on the outside of the cytoplasmic membrane. This interaction blocks formation of mature peptidoglycan, principally by denying transpeptidase access to its substrate. B) Vancomycin resistant bacteria reprogramme cell wall biosynthesis such that the stem pentapeptide terminates in D-alanyl-D-lactate (D-Ala-D-Lac). The affinity of vancomycin for D-Ala-D-Lac is ~1000-fold lower than for D-Ala-D-Ala, allowing transpeptidation to occur. Note that the peptidoglycan precursor shown is the one present in *Streptomyces*, but the exact nature of the precursor varies from genus to genus; in *Streptomyces* it has LL-diaminopimelic acid (LL-dpm) at position 3 of the stem pentapeptide and the branch is a single glycine.

## A Range of Different VanS/VanR Systems

In the context of this book, the key point about the vancomycin resistance (*van*) genes is that they are expressed only in the presence of extracellular glycopeptides and that signal transduction is mediated by a two-component system consisting of a sensor kinase (VanS) and a response regulator (VanR). Five different VanS/VanR two-component systems have been examined, albeit to very differing extents. Most effort has been concentrated on the VanS/VanR systems associated with the clinically important VanA ($VanS_A/VanR_A$) and VanB ($VanS_B/VanR_B$) enterococcal strains that first appeared in hospitals in the late 1980s.[11] More recently, there has been analysis of VanS/VanR systems found in actinomycetes, the order of bacteria that make all of the known glycopeptides. In every case where it has been studied, the *vanS/vanR* genes are themselves under VanS/VanR control, creating an auto-amplification loop in the presence of inducer (Fig. 2).

### *Enterococcal VanA and VanB Strains*

Enterococcal VanA strains are resistant to vancomycin itself and also to the lipidated glycopeptide teicoplanin (Fig. 3), while VanB strains are resistant to vancomycin but sensitive to teicoplanin. The *van* genes in VanA strains are carried on the transposon Tn*1546* and there is very little sequence variation between the *van* genes in VanA isolates. The first isolates of the new *S. aureus* hospital 'superbug,' VRSA, arose from intergeneric transfer of Tn*1546* from a co-isolate of *E. faecalis*.[7,10] The *van* genes in VanB strains are chromosomally encoded and are more diverse in sequence than their VanA equivalents.

In contrast to these VanA and VanB resistant strains, the comparatively rare VanC, VanE and VanG isolates of *Enterococci* have a D-alanyl-D-serine (D-Ala-D-Ser) ligase instead of a D-Ala-D-Lac ligase.[6,12] The substitution of D-Ser for D-Ala results in a ~6-fold decrease in affinity for vancomycin and therefore low-level resistance.[13] The D-Ser-based systems will not be considered further here.

### *Glycopeptide-Resistant Actinomycetes*

Glycopeptide resistance has been explored in three different actinomycetes: *Streptomyces coelicolor*, which does not make a glycopeptide and *Streptomyces toyocaensis* and *Actinoplanes teichomyceticus*, which do make glycopeptides. Glycopeptide resistance has an additional significance

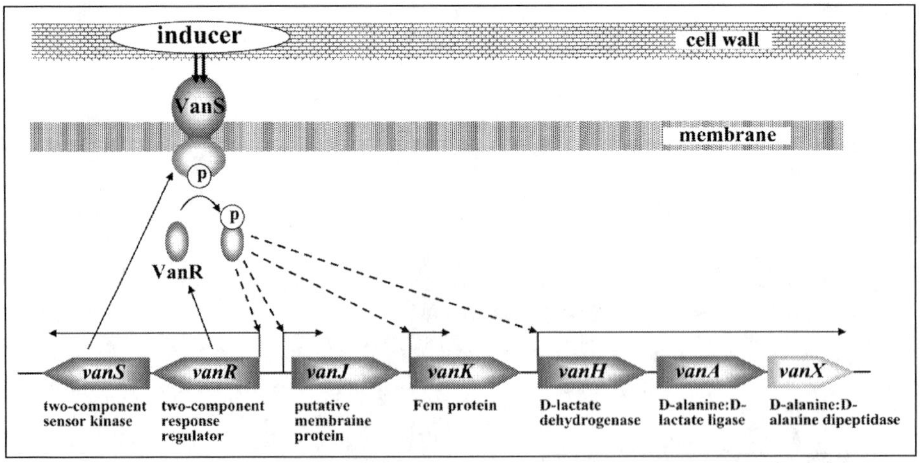


Figure 2. Organisation and regulation of the vancomycin resistance (*van*) gene cluster of *S. coelicolor*. The genes are organized into four transcription units, *vanS/vanR*, *vanJ*, *vanK* and *vanHAX* and these transcripts are induced by vancomycin in a *vanR*-dependent manner. Reproduced with permission from reference 23.

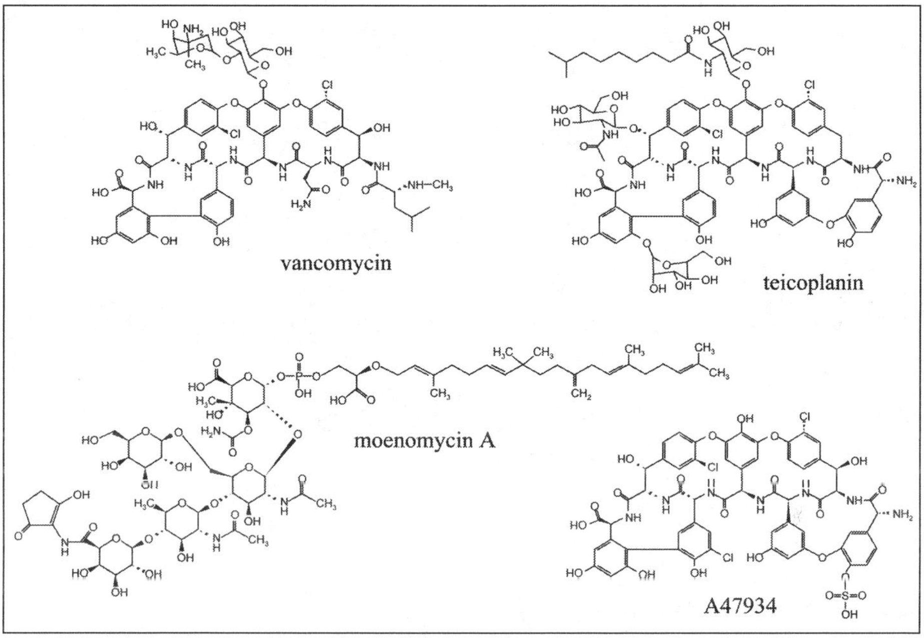


Figure 3. Structures of the glycopeptide antibiotics vancomycin, teicoplanin and A47934 and the cell wall-specific, nonglycopeptide antibiotic, moenomycin A.

in glycopeptide producers, where activation of the resistance genes by the endogenously produced antibiotic prevents suicide (auto-toxicity).

*S. coelicolor* is genetically the model species of a genus of Gram-positive, mycelial soil bacteria responsible for the production of two-thirds of the commercially important antibiotics. Like most other nonpathogenic actinomycetes, *S. coelicolor* lives in the soil and it seems likely that it encounters glycopeptide producers such that the *van* gene cluster (Figs. 2 and 4) confers a selective advantage. Further, it is widely believed that all glycopeptide resistance genes are ultimately derived from actinomycete glycopeptide producers.[14] Consistent with this idea, the *S. coelicolor* resistance genes are clearly associated with a laterally acquired DNA element (G. Chandra and H.-J. Hong, unpublished).

*S. toyocaensis* is the producer of the 'sugarless' glycopeptide A47934 (Fig. 3) and the *van* resistance genes in this organism (Fig. 4) are associated with the A47934 biosynthetic cluster.[15] *S. toyocaensis* is resistant to A47934 but sensitive to both vancomycin and teicoplanin.

*A. teichomyceticus* is the producer of teicoplanin (Fig. 3) and carries a *van* cluster (Fig. 4), including *vanS/vanR*, associated with the teicoplanin biosynthetic genes.[16,17] *A. teichomyceticus* is resistant to all glycopeptides tested, but it now seems clear that this 'pan-glycopeptide resistance' does not arise from pan-glycopeptide induction of the *van* genes but rather because the *van* genes are expressed constitutively, even in the absence of antibiotic.[18] The cause of the constitutive expression of the *van* genes is unknown, but one possibility is that the VanS sensor kinase is locked in the 'on' state in this organism.

## What Do the *van* Genes Encode?

The number of genes present in the *van* cluster varies (Fig. 4), but the 'core' cluster consists of five genes—*vanS/vanR*, plus a *vanHAX* operon encoding the three enzymes required for remodelling cell wall precursors: VanH, which converts pyruvate to D-lactate; VanA, a D-Ala-D-Lac ligase; and VanX, a D-Ala-D-Ala dipeptidase that cleaves any residual D-Ala-D-Ala dipeptide, ensuring

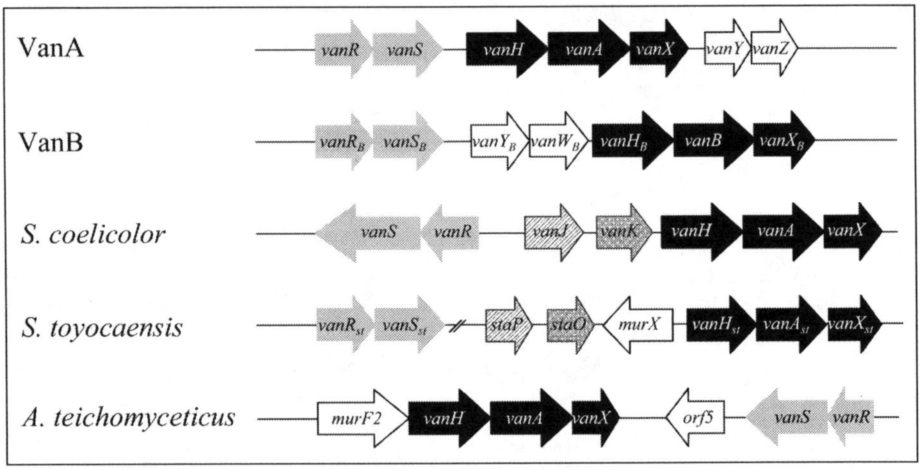


Figure 4. Comparison of the *van* gene clusters from enterococcal VanA and VanB strains, *S. coelicolor, S. toyocaensis* and *A. teichomyceticus*.

that peptidoglycan precursors terminate uniformly in D-Ala-D-Lac. In addition to this minimal set, other genes are sometimes present. The VanA transposon Tn*1546* encodes two accessory proteins, VanY and VanZ, which are not required for, but can contribute to, high level resistance to vancomycin and teicoplanin. VanY is a D,D-carboxypeptidase that can cleave the C-terminal D-Ala of peptidoglycan precursors (but has no activity against free D-Ala-D-Ala dipeptide, the VanX substrate).[19,20] VanZ confers low level teicoplanin resistance in the absence of the other resistance proteins by an unknown mechanism.[21]

The *S. coelicolor* cluster consists of seven genes, *vanSRJKHAX*, divided into four transcription units and carries two genes, *vanJ* and *vanK*, not found in enterococcal VanA and VanB strains (Figs. 2 and 4).[22] *vanJ*, encoding a predicted membrane protein of unknown function, is not required for vancomycin resistance, but *vanK* is essential for resistance.[22] VanK is a member of the Fem family of enzymes, which add the 'branch' amino acid(s) to the stem pentapeptide of peptidoglycan precursors. In *S. coelicolor*, the branch is a single glycine residue (Fig. 1) and, in the absence of vancomycin, this residue is added by an enzyme called FemX.[23] However, the constitutive FemX activity of *S. coelicolor* can recognise only precursors that terminate in D-Ala-D-Ala as a substrate. VanK is required for vancomycin resistance because it is the only enzyme that can add the Gly branch to precursors terminating in D-Ala-D-Lac (production of precursors lacking the Gly branch is lethal in *Streptomyces* because it prevents cross-linking of the peptidoglycan by transpeptidase, leading to cell lysis).[23] The absence of orthologues of *vanK* in the vancomycin-resistance gene clusters of pathogenic enterococci implies that FemX of enterococci can recognise precursors terminating in either D-Ala-D-Lac or D-Ala-D-Ala.

## VanS/VanR Biochemistry

VanS/VanR systems from enterococci and *S. coelicolor* have been characterised in vitro. Wright et al[24] demonstrated that a fusion protein consisting of maltose binding protein (MBP) and the cytosolic domain of enterococcal $VanS_A$ could catalyse both autophosphorylation and rapid phosphotransfer to purified $VanR_A$. Incubation of MBP-$VanS_A$ with phosphorylated $VanR_A$ ($VanR_A \sim P$) increased its dephosphorylation approximately 6-fold, suggesting that VanS can also act as a $VanR_A$-specific phosphatase.[24] Similar experiments showed that enterococcal $VanS_B$ also possesses both $VanR_B$ kinase and phosphatase activity.[25] Further, Hutchings et al[26] showed that the cytosolic domain of *S. coelicolor* VanS can autophosphorylate and catalyse both phosphorylation and dephosphorylation of *S. coelicolor* VanR in vitro. Thus, the in vitro biochemical evidence suggests that VanS is a bifunctional protein that can switch between kinase and phosphatase activities.

Using gel shift assays and a DNA fragment carrying the *vanH*$_A$ promoter region, Holman et al[27] showed that phosphorylation of enterococcal VanR$_A$ results in a 500-fold increase in DNA-binding activity. Similarly, VanR$_B$ ~ P was found to bind target promoters more tightly than unphosphorylated VanR$_B$ and to be more efficient in promoting open complex formation by RNA polymerase.[28] DNaseI footprinting experiments suggested that phosphorylation of VanR$_A$ resulted in oligomerisation of the protein at the *vanH*$_A$ promoter. Unphosphorylated VanR$_A$, or a D53A variant which cannot be phosphorylated, exhibited lower DNA binding-affinity and a smaller footprint at the *vanH*$_A$ promoter.[27] Investigation of VanR$_B$ oligomerization using gel filtration suggests that enterococcal VanR$_B$ is converted from monomer to dimer on phosphorylation.[28] The intrinsic in vitro stability of phosphorylated response regulators varies widely, perhaps reflecting their physiological roles, with isolated proteins displaying half-lives ranging from 23 s for CheY ~ P, involved in chemotaxis, to 180 min for Spo0F ~ P, involved in *Bacillus* sporulation.[25] The half-life of VanR$_B$ ~ P is ~150 min.[25]

## VanS/VanR and Acetyl Phosphate

In many two-component systems, loss of the sensor kinase or loss of the response regulator leads to the same phenotype—loss of expression of the target genes. However, in both *S. coelicolor* and enterococci, deletion of *vanS* results in constitutive expression of the vancomycin resistance genes, suggesting that VanS negatively regulates VanR function in the absence of antibiotic. In other words, VanR ~ P can be generated in a VanS-independent manner and VanS acts as a VanR ~ P phosphatase in the absence of vancomycin. In *S. coelicolor*, VanS-independent synthesis of VanR ~ P appears to arise because VanR can be activated in vivo by the small molecule phosphodonor acetyl phosphate. Deletion of *vanS* in *S. coelicolor* results in constitutive expression of the *van* genes but a *vanS pta ackA* triple mutant, which should not be able to synthesise acetyl phosphate, fails to express the *van* genes, whereas a *pta ackA* double mutant shows wild-type, regulated induction of the *van* genes.[26] These results suggest that in the absence of vancomycin, acetyl phosphate phosphorylates VanR and VanS acts as a phosphatase to suppress the levels of VanR~P. On exposure to vancomycin, VanS activity switches from a phosphatase to a kinase and vancomycin resistance is induced (Fig. 5). It should be noted that transcription of the *S. coelicolor vanS/vanR* operon is itself under VanS/VanR control (Fig. 2)[22] and so there will be very little VanR or VanS protein in *S. coelicolor* growing in the absence of vancomycin. Thus, the 'futile cycle' of VanR phosphorylation and dephosphorylation shown to occur in the absence of vancomycin in Fig. 5 will occur at a significant level only after the organism has been transiently exposed to the antibiotic. Similar results have been obtained in enterococcal VanA strains. Arthur et al[29] showed that the *van* promoters of an *E. faecium* VanA strain were constitutively activated by VanR$_A$ in the absence of VanS$_A$ and concluded that VanS$_A$ negatively controls VanR$_A$ in the absence of glycopeptide inducer, presumably by dephosphorylation. Further, Haldimann et al[30] introduced a *vanH*$_A$*-lacZ* fusion into an *ackA* strain of *E. coli*, which overproduces acetyl phosphate. Heterologous expression of enterococcal VanR$_A$ in this strain stimulated high levels of β-galactosidase production, suggesting that acetyl phosphate could act as an in vivo phosphodonor to the *E. faecium* VanR$_A$ protein in *E. coli*.

## 'Crosstalk' with Other Two-Component Systems

In an elegant study using flow cytometry, Baptista et al[31] took advantage of a *vanY*$_B$*-gfp* transcriptional fusion to examine induction of *van* gene expression in single cells of an enterococcal VanB strain. In enterococcal VanB strains, null mutations in *vanS*$_B$ lead to a phenotype termed 'heterogeneous,' in which, in the absence of antibiotic, only a minority of the bacteria express the *van* genes.[31] Further, addition of antibiotic leads to uniform induction of the whole population, rather than selection of the subpopulation initially expressing resistance under non-inducing conditions. They concluded that a heterologous kinase activated VanR$_B$ in the absence of VanS$_B$. Interestingly, this putative kinase was stimulated by vancomycin, teicoplanin and the nonglycopeptide cell wall inhibitor moenomycin (Fig. 3), suggesting that it might respond to the same indirect signal as

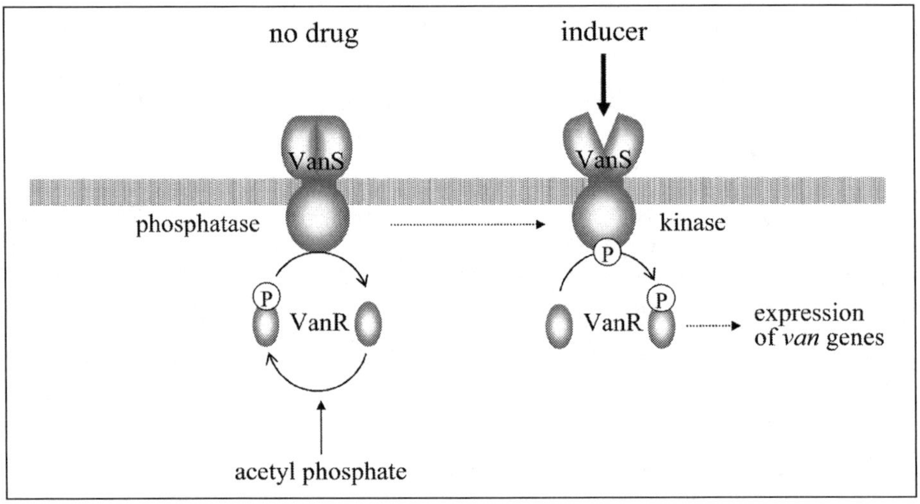


Figure 5. A model for the function of the vancomycin resistance VanS/VanR two-component signal transduction system in *S. coelicolor.* In the absence of antibiotic, acetyl phosphate phosphorylates D51 of VanR and VanS acts as a phosphatase to suppress the levels of VanR ~ P. In the presence of antibiotic, VanS is converted from a phosphatase into a kinase, leading to the accumulation of VanR ~ P and activation of the four promoters of the *van* gene cluster. Transcription of the *vanS/vanR* operon is itself under VanS/VanR control and so there will be very little VanR or VanS protein in *S. coelicolor* growing in the absence of vancomycin. Thus, the 'futile cycle' of VanR phosphorylation and dephosphorylation shown in the absence of vancomycin will occur at a significant level only after the organism has been transiently exposed to the antibiotic.

$VanS_A$ from VanA-type enterococci (see below). Presumably, in wild-type enterococcal VanB strains, the phosphatase activity of $VanS_B$ keeps $VanR_B$ in the unphosphorylated state in the presence of teicoplanin and moenomycin, preventing the putative heterologous kinase from activating *van* gene expression. The putative heterologous kinase has not been identified but a possible candidate is CroS, since it is known to be induced by vancomycin, teicoplanin and moenomycin A.[32] CroS is required for intrinsic β-lactam resistance in *E. faecalis* but the target genes of the CroRS two-component system involved in this resistance have not been identified.[32]

## Relationships Between VanS Proteins of Different Origin

In considering the nature of the effector ligand(s) that activate VanS, it is important to keep in mind the relationships between VanS proteins of different origin. First, the differences in the sizes of the extracytoplasmic sensor domains are striking. The putative extracytoplasmic sensor domain of $VanS_A$ is 103 amino acids long, the equivalent domain of $VanS_B$ consists of 37 amino acids, whereas the putative extracytoplasmic sensor domains of the three actinomycete VanS proteins contain only 26-27 amino acids. These sensor domains are sufficiently small for the actinomycete VanS proteins to have been included in a review of 'intramembrane-sensing' sensor kinases.[33] The VanS proteins from enterococcal VanA and VanB strains are only distantly related (16% overall identity) and the putative $VanS_A$ and $VanS_B$ sensor domains are not related in amino acid sequence. The enterococcal $VanS_A$ and $VanS_B$ proteins are also very diverged from their actinomycete equivalents (~15% overall identity in pairwise comparisons). In contrast, the VanS proteins from the three actinomycetes strongly resemble each other (65-77% overall identity in pairwise comparisons). However, in comparing VanS from *S. coelicolor* and *S. toyocaensis*, it is clear that this high similarity breaks down in the 26-27-residue stretch between the two predicted transmembrane helices, corresponding to the putative VanS sensor domain (Fig. 6). It now seems clear that the *van* genes

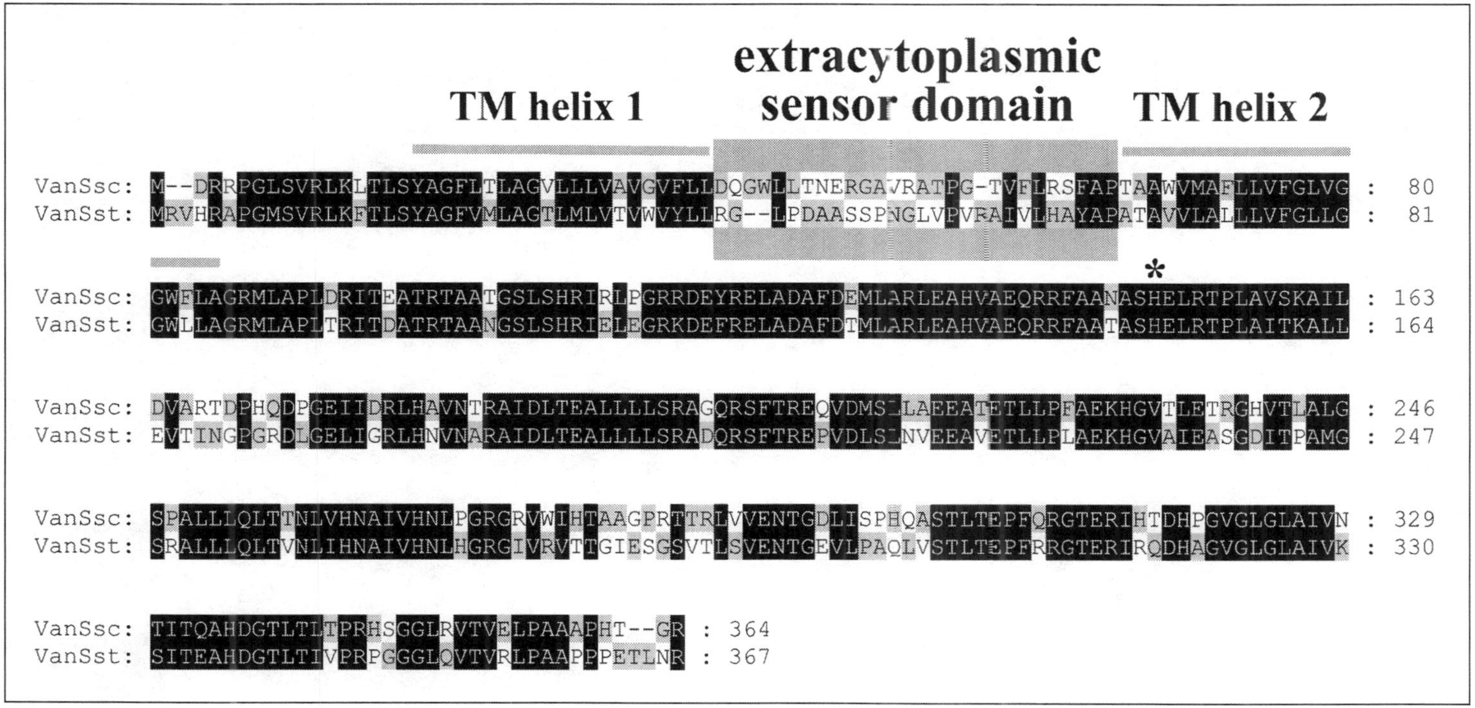


Figure 6. Alignment of the VanS proteins from *S. coelicolor* (VanSsc) and *S. toyocaensis* (VanSst). The two predicted transmembrane helices and the 26-27-amino acid putative extracytoplasmic sensor domain lying in between are highlighted. The histidine that is the putative site of autophosphorylation is marked with an asterisk.

of *A. teichomyceticus* are not inducible but constitutively active,[18] and so it is hard to know if *A. teichomyceticus* VanS binds effector molecule(s). Nevertheless, it is interesting to note that the putative 26-residue extracytoplasmic sensor domain of *A. teichomyceticus* VanS differs from that of *S. toyocaensis* at only 4 residues.

## What Is the Effector Ligand Recognised by VanS?

The nature of the direct molecular ligand that activates VanS has not been determined for any VanS/VanR signal transduction system. Two distinct models exist: direct induction, in which the sensor kinase is activated by direct binding of antibiotic to the sensor domain and indirect induction, in which the sensor kinase is activated by binding an intermediate in cell wall biosynthesis or degradation that accumulates as a result of antibiotic action. These two models are not mutually exclusively, since a further possibility is that the VanS inducer is the antibiotic bound to a D-Ala-D-Ala-containing cell wall precursor, such as lipid II. Given that the sensor domains of VanS proteins are not homologous, it is possible or even likely that some VanS proteins respond directly to the antibiotic while others respond indirectly. A summary of the genetic evidence that addresses this question is presented below, but it seems unlikely that genetics alone can identify the nature of the inducer and that biochemical (in vitro reconstitution studies; in vivo cross-linking) or structural studies will be required to provide a definitive answer.

### *Induction in Enterococcal VanA Strains*

Screens for inducers of $VanS_A$ have been established by coupling a promoter under the control of VanS/VanR to suitable reporter genes,[34-38] by assaying VanX activity in cell extracts,[39] by monitoring induction of Lac-containing precursors,[40] or by looking for induced vancomycin resistance in pretreated cultures.[40,41] All these reports agree that VanA strains are induced by vancomycin and teicoplanin. However, the most interesting results from these papers concern the potential for nonglycopeptide cell wall-specific antibiotics to induce VanS. All reports agree that VanA strains are inducible by the nonglycopeptide moenomycin A.[36-41] The experiments of Baptista et al[39] are particularly compelling since they assayed induction of VanX enzymatic activity in cell extracts and did not rely on multicopy plasmids or reporter genes. Since moenomycin A is not structurally related to glycopeptides it seems unlikely that the sensor domain of $VanS_A$ could bind both glycopeptides and moenomycin directly. The general conclusion has therefore been that $VanS_A$ must be activated by an intermediate in cell wall biosynthesis that accumulates in response to both glycopeptides and moenomycin A. Because moenomycin A inhibits transglycosylase,[42,43] both glycopeptides and moenomycin A are likely to lead to accumulation of lipid II (a lipid-anchored cell wall precursor) on the external face of the cytoplasmic membrane and it has been speculated that lipid II might be the direct effector ligand of $VanS_A$.[31]

### *Induction in Enterococcal VanB Strains*

Induction of VanB strains has been also been addressed.[31,39,44] In contrast to VanA strains, all the nonglycopeptides tested, including moenomycin A, failed to induce $VanS_B$. Since all $VanS_B$ inducers identified are structurally related glycopeptides, the simplest interpretation of the data is that $VanS_B$ is likely to be induced directly by the drug itself.[31] VanB strains are sensitive to the lipidated antibiotic teicoplanin because the $VanS_B$/$VanR_B$ signal transduction system is not induced by teicoplanin.[31,39,44] In further experiments, Baptista et al[31,39,44] isolated teicoplanin-resistant mutants of VanB strains, six of which showed induction of the *van* genes by teicoplanin (but not by the nonglycopeptide moenomycin A). These six mutants all carried single amino acid substitutions in the N-terminal half of $VanS_B$. How to interpret these gain-of-function mutations is not clear. Two were in the predicted extracytoplasmic sensor domain where they could potentially directly improve interaction with an extracellular ligand, such as teicoplanin. However, the remaining four were in the cytoplasmic linker domain that connects the sensor and kinase domains. It is possible that wild-type $VanS_B$ binds teicoplanin unproductively and that these four amino acid substitutions affect propagation of the induction signal through the membrane such that signal transduction now occurs. However, it should be noted that in the case examined in detail (an A167S substitution in the linker domain), the substitution conferring teicoplanin inducibility also conferred hyper-inducibility by vancomycin. Again, it is possible that wild-type $VanS_B$ binds teicoplanin unproductively and that the A167S mutation makes $VanS_B$ hypersensitive to inducers. Thus, the teicoplanin-inducible $VanS_B$ mutations may be qualitative and involve a change in induction specificity, or they may be quantitative and involve an increase in the sensitivity of the protein to inducers.[31,39,44]

### *Induction in Actinomycete Species*

Inducers of VanS in *S. coelicolor* were identified using a bioassay. *S. coelicolor femX* null mutants are viable only in the presence of compounds that activate the VanS/VanR signal transduction system, because they rely on expression of VanK for survival. Hutchings et al[26] took advantage of this antibiotic-dependent phenotype to create a simple bioassay for inducers of the *van* genes in *S. coelicolor*. The structurally closely related glycopeptide antibiotics vancomycin, ristocetin, chloroeremomycin and A47934 all acted as inducers of the VanS/VanR system, but the lipidated glycopeptide teicoplanin and the nonglycopeptide moenomycin A did not.

To address the effector ligand issue further, Hutchings et al[26] carried out a "VanS/VanR swap" experiment between two glycopeptide-resistant *Streptomyces* species with differing spectra of inducer molecules, to see if inducer specificity was determined by VanS/VanR itself or by the host background. In *S. coelicolor*, the *van* genes are induced by both A47934 and vancomycin, while in *S. toyocaensis*, resistance is induced by A47934 but not by vancomycin.[45] Introduction of the *S. toyocaensis* VanS/VanR signal transduction system into an *S. coelicolor vanS/vanR* null mutant switched inducer specificity to that of *S. toyocaensis*. Thus, inducer specificity is determined by the origin of VanS/VanR. There are two potential explanations for this observation. If *Streptomyces* VanS is activated by accumulation of a cell wall intermediate, vancomycin must induce a radically different spectrum of cell wall intermediates in *S. coelicolor* and *S. toyocaensis*, which seems unlikely. The more likely alternative is that *Streptomyces* VanS is directly activated by binding antibiotic (or possibly antibiotic bound to D-Ala-D-Ala-containing cell wall precursors, such as lipid II) and that *S. toyocaensis* VanS interacts productively with A47934 but not with vancomycin, whereas *S. coelicolor* VanS interacts productively with both antibiotics. This would also be consistent with the fact that the nonglycopeptide moenomycin is not an inducer of VanS in *Streptomyces*.

Whether the *Streptomyces* VanS effector ligand is a cell wall intermediate or the antibiotic itself, the ability to respond differentially to vancomycin and A47934 must reside in differences between the sensor domains of the *S. toyocaensis* and *S. coelicolor* VanS proteins. The VanS proteins from these two species are very similar, with 65% identity overall (Fig. 6). However, as noted above, it is striking that this high level of identity breaks down in the 26-27-residue putative sensor domain lying between the two predicted transmembrane helices (Fig. 6).

## Functional Differences between Vancomycin and Teicoplanin

Enterococcal VanA strains are resistant to teicoplanin, whereas VanB strains are sensitive because teioplanin fails to induce $VanS_B$. Vancomycin and teicoplanin (Fig. 3) differ in the structure of the aglycone (the peptide part of the molecule), the glycosylation pattern and in the presence of a fatty-acid chain attached to teicoplanin that is absent in vancomycin. Through the chemo-enzymatic synthesis of a spectrum of vancomycin and teicoplanin deriviatives, Kahne and colleagues showed definitively that the key functional difference between these two antibiotics is the presence or absence of the lipid: removal of the lipid from teicoplanin prevents it from killing VanB strains and addition of a lipid to vancomycin makes it an effective antibiotic against VanB strains.[46] Taking this a stage further, using the same range of vancomycin and teicoplanin derivatives, it has been shown that it is the presence or absence of the lipid and not the differences in aglycone structure or glycosylation pattern, that is the key difference between the two antibiotics in determining *van* gene inducer activity in *S. coelicolor* (M. Oberthür, H.-J. Hong, C. Leimkuhler, B. Falcone, C. Walsh, M. Buttner and D. Kahne unpublished). These observations raise interesting questions about the evolution of teicoplanin. Perhaps addition of the lipid was selected during evolution of the producing organism, *A. teichomyceticus*, at least in part because it prevents competing soil bacteria like *S. coelicolor* from sensing the antibiotic and generating a resistance response.

The lipid moiety can serve to anchor teicoplanin in the membrane,[47-49] and advocates of direct induction of $VanS_B$ have proposed that membrane anchoring prevents teicoplanin from interacting productively with the $VanS_B$ sensor domain. However, it should be noted that this lipid moiety is relatively short and that teicoplanin is water soluble, implying that capture of teicoplanin by the membrane would not be as complete as for other molecules carrying longer lipid tails such as, for example, lipoproteins. A further issue concerns the mode of action of these drugs. Both teicoplanin and vancomycin bind to D-Ala-D-Ala and inhibit both transpeptidation and transglycosylation. However, vancomycin exerts its major effect on transpeptidation whereas lipidated glycopeptides inhibit transglycoslylation more strongly.[43,50-52] These observations suggest that the actions of vancomycin and teicoplanin will lead to the accumulation of somewhat different spectra of cell wall intermediates, leaving open the possibility that the enterococcal VanB phenotype could be accounted for through an indirect induction mechanism.

## Evolution of the *van* Cluster

An important unanswered question is whether there is selective pressure against the evolution of constitutive expression of the *van* genes. Such pressure might arise from the relative thermodynamic instability of the D-Ala-D-Lac ester linkage. The D-Ala-D-Ala peptide bond is more stable than the D-Ala-D-Lac ester linkage and spontaneous hydrolysis of lactate from cell wall precursors would yield molecules incapable of supporting cell wall crosslinking (because the D-Ala-D-Lac or D-Ala-D-Ala bond is cleaved during transpeptidation and the energy of the bond is conserved to form the peptide crosslink with the pendant peptide of an adjacent polysaccharide chain; Fig. 1). Potentially consistent with this logic, *vanS* null mutants of *S. coelicolor*, which express D-Ala-D-Lac precursors constitutively, suffer, for unknown reasons, a growth rate disadvantage relative to the wild type expressing D-Ala-D-Ala precursors (Hong, Hutchings and Buttner, unpublished). However, constitutive expression of D-Ala-D-Lac precursors does not seem to cause a growth rate disadvantage in enterococci (Michel Arthur, pers. comm.) and, most importantly, members of the genera *Leuconostoc*, *Lactobacillus* and *Pediococcus* are all naturally resistant to vancomycin because they constitutively express cell wall precursors terminating in D-Ala-D-Lac.[53-56]

Finally, it should be noted that expression of vancomycin resistance can bring into play other selective pressures that have very important clinical consequences. For example, enterococcal strains expressing vancomycin resistance become sensitive to third generation cephalosporins, like ceftriaxone, because the ceftriaxone-resistant penicillin-binding protein (called PBP5) cannot recognise cell wall precursors terminating in D-Ala-D-Lac as substrates.[57] Likewise, the recent intergeneric transfer of Tn*1546* from enterococci into *S. aureus* to create VRSA strains has similar interesting consequences: VRSA strains expressing vancomycin resistance become sensitive to β-lactams, because PBP2A (encoded by *mecA*), which confers β-lactam resistance, cannot recognise cell wall precursors terminating in D-Ala-D-Lac as substrates.[58] Thus, while VRSA strains are highly resistant to vancomycin (MIC = 512 μg/ml) or β-lactams such oxacillin (MIC = 800 μg/ml) when applied individually, they are very effectively killed by low concentrations of these two drugs in combination (for example 40 μg/ml oxacillin with 12 μg/ml vancomycin is lethal).[58,59]

### *Acknowledgements*

We are very grateful to Michel Arthur, Daniel Kahne, Chris Walsh, David Hopwood and Flavia Marinelli for helpful discussion, communication of unpublished data and comments on the manuscript. The authors' work on vancomycin resistance in *Streptomyces* was funded by the Biotechnology and Biological Sciences Research Council (BBSRC) of the United Kingdom.

## References


1. Williams DH, Williamson MP, Butcher DW et al. Detailed binding sites of the antibiotics vancomycin and ristocetin A: determination of intermolecular distances in antibiotic/substrate complexes by use of the time-dependent NOE. J Am Chem Soc 1983; 105:1332-1339.
2. Barna JCJ, Williams DH. The structure and mode of action of glycopeptide antibiotics of the vancomycin group. Annu Rev Microbiol 1984; 38:339-57.
3. Bugg TDH, Wright GD, Dutka-Malen S et al. Molecular basis for vancomycin resistance in Enterococcus faecium BM4147: biosynthesis of a depsipeptide peptidoglycan precursor by vancomycin resistance proteins VanH and VanA. Biochemistry 1991; 30:10408-10415.
4. Healy VL, Lessard IA, Roper DI et al. Vancomycin resistance in enterococci: reprogramming of the D-Ala-D-Ala ligases in bacterial peptidoglycan biosynthesis. Chem Biol 2000; 7:R109-119.
5. Walsh CT, Fisher SL, Park IS et al. Bacterial resistance to vancomycin: five genes and one missing hydrogen bond tell the story. Chem Biol 1996; 3:21-8.
6. Pootoolal J, Neu J, Wright GD. Glycopeptide antibiotic resistance. Annu Rev Pharmacol Toxicol 2002; 42:381-408.
7. Weigel, LM, Clewell DB, Gill SR et al. Genetic analysis of a high-level vancomycin-resistant isolate of Staphylococcus aureus. Science 2003; 28:1569-1571.
8. Pearson, H. 'Superbug' hurdles key drug barrier. Nature 2002; 418:469-470.
9. Chang S, Sievert DM, Hageman JC et al. Infection with vancomycin-resistant Staphylococcus aureus containing the vanA resistance gene. N Engl J Med 2003; 348:1342-1347.
10. Tenover FC, Weigel LM, Appelbaum PC et al. Vancomycin-resistant Staphylococcus aureus isolate from a patient in Pennsylvania. Antimicrob Agents Chemother 2004; 48:275-280.
11. Arthur M, Quintiliani R. Regulation of VanA- and VanB-type glycopeptide resistance in enterococci. Antimicrob Agents Chemother 2001; 45:375-81.
12. Reynolds PE, Courvalin P. Vancomycin resistance in enterococci due to synthesis of precursors terminating in D-alanyl-D-serine. Antimicrob Agents Chemother 2005; 49:21-5.
13. Billot-Klein D, Blanot D, Gutmann L et al. Association constants for the binding of vancomycin and teicoplanin to N-acetyl-D-alanyl-D-alanine and N-acetyl-D-alanyl-D-serine. Biochem J 1994; 304:1021-1022.
14. Davies J. Inactivation of antibiotics and the dissemination of resistance genes. Science 1994; 264:375-382.
15. Pootoolal J, Thomas MG, Marshall CG et al. Assembling the glycopeptide antibiotic scaffold: the biosynthesis of A47934 from Streptomyces toyocaensis. Proc Natl Acad Sci USA 2002; 99:8962-8967.
16. Sosio M, Kloosterman H, Bianchi A et al. Organization of the teicoplanin gene cluster in Actinoplanes teichomyceticus. Microbiology 2004; 150:95-102.
17. Serina S, Radice F, Maffioli S et al. Glycopeptide resistance determinants from the teicoplanin producer Actinoplanes teichomyceticus. FEMS Microbiol Lett 2004; 240:69-74.
18. Beltrametti F, Consolandi A, Carrano L et al. Resistance to glycopeptide antibiotics in the teicoplanin producer is mediated by van-gene homologue expression directing the synthesis of a modified cell wall peptidoglycan. Antimicrob Agents Chemother 2007; 51:1135-41.
19. Arthur M, Depardieu F, Cabanie L et al. Requirement of the VanY and VanX D,D-peptidases for glycopeptide resistance in enterococci. Mol Microbiol 1998; 30:819-30.
20. Wright GD, Molinas C, Arthur M et al. Characterization of VanY, a DD-carboxypeptidase from vancomycin-resistant Enterococcus faecium BM4147. Antimicrob Agents Chemother 1992; 36:1514-8.
21. Arthur M, Depardieu F, Molinas C et al. The vanZ gene of Tn1546 from Enterococcus faecium BM4147 confers resistance to teicoplanin. Gene 1995; 154:87-92.
22. Hong H-J, Hutchings MI, Neu JM et al. Characterisation of an inducible vancomycin resistance system in Streptomyces coelicolor reveals a novel gene (vanK) required for drug resistance. Mol Microbiol 2004; 52:1107-1121.
23. Hong H-J, Hutchings MI, Hill LM et al. The role of the novel Fem protein VanK in vancomycin resistance in Streptomyces coelicolor. J Biol Chem 2005; 280:13055-13061.
24. Wright GD, Holman TR, Walsh CT. Purification and characterization of VanR and the cytosolic domain of VanS: a two-component regulatory system required for vancomycin resistance in Enterococcus faecium BM4147. Biochemistry 1993; 32:5057-63.
25. Depardieu F, Courvalin P, Msadek T. A six amino acid deletion, partially overlapping the $VanS_B$ G2 ATP-binding motif, leads to constitutive glycopeptide resistance in VanB-type Enterococcus faecium. Mol Microbiol 2003; 50:1069-1083.
26. Hutchings MI, Hong H-J, Buttner MJ. The vancomycin resistance VanS/VanR two-component signal transduction system of Streptomyces coelicolor. Mol Microbiol 2006; 59:923-935.

27. Holman TR, Wu Z, Wanner BL et al. Identification of the DNA-binding site for the phosphorylated VanR protein required for vancomycin resistance in Enterococcus faecium Biochemistry 1994; 33:4625-31.
28. Depardieu F, Courvalin P, Kolb A. Binding sites of $VanR_B$ and $\sigma^{70}$ RNA polymerase in the vanB vancomycin resistance operon of Enterococcus faecium BM4524. Mol Microbiol 2005; 57:550-64.
29. Arthur M, Depardieu F, Gerbaud G et al. The VanS sensor negatively controls VanR-mediated transcriptional activation of glycopeptide resistance genes of Tn1546 and related elements in the absence of induction. J Bacteriol 1997; 179:97-106.
30. Haldimann A, Fisher SL, Daniels LL et al. Transcriptional regulation of the Enterococcus faecium BM4147 vancomycin resistance gene cluster by the VanS-VanR two-component regulatory system in Escherichia coli K-12. J Bacteriol 1997; 179:5903-5913.
31. Baptista M, Rodrigues P, Depardieu F et al. Single-cell analysis of glycopeptide resistance gene expression in teicoplanin-resistant mutants of a VanB-type Enterococcus faecalis. Mol Microbiol 1999; 32:17-28.
32. Comenge Y, Quintiliani R Jr, Li L et al. The CroRS two component regulatory system is required for intrinsic beta-lactam resistance in Enterococcus faecalis. J Bacteriol 2003; 185:7184-92.
33. Mascher T. Intramembrane-sensing histidine kinases: a new family of bacterial cell envelope stress sensors. FEMS Microbiol Lett 2006; 264:133-144.
34. Ulijasz AT, Grenader A, Weisblum B. A vancomycin-inducible lacZ reporter system in Bacillus subtilis: induction by antibiotics that inhibit cell wall synthesis and by lysozyme. J Bacteriol 1996; 178:6305-6309.
35. Arthur M, Depardieu F, Courvalin P. Regulated interactions between partner and nonpartner sensors and response regulators that control glycopeptide resistance gene expression in enterococci. Microbiology 1999; 145:1849-58.
36. Lai MH, Kirsch DR. Induction signals for vancomycin resistance encoded by the vanA gene cluster in Enterococcus faecium. Antimicrob Agents Chemother 1996; 40:1645-1648.
37. Mani N, Sancheti P, Jiang ZD et al. Screening systems for detecting inhibitors of cell wall transglycosylation in Enterococcus. Cell wall transglycosylation inhibitors in Enterococcus. J Antibiot 1998; 51:471-479.
38. Grissom-Arnold J, Alborn WE, Nicas TI et al. Induction of VanA vancomycin resistance genes in Enterococcus faecalis: use of a promoter fusion to evaluate glycopeptide and nonglycopeptide induction signals. Microbial Drug Resistance 1997; 3:53-64.
39. Baptista M, Depardieu F, Courvalin P et al. Specificity of induction of glycopeptide resistance genes in Enterococcus faecalis. Antimicrob Agents Chemother 1996; 40:2291-2295.
40. Allen NE, Hobbs JN. Induction of vancomycin resistance in Enterococcus faecium by nonglycopeptide antibiotics. FEMS Microbiol Lett 1995; 132:107-114.
41. Handwerger S, Kolokathis A. Induction of vancomycin resistance in Enterococcus faecium by inhibition of transglycosylation. FEMS Microbiol Lett 1990; 58:167-170.
42. van Heijenoort J. Formation of the glycan chains in the synthesis of bacterial peptidoglycan. Glycobiology 2001; 11:25R-36R.
43. Chen L, Walker D, Sun B. et al. Vancomycin analogues active against vanA-resistant strains inhibit bacterial transglycosylase without binding substrate. Proc Natl Acad Sci USA 2003; 100:5658-5663.
44. Baptista M, Depardieu F, Reynolds P et al. Mutations leading to increased levels of resistance to glycopeptide antibiotics in VanB-type enterococci. Mol Microbiol 1997; 25:93-105.
45. Neu JM, Wright GD. Inhibition of sporulation, glycopeptide antibiotic production and resistance in Streptomyces toyocaensis NRRL 15009 by protein kinase inhibitors. FEMS Microbiol Lett 2001; 199:15-20.
46. Dong SD, Oberthur M, Losey HC et al. The structural basis for induction of VanB resistance. J Am Chem Soc 2002; 124:9064-5.
47. Sharman GJ, Try AC, Dancer RJ et al. The roles of dimerization and membrane anchoring in activity of glycopeptide antibiotics against vancomycin-resistant bacteria. J Am Chem Soc 1997; 119:12041-12047.
48. Cooper MA, Williams, DH. Binding of glycopeptide antibiotics to a model of a vancomycin-resistant bacterium. Chem Biol 1999; 6:891-899.
49. Beauregard DA, Williams DH, Gwynn MN et al. Dimerization and membrane anchors in extracellular targeting of vancomycin group antibiotics. Antimicrob Agents Chemother 1995; 39:781-785.
50. Ge M, Chem Z, Onishi HR et al. Vancomycin derivatives that inhibit peptidoglycan biosynthesis without binding D-Ala-D-Ala. Science 1999; 284:507-511.
51. Sinha Roy R, Yang P, Kodali S et al. Direct interaction of a vancomycin derivative with bacterial enzymes involved in cell wall biosynthesis. Chem Biol 2001; 8:1095-1106.
52. Kerns, R. Dong, S.D. Fukuzawa S et al. The Role of hydrophobic substituents in the biological activity of glycopeptide antibiotics J Am Chem Soc 2000; 122:12608-12609.

53. Billot-Klein D, Gutmann L, Sable S et al. Modification of peptidoglycan precursors is a common feature of the low-level vancomycin-resistant VANB-type Enterococcus D366 and of the naturally glycopeptide-resistant species Lactobacillus casei, Pediococcus pentosaceus, Leuconostoc mesenteroides and Enterococcus gallinarum. J Bacteriol 1994; 176:2398-405.
54. Billot-Klein D, Legrand R, Schoot B et al. Peptidoglycan structure of Lactobacillus casei, a species highly resistant to glycopeptide antibiotics. J Bacteriol 1997; 179:6208-12.
55. Handwerger S, Pucci MJ, Volk KJ et al. Vancomycin-resistant Leuconostoc mesenteroides and Lactobacillus casei synthesize cytoplasmic peptidoglycan precursors that terminate in lactate. J Bacteriol 1994; 176:260-4.
56. Nicas TI, Cole CT, Preston DA et al. Activity of glycopeptides against vancomycin-resistant Gram-positive bacteria. Antimicrob Agents Chemother 1989; 33:1477-81.
57. Gutmann L, al-Obeid S, Billot-Klein D et al. Synergy and resistance to synergy between β-lactam antibiotics and glycopeptides against glycopeptide-resistant strains of Enterococcus faecium. Antimicrob Agents Chemother 1994; 38:824-9.
58. Severin A, Wu SW, Tabei K et al. Penicillin-binding protein 2 is essential for expression of high-level vancomycin resistance and cell wall synthesis in vancomycin-resistant Staphylococcus aureus carrying the enterococcal vanA gene complex. Antimicrob Agents Chemother 2004; 48:4566-73.
59. Severin A, Tabei K, Tenover F et al. High level oxacillin and vancomycin resistance and altered cell wall composition in Staphylococcus aureus carrying the staphylococcal mecA and the enterococcal vanA gene complex. J Biol Chem 2004; 279:3398-3407.

# Chapter 15

# Tearing Down the Wall:

## Peptidoglycan Metabolism and the WalK/WalR (YycG/YycF) Essential Two-Component System

**Sarah Dubrac and Tarek Msadek***


## Abstract

In order to survive, bacteria have developed a variety of highly sophisticated and sensitive signal transduction pathways with which they adapt their genetic expression to meet the challenges of their ever-changing surroundings. These mechanisms enable bacterial cells to communicate with their environment, their hosts and each other, allowing them adopt specific responses, or develop specialised structures such as biofilms or spores to ensure survival, colonization of their ecological niches and dissemination. As highlighted in this book, the so-called two-component systems (TCSs) are one of the most widespread and efficient strategies used for this purpose, where signal acquisition involves autophosphorylation of a sensor histidine kinase and transduction takes place when the kinase phosphorylates its cognate response regulator protein, leading in turn to specific alteration of gene expression.

In their simplest form, TCSs elegantly combine sensing, transducing and transcription activation modules within two proteins, effectively coupling external signals to genetic adaptation. The high degree of conservation among TCS phosphotransfer domains, their ubiquitous nature and the fact that several are essential for cell viability has made them an attractive target for novel classes of antimicrobial compounds. The WalK/WalR (aka YycG/YycF) two-component system, originally identified in *Bacillus subtilis*, is very highly conserved and specific to low G + C Gram-positive bacteria, including several pathogens such as *Staphylococcus aureus*. While this system is essential for cell viability, both the nature of its regulon and its physiological role had remained mostly uncharacterized. A number of recent studies have now unveiled a conserved function for this system in different bacteria, defining this signal transduction pathway as a master regulatory system for cell wall metabolism, which we have accordingly renamed WalK/WalR. This review will focus on the cellular function of the WalK/WalR TCS in different bacterial species and the attractive target it constitutes for novel classes of antimicrobial compounds.


## Introduction

Bacterial survival in the environment is as delicate a balancing act as swordplay on a tightrope, where the slightest misstep is fatal. Gene expression must therefore be swiftly and accurately fine-tuned, with each detrimental environmental thrust instantly met by a successful bacterial parry. The uncanny adaptability displayed by microorganisms relies on their expert swordsmanship, where feints and ripostes are replaced by highly efficient signal transduction pathways that can rapidly be brought into play at the appropriate time, allowing them to weather a broad range

*Corresponding Author: Tarek Msadek—Biology of Gram Positive Pathogens, Department of Microbiology, Institut Pasteur, 25, Rue Du Dr. Roux 75724 Paris Cedex 15, France.
Email: tmsadek@pasteur.fr

*Bacterial Signal Transduction: Networks and Drug Targets*, edited by Ryutaro Utsumi.

of inhospitable conditions. As reviewed in this book, two-component systems (TCS) are the most widespread of these extremely sophisticated signaling mechanisms, transducing external signals across the cell membrane to effect specific changes in genetic expression or bacterial behavior. This occurs when the first component, a histidine kinase, usually membrane-bound, receives a signal from the external milieu through its variable input domain. The signal is then transduced to the conserved histidine kinase domain, leading, through interaction of two monomers with ATP, to autophosphorylation of the kinase on a histidine residue.[1] The phosphate moiety is subsequently transferred to an invariant aspartate residue in the conserved receiver domain of the second protein, the response regulator. This entails conformational changes that allow its variable output domain, usually involved in DNA-binding and transcription activation, to become active, accordingly modifying gene expression in response to a specific signal.[2]

As a rule, these signaling pathways are thus typically geared towards adaptation, with few being involved in cell viability. The rare exceptions where a TCS has been reported to be essential include the extensively studied CckA/CtrA, DivJ/DivK and CenK/CenR systems of *Caulobacter crescentus*, involved in cell division, morphological differentiation and DNA replication,[3,4] as well as the MtrB/MtrA TCS of *Mycobacterium tuberculosis* which controls the expression of the essential replication gene *dnaA*.[5,6] By far the most widely distributed essential TCS, ubiquitous among low G + C% Gram-positive bacteria and specific to the Firmicutes phylum, is the WalK/WalR system, (aka YycG/YycF), the only essential TCS described in this clade. Originally identified in *Bacillus subtilis*,[7,8] this system has since been extensively studied and described as essential in several closely related pathogens (*Staphylococcus aureus*, *Streptococcus pneumoniae*, *S. mutans*, *S. pyogenes*), where it has been referred to under various designations (YycG/YycF, VicK/VicR, MicA/MicB).[9-15] Initial attempts to inactivate the *walRK* locus in *Listeria monocytogenes*, *Enterococcus faecalis* and *Staphylococcus epidermidis* were unsuccessful, suggesting that it is also essential in these bacteria although this was not formally demonstrated.[16-18]

While the function of this TCS had remained obscure for the past ten years, a number of recent reports have now revealed a key role for this system in cell wall metabolism.[10,19-22] In an effort to standardize the nomenclature for this system in different bacteria and better reflect its true function, we have recently proposed that it be henceforth referred to as WalK (histidine kinase) and WalR (response regulator).[19,23] This review will focus on the physiological role of the WalKR system and the potential it holds as a target for designing new types of antibiotics.

## *walRK* Operon Structure

Genes encoding the WalKR TCS are located within an operon encompassing between 3 to 6 cistrons among the Bacilli class of Firmicutes, with three types of structural organization (see Fig. 1). Within the Bacillale order, the Bacillaceae and Listeriaceae families have a 6-gene operon (*walR*, *walK*, *yycH*, *yycI*, *yycJ*, *yycK*), whereas only five are found in Staphylococcaceae, where the last gene, *yycK*, encoding an HtrA-like protease, is missing (Fig. 1). Among Lactobacillales, the Enterococcaceae and Lactobacillaceae families also have this conserved five-gene operon, whereas Streptococcaceae (Streptococci, Lactococci) have a further reduced 3-gene operon (*walR*, *walK* and *yycJ*) with the *yycH* and *yycI* genes missing from the genome (Fig. 1). Interestingly, the only other bacterium with a WalKR ortholog is *Clostridium thermocellum*, unique among Firmicutes of the Clostridia class. Both *yycH* and *yycI* are missing from the *C. thermocellum* genome and there is no proximal *yycJ* gene, with the closest ortholog located instead five genes downstream from the *walR* and *walK* genes, further emphasizing the unique architecture of the locus in this bacterium (Fig. 1).

Apart from their orthologs, YycH and YycI share no similarities with other known proteins. Deletion of *yycH* or *yycI* led to overexpression of a WalKR-dependent gene in *B. subtilis*, suggesting that the encoded proteins negatively regulate the activity of this TCS and also affected growth and cell wall integrity.[24,25] YycH and YycI are localized outside the cell, anchored to the membrane by an N-terminal transmembrane domain and bacterial two-hybrid assays indicate that they form a ternary complex together with the WalK kinase.[24,25] Taken together, these results suggest that

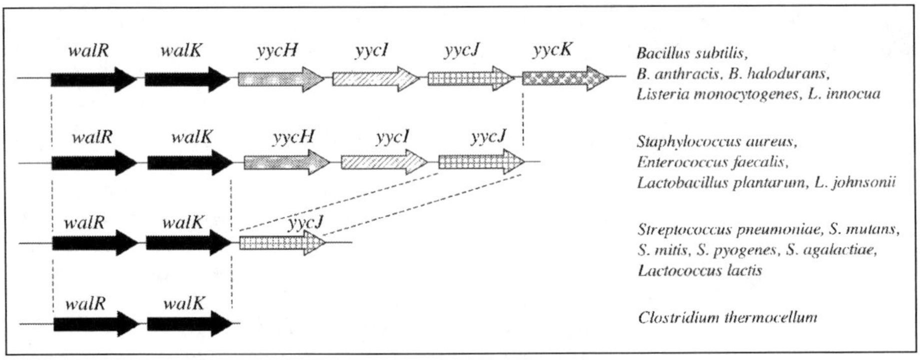


Figure 1. Genetic map of the *walRK* operon displaying the four types of structural organization found in low GC% Gram positive bacteria. Major bacterial species for each type of organization are indicated on the right.

YycH and YycI interact with WalK to negatively control its activity, through their transmembrane or extracellular domains. Although they share no significant amino acid sequence similarities, the crystal structures of YycH and YycI have very closely related structural domains with a common fold, suggesting they originated from a shared ancestor and have since diverged significantly.[26,27]

Except for the so far unique case of *C. thermocellum*, *yycJ* is always cotranscribed with *walR* and *walK*, constituting the minimal *walRK* operon present in Streptococcaceae. Although its function is unknown, Hidden Markov Model (HMM) predictions indicate that YycJ has a domain shared with the metallo-β-lactamase superfamily, containing a predicted HxHxDH zinc or iron-binding site, suggesting a potential role for this protein in cell wall metabolism.[28,29]

Inactivation of *yycJ* in *B. subtilis* or *Streptococcus pneumoniae* did not lead to any significant phenotype under laboratory growth conditions, apart from a modification of colony morphology in *Bacillus*.[8,25,30] In *S. mutans*, inactivation of *yycJ* (aka *vicX*) showed a more pronounced effect. Compared to the wild type isogenic strain, the mutant showed a slight increase in doubling-time, an increase in biofilm formation, drastically compromised natural competence and a large decrease in oxidative stress tolerance.[31] Phenotypes of a *S. mutans* or *S. pneumoniae walK* mutant also included effects on genetic competence development and biofilm formation, strongly suggesting that YycJ also participates in the WalKR signal transduction pathway.[13,14,30] In contrast, *S. aureus* cells lacking *yycJ* did not display any distinguishing phenotypes and cell morphology and oxidative stress resistance were not affected (O. Poupel and S. Dubrac, unpublished results).

The last gene of the operon, *yycK* (aka *yyxA* in *B. subtilis*), appears to be specific to Bacillaceae and Listeriaceae. It encodes a serine protease closely related to HtrA (35% identity with HtrA of *E. coli*).[32] HMM domain prediction suggests that YycK is anchored in the bacterial membrane by a single N-terminal transmembrane segment, followed by trypsin-like catalytic and PDZ protein-protein interaction domains, which are common to all HtrA family members.[33] HtrA proteases are involved in proteolysis of abnormal proteins and have been shown to be required for resistance to oxidative and heat stress.[33] In *Listeria monocytogenes*, YycK (Lmo 0292) was shown to be involved in stress response and pathogenesis in a murine infection model.[34]

## The WalK Histidine Kinase

WalK is a membrane-linked kinase, with a conserved C-terminal cytoplasmic histidine kinase region, including ATP binding and phosphoacceptor domains (Fig. 2). Sequence similarities indicate that WalK phosphorylation occurs on conserved residue H387 (all amino acid sequence coordinates are those of the prototypical *B. subtilis* proteins), although the aminoterminal domain of the WalK histidine kinase displays considerable variation among species (45% sequence identity on average over the first 300 amino acids). A PAS (PER-ARNT-SIM) sensor domain is also present in the

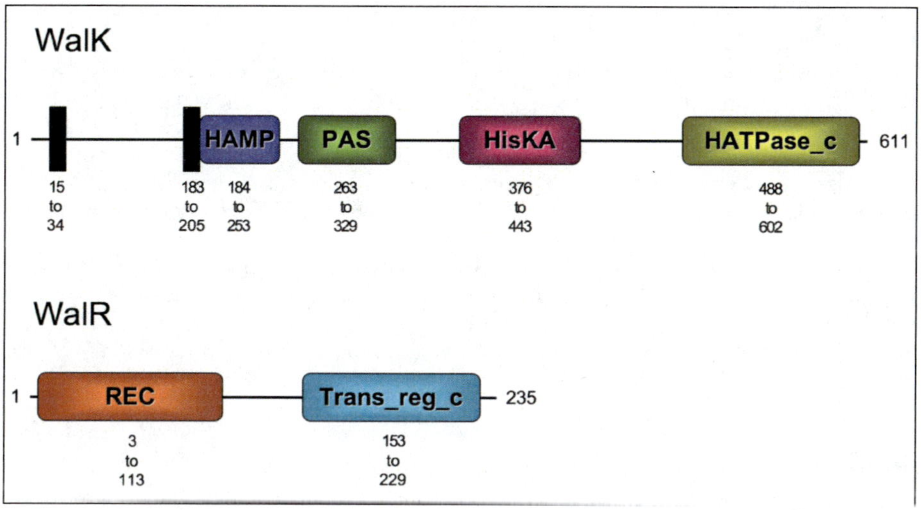


Figure 2. Domain architecture of the WalK histidine kinase and its cognate WalR regulator. Coordinates for each domain are indicated based on the WalK and WalR proteins of *Bacillus subtilis*. The graphical representation is derived from data from the SMART web interface at http://smart.embl-heidelberg.de

cytoplasmic part of the protein. PAS domains have been described as structures that adopt typical α/β folds and are involved in sensing signals such as oxygen, light, redox potential or the presence of specific ligands.[35] Although infrequent, these domains are found in all kingdoms of life. It is interesting to note that in *S. aureus* the only two proteins with PAS domains are the PhoR and WalK histidine kinases, whereas in *S. pneumoniae* WalK is the only protein containing a PAS domain.[36] The role of this domain in WalK is still not clear, but in *S. pneumoniae* it has been reported to be necessary for repression of competence development by WalKR under microaerobiosis,[13] although it is not required for autophosphorylation of the cytoplasmic WalK kinase domain in vitro.[37]

Several members of the histidine protein kinase family are also known to act as phosphoprotein phosphatases, catalyzing the dephosphorylation of the associated response regulator. Regulation by environmental signals may therefore take place by varying either the kinase or the phosphatase activity of the sensor. Accumulation of the phosphorylated form of the response regulator is thus the end result of a series of reversible enzymatic reactions: autophosphorylation of the kinase, autophosphatase activity of the kinase, phosphotransfer to the regulator, backtransfer to the kinase, autophosphatase activity of the regulator and phosphoprotein phosphatase activity of the kinase towards the regulator. The time-scale response of the signal transduction system can thus be modulated through the combined kinase/phosphatase activities of the histidine kinases controlling the amount of the phosphorylated form of the response regulator present in the cell in response to environmental signals. In many cases, regulation occurs essentially through the control of the phosphatase activity of the kinase, such as the VanS/VanR system.[38] Phosphatase activity is particularly critical in the case of response regulators whose phosphorylated form is highly stable, to ensure that the protein is not permanently activated. It is thus interesting to note that a structure- and mathematical modeling-based comparative analysis of TCSs led to the prediction that YycG should act as a so-called bifunctional sensor, i.e., one endowed with both kinase and phosphatase activities,[39] although this has yet to be demonstrated.

The cytoplasmic domains of WalK from *B. subtilis*, *S. aureus* and *S. pneumoniae* have been purified, shown to be autophosphorylated and to phosphorylate their cognate response regulators in vitro.[10,13,40] In addition, noncognate phosphotransfer was shown between the WalKR systems of *S. aureus* and *S. pneumoniae*, underlining how closely related the systems are between species even though the primary structures of the kinases are not highly conserved (45% amino acid sequence identity between WalK of *S. aureus* and *S. pneumoniae*).[37]

Despite the presence of conserved domains, there are clear structural differences among WalK orthologs, namely the number of transmembrane domains and the presence and length of the extra-cytoplasmic loop. Most WalK orthologs contain a large extra-cytoplasmic loop, 142 to 154 amino acids long, flanked by two transmembrane domains. One exception is represented by WalK of *Lactococcus lactis,* where the loop is only 4 amino acids long.[41] Furthermore, WalK orthologs in Streptococci lack an extracellular loop and contain a single transmembrane domain, with only 4 to 12 amino acids protruding outside the cytoplasm.[29]

The extra-cytoplasmic loop is known to play a crucial role in detecting the external stimulus in several histidine kinases, while the transmembrane domains function as an anchor to the cytoplasmic membrane, but can also be involved in signal perception.[42] Structural differences among WalK orthologs are thus likely to reflect a variety of signaling mechanisms: WalK might sense different signals in various organisms, according to their specific metabolism, physiology and ecological niche.

## The WalR Response Regulator

WalR is a typical response regulator with an N-terminal receiver domain and a C-terminal DNA binding domain (Fig. 2). The receiver domain has the invariant active site residues, including a conserved aspartate residue (D54), which is the likely phosphorylation site (Fig. 3).[7] In contrast to the lower conservation observed for WalK, the WalR primary structure is very highly conserved (more than 70% amino acid sequence identity with WalR of *B. subtilis* on average, with 60 to 70% identity for Streptococci orthologs) (see Fig. 3).

As is typical for TCS response regulators, the structural organization of WalR is bipartite. The N-terminal part (positions 3 to 113, Figs. 2,3) is involved in signal reception from the kinase, whereas the C-terminal domain (positions 153 to 229, Figs. 2,3) is responsible for DNA-binding and activation of gene transcription. WalR belongs to the OmpR/PhoB subfamily of response regulators and as such contains a characteristic winged helix-turn-helix DNA-binding domain.[43,44] Response regulators of the OmpR/PhoB subfamily control transcription by binding as dimers to directly repeated DNA sequences. The crystal structure of the PhoB-DNA complex reveals that the two alpha helices are in contact with DNA (the first, a structural helix, for positioning in the major groove of the DNA helix and the second, the DNA recognition helix, involved in specific binding to the target site) whereas the alpha loop separating the two helices is responsible for interactions of PhoB and OmpR with the sigma and alpha subunits of RNA polymerase, respectively.[45-47] The crystal structure of the C-terminal domain of WalR from *Enterococcus faecalis* has been solved recently,[48] suggesting a role for the two helices in contacting DNA and the alpha loop in interactions with RNA polymerase (see Fig. 3) and indicating that the alpha-loop is similar to that of PhoB but differs from that of OmpR. The loop following the helix-turn-helix motif on the other hand is more closely related to that of OmpR than PhoB.[48]

## A Matter of Life and Death: To Be or Not to Be Essential

As mentioned above, the WalKR TCS has been shown or is thought to be essential in several bacteria including *B. subtilis, S. aureus, E. faecalis, L. monocytogenes* and *S. pneumoniae*.[7-12,16,17] While it has been clearly shown that both the WalK kinase and the WalR regulator are essential in *B. subtilis* and *S. aureus*, the situation differs in *S. pneumoniae* since the regulator is essential but not the kinase.[8,9,11] Although *walK* of *S. pneumoniae* can be disrupted, the growth rate of the resulting strain was lowered to about 30% of that of the parental strain.[30] Likewise, growth of a *walK* inactivation mutant of *Lactococcus lactis* was reported to be impaired, with cells displaying clumping.[41] Recent in vitro data have shown that acetyl phosphate is a phosphoryl group donor for *S. pneumoniae* WalR.[49] Acetyl phosphate is an important signaling molecule in many bacteria and its capacity to phosphorylate WalR could partly account for the fact that the cognate WalK histidine kinase is not essential in *S. pneumoniae*. However, interactions between WalR and noncognate histidine kinases

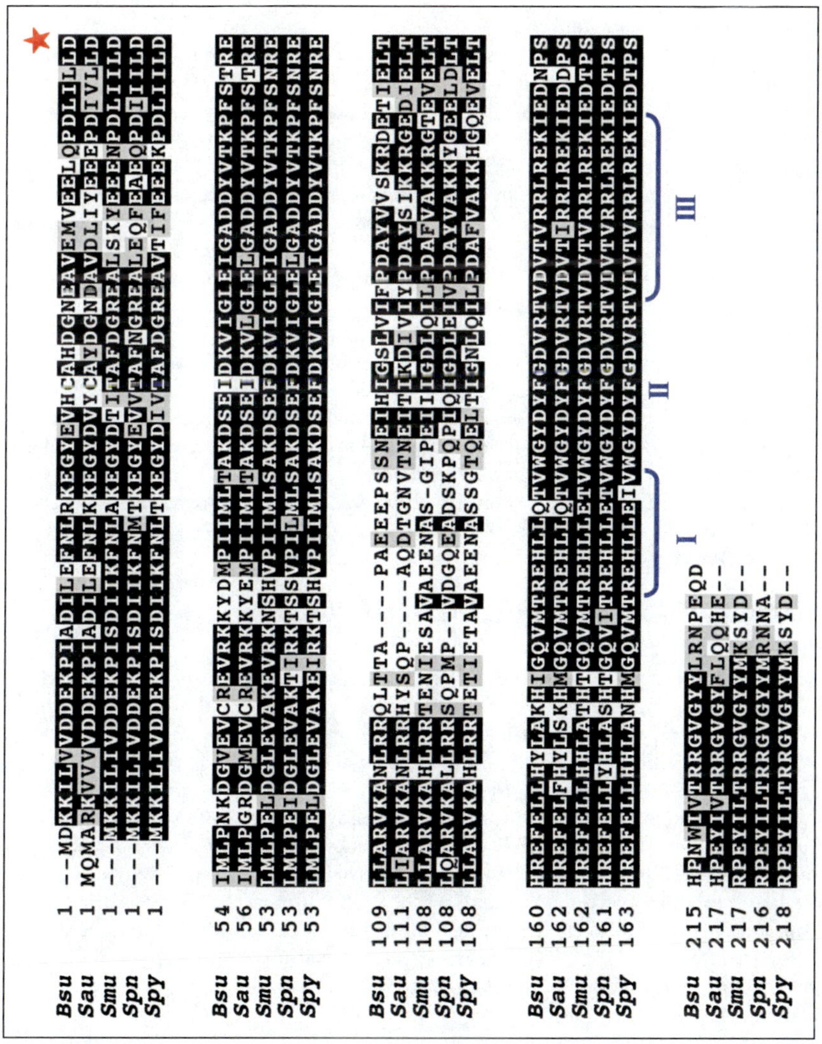


Figure 3. Clustal W[72] alignment of WalR orthologs from *Bacillus subtilis (Bsu)*, *Staphylococcus aureus (Sau)*, *Streptococcus mutans (Smu)*, *Streptococcus pneumoniae (Spn)* and *Streptococcus pyogenes (Spy)*. The star indicates the phosphorylated aspartate residue. The three segments of the DNA binding domain as defined by Trinh et al[48] are indicated: I: positioning helix; II: transactivation loop; III: DNA binding helix.

in *S. pneumoniae* cannot be ruled out since it has been shown that it can be phosphorylated by the heterologous VanS histidine kinase from *Enterococcus faecalis* in vitro.[30] In *B. subtilis*, the *walK* gene can only be inactivated when the strain contains a point mutation in *walR*, leading to a constitutively active form of the response regulator (WalRD54H).[7] Interestingly, purified WalR of *B. subtilis* or *S. aureus* could not be phosphorylated using radiolabeled acetyl phosphate (S. Dubrac, unpublished results). It is also worth noting that the WalKR system is one of the rare examples where physiologically relevant in vivo crosstalk has been demonstrated, since it has been shown to interact with the PhoPR system in *B. subtilis*, where the PhoR kinase is able to phosphorylate the noncognate WalR response regulator upon phosphate limitation.[50]

In *S. pyogenes* (group A *Streptococcus*), an insertional mutant of *walR* was reported to be viable, with no major growth defect in rich medium except a higher rate of mortality in stationary phase.[15] While the authors showed that the strain carrying the insertional *walR* mutation did not product the WalR regulator, they failed to obtain a *walR* deletion mutant. In *S. mutans*, two contradictory reports have appeared regarding the essential nature of WalR. In 2004, Lee et al claimed they had inactivated *walR*, which was erroneously referred to as *covR*.[51] A later paper published in 2005 by Senadheera et al reported that inactivation of *walR* in *S. mutans* was lethal.[14] These conflicting reports do not allow one to firmly conclude that the WalKR TCS is not essential in *S. pyogenes* or *S. mutans*. To date, the only organism in which the WalR regulator has been reported not to be essential is *Lactococcus lactis*, where an inactivation mutant of the gene was described several years ago.[41]

There is an interesting correlation between the type of *walKR* operon organization and the features of the WalK histidine kinases. Bacterial species that encode YycH and YycI orthologs have an essential WalK kinase, with two transmembrane domains flanking an extra-cytoplasmic loop, while in species that lack these two proteins the WalK kinase is not essential and has a single transmembrane domain.[29] The exception again is *Lactococcus lactis,* which lacks the *yycH* and *yycI* genes but encodes a WalK kinase with two transmembrane domains. However in this species both WalR and WalK are reported to be dispensable and the extracytoplasmic loop of the kinase is unusually short, containing only 4 amino acids, as mentioned above.[41]

## Global Analyses of WalKR-Regulated Genes Reveal a Major Role in Cell Wall Homeostasis

*Bacillus subtilis* cell morphology is altered under conditions of depletion or overproduction of the WalR regulator, leading respectively to very long cells or mini cells.[7] This allowed the first identification of a WalKR-regulated gene, the *ftsAZ* operon involved in cell division, though the system only controls expression from a secondary non-essential promoter.[7,40] Expression profiling experiments carried out in *Bacillus subtilis* using either a hybrid regulator approach involving a chimeric PhoP'-'WalR protein or a *walRK* inducible strain under conditions of WalKR depletion allowed identification of several other WalKR-dependent genes. Several appear to be involved in cell wall metabolism, namely *yocH*, encoding a potential autolysin, *lytE* and *yvcE*, encoding two endopeptidase-like autolysins involved in cell wall synthesis and turn over respectively, the *tagAB* and *tagDEF* operons, involved in teichoic acid biosynthesis and *yjeA*, encoding a putative peptidoglycan deacetylase (see Table 1).[20,40,50] The purified WalR regulator was shown to bind directly to the promoter regions of these genes, indicating that WalR activates or represses expression of these genes directly.[20,40,50]

The WalR regulators of *B. subtilis* and *S. aureus* are highly similar, sharing 75% amino acid sequence identity, particularly in the winged helix-turn-helix DNA-binding domains with 30 identical amino acids out of 31 (Fig. 3 and next section). This allowed the identification of several potential WalKR-regulated genes in *S. aureus* by genome mining, using the previously established consensus WalR recognition sequence (5'-TGTWAH-$N_5$-TGTWAH-3').[10] Among these, transcriptional analysis showed that WalKR activates the expression of nine cell wall metabolism genes, encoding the two major *S. aureus* autolysins (AtlA and LytM), two proteins with a transglycosylase domain (IsaA and SceD) and five proteins with a CHAP amidase domain (SsaA, SA0620, SA2097,

***Table 1.** Genes involved in cell wall metabolism, fatty acid metabolism and virulence shown to be WalKR-regulated in* **Bacillus subtilis, Staphylococcus aureus** *or* **Streptococcus pneumoniae**

| | *Bacillus subtilis* | | | *Staphylococcus aureus* | | | *Streptococcus pneumoniae* | | |
|---|---|---|---|---|---|---|---|---|---|
| | **Regulated Genes** | **Direct Effect** | **Reference** | **Regulated Genes** | **Direct Effect** | **Reference** | **Regulated Genes** | **Direct Effect** | **Reference** |
| **Cell wall metabolism** | *yocH* (+) | Yes | 40 | *atlA* (+) | ND | 19 | *pcsB* (+) | Yes | 21,49 |
| | *lytE* (+) | Yes | 20 | *lytM* (+) | Yes | 10,19 | *lytB* (+) | ND | 21 |
| | *yvcE* (+) | Yes | 20 | *isaA* (+) | Yes | 10,19 | | | |
| | *tagAB* (–) | Yes | 40,50 | *sceD* (+) | ND | 19 | | | |
| | *tagDEF* (–) | Yes | 40,50 | *ssaA* (+) | Yes | 10,19 | | | |
| | *yjeA* (–) | Yes | 20 | *SA0620* (+) | ND | 19 | | | |
| | *ykvT* (ND) | Yes | 40 | *SA2097* (+) | ND | 19 | | | |
| | | | | *SA2353* (+) | ND | 19 | | | |
| | | | | *SA0710* (+) | ND | 19 | | | |
| **Fatty acid metabolism** | *des* (–) | No | 20 | | | | *fabK* op. (+) | ND | 54 |
| | | | | *ebpS* (+) | ND | (S. Dubrac, unpublished data) | *pspA* (+) | Yes | 21,49 |
| **Virulence** | | | | *sdrD* (+) | ND | (S. Dubrac, unpublished data) | | | |

The effect of WalKR on gene expression is indicated under the name of the gene: (+) for activated gene expression, (–) for repressed genes and (ND) when the effect of the WalKR system has not been determined other than in vitro. Direct effects of WalKR on gene expression were shown by in vitro protein/DNA interaction assays, ND: not determined.

SA2353, SA0710) (see Table 1).[19] IsaA and SceD have both been shown to have autolytic activity.[52] WalR of *S. aureus* was shown to bind directly to the promoter regions of *ssaA*, *isaA* and *lytM*.[10] In agreement with its major role in controlling peptidoglycan metabolism, a reduced rate of cell wall biosynthesis and turnover was measured in WalKR-depleted cells.[19]

As shown for *B. subtilis* and *S. aureus*, the WalKR TCS also regulates cell wall metabolism in *S. pneumoniae* and *S. mutans*. Transcriptome analyses in *S. pneumoniae* unveiled WalKR-mediated activation of two genes encoding cell wall hydrolases: *pcsB* and *lytB*, as well as several genes encoding proteins with LysM cell wall binding domains.[21] Depletion for either WalKR or PcsB in *S. pneumoniae* was shown to cause defects in cell morphology and murein synthesis.[21,22] The *pcsB* gene has been shown to be essential in *S. pneumoniae* and constitutive expression of *pcsB* can suppress the WalKR requirement for viability.[21] This is the only instance so far where essentiality of the WalKR TCS can be assigned to regulation of a single essential gene. In *S. mutans*, the WalKR system has been shown to be involved in controlling expression of the *atlA* gene, as well as autolysis.[53]

While the major role of the WalKR TCS is clearly linked to the regulation of cell wall metabolism, expression profiling studies in *B. subtilis* and *S. pneumoniae* have also shown a role in regulating fatty acid metabolism. In *S. pneumoniae* expression of a cluster of 12 genes involved in fatty acid biosynthesis was significantly altered by WalKR levels in the cell, resulting in increased fatty acid chain lengths and in lower cell membrane fluidity (Table 1).[54] However, this effect is likely indirect since the *S. pneumoniae* WalR regulator does not bind to the *fabR* promoter region.[49] In *B. subtilis*, expression of the *des* gene encoding fatty acid desaturase appeared to be indirectly repressed by WalKR.[20] Expression of the *des* gene is induced in response to reduced cell membrane fluidity via the DesKR TCS.[55] Up-regulation of *des* upon depletion for WalKR could indicate membrane fatty acid perturbation, which could be caused by misregulation of cell membrane metabolism.

Several transcriptome analyses have shown that many *S. pneumoniae* stress-induced genes are activated in a WalKR depleted strain, probably reflecting the cell stress this causes.[21] In *S. mutans* the system has also been reported to be involved in resistance to oxidative stress, in agreement with the presence of a PAS domain in WalK.[53,56] It is interesting to note that except in *S. pneumoniae*, where essentiality of the WalKR TCS is linked to expression of the essential *pcsB* gene, none of the genes identified as regulated by the WalKR TCS in *B. subtilis* or *S. aureus* appear to be required for cell viability. This suggests that the essential nature of this system would be multifactorial in these cases, involving expression of several genes, none of which are essential on their own.

## *walRK* Operon Expression

In *B. subtilis* a polycistronic transcript covering the six genes of the operon (*walRKyycHIJK*), as well as a shorter transcript corresponding to *walR*, were detected during early exponential growth phase, suggesting that *walRK* operon transcription is driven by a typical $\sigma^A$-type promoter.[7] In *Bacillus subtilis*, although *yycK* was shown to be cotranscribed with the *walRKyycHIJ* genes, it is also transcribed independently during sporulation from a likely $\sigma^G$-dependent promoter, suggesting it may play a role in this developmental process, although inactivation of *yycK* did not affect spore formation.[7,8] Transcriptional *lacZ* fusions with the *walRK* promoter region showed expression during early exponential phase and a rapid shut off at the transition phase,[8] although Western blot analyses indicate that WalK levels do not vary throughout the growth curve in *B. subtilis*.[24,50] Lowered *walRK* transcription does not correspond to diminished WalK protein levels, most likely due to slower cell division upon entry into stationary phase, allowing accumulation of the protein within the cells. However, the expression pattern of several WalKR dependent genes mirrors that of the *walRK* operon, suggesting the system is only active and required during exponential growth. Indeed, in *B. subtilis* expression of several members of the WalKR regulon occurs mainly during the exponential growth phase and is shut off as cells enter stationary phase, including *yocH*, *ykvT*, the *tagDA* divergon as well as the *ydjM* and *yvcE* genes.[20,40] This is also the case in *Staphylococcus aureus* where the expression profiles of the WalKR dependent *isaA* and *lytM* genes follow the same pattern as well, again suggesting that the WalKR system might also be inactive during the stationary growth phase.[10,19,57,58] In *S. mutans*, *walR* expression was reported to be induced in the presence of xylitol.[59]

Little is known about *walRK* operon transcription regulation. In *B. subtilis*, in vitro protein/DNA interaction assays have shown that WalR does not bind to its own promoter region, which does not contain a potential WalR binding site (S. Dubrac, unpublished data) and expression of the *walRK* operon is not regulated by WalKR levels in the cell.[8] In *S. pneumoniae*, two studies based on transcriptome data led to contradictory conclusions about autoregulation,[49,54] yet in vitro gel mobility shift assays failed to detect any interactions between purified WalR of *S. pneumoniae* and the upstream region of the *walRK* operon[49] strongly suggesting that the WalKR system does not regulate its own expression.

## DNA Sequence Targeted by the WalR Regulator

As mentioned earlier, the WalR regulator belongs to the OmpR/PhoB subfamily of response regulators that are classified by their winged helix-turn-helix DNA-binding domains.[43,44] Response regulators of this subfamily control gene transcription by binding as dimers to directly repeated DNA sequences.[46] In vitro interactions using DNase I footprinting between the purified *B. subtilis* WalR regulator and the promoter regions of two members of the WalKR regulon, *ftsAZ* and *yocH*, allowed the identification of a conserved directly repeated hexanucleotide DNA sequence targeted by the WalR regulator: 5'-TGTWAH-$N_5$-TGTWAH-3'.[10,40] This sequence has been found in the upstream regions of all WalKR-regulated genes in *B. subtilis* and *S. aureus* and a derived consensus albeit slightly modified has also been defined for WalKR-regulated genes in *S. pneumoniae* (5'-TGTNAN-N5-NGTNANA-3').[10,40,49] Site-directed mutagenesis of the WalR target sequence followed by in vitro binding assays and in vivo transcriptional studies confirm the role of this DNA sequence in WalKR-mediated regulation with some exceptions, especially in *S. pneumoniae* where it appears that some genes preceded by this sequence are not regulated by the WalKR TCS and some WalKR-regulated genes do not have this sequence in their upstream region.[10,40,49] While the WalR target sequence needs to be defined more precisely, the characterized direct repeat consensus sequence has already proven to be invaluable as a genome-mining tool to identify WalKR-regulated genes.[19]

## Phenotypes Associated with a Defect in WalKR Activity

Many WalKR associated phenotypes have been described since this TCS was first identified, but the increased interest this system has received over the last few years has revealed the molecular basis for most of these phenotypes.

As mentioned above, since the WalKR TCS is essential to bacterial survival, phenotypes associated with a WalKR defect are usually studied by comparing conditions where either the expression of the system is modulated by an inducible promoter or the activity of the system is controlled by a thermosensitive mutation of the WalR regulator or, in the case of streptococcal species, by inactivating the *walK* gene encoding the sensor protein. In *B. subtilis* as well as in *S. pneumoniae*, WalKR depletion leads to abnormal cell morphology, i.e., large empty cells, that has been correlated to aberrant bacterial division.[7,8,21,22] In *B. subtilis* WalKR-dependent up-regulation of *ftsAZ* expression is thought to be responsible for minicell formation when the system is overexpressed,[7] but in *S. pneumoniae* expression of the *ftsAZ* ortholog does not appear to be regulated by WalKR. Instead, WalKR-related deficiency of bacterial division is linked to expression of *pcsB*, encoding a cell wall amidase involved in the dynamics of cell wall metabolism, a crucial event during cell septation.[22] Several WalKR-associated phenotypes can be correlated to regulation of cell wall metabolism genes, which now appears to be the main function controlled by the WalKR system (see Table 1). In *B. subtilis*, it has recently been shown that cell walls extracted from WalKR depleted cells had a much lower autodigestive capacity than those extracted from WalKR replete cells.[20] Likewise, *S. aureus* WalKR depleted cells display increased resistance to Triton X-100 induced lysis.[19] WalKR-dependent activation of autolytic activities is directly correlated to transcriptional regulation of genes involved in cell wall degradation.[19,20] Several recent reports describe the impact of the WalKR TCS on antibiotic resistance. The first *walR* conditional mutant isolated in *S. aureus* was reported to be more susceptible to macrolides and lincosamides.[9] This hypersensitive phenotype was reversed by overexpression of the *ssaA* gene, which has been shown to be activated by the WalKR system.[10,19,60] However, the link between the *ssaA* gene, which encodes a protein with a CHAP-type amidase domain, suggesting a role in cell wall degradation and resistance to macrolides and lincosamides is still not clear.

It has also been reported that the WalKR system could be involved in resistance to daptomycin, a lipopeptide antibiotic, since daptomycin *S. aureus* resistant strains, either selected in vitro or clinical isolates, were shown to have mutations in the *walK* gene.[61] Considering the the major role of WalKR in controlling cell wall metabolism, an impact on vancomycin resistance in *S. aureus* could be predicted. Two recent publications report a link between the WalKR system and vancomycin intermediate resistance (VISA strains). It has been shown that in a VISA clinical isolate, an insertion in the promoter of the *walKR* operon resulted in up-regulation of its transcription (through generation of an hybrid promoter which is much more efficient than the natural *walKR* promoter).[62] Whole-genome sequencing of *S. aureus* isogenic clinical isolates with increasing MICs of vancomycin allowed the identification of a mutation in the *yycH* gene,[63] whose product has been shown to down regulate the activity of the WalKR system.[25] These data indicate that changes in WalKR activity levels (either by regulation of its synthesis or by posttranslational modulation of its activity) could be directly correlated to the level of vancomycin susceptibility, probably through regulation of cell wall metabolism.

## Impact of the WalKR System on Virulence

A link between expression of genes encoding virulence factors and the WalKR TCS has been established in both *S. aureus* and *S. pneumoniae*. Most of the available data concerning the impact of WalKR on host infection and virulence concerns the streptococcal species in which the kinase of the system is not essential and *walK* mutants can be easily used to test virulence. While the data concerning *S. pneumoniae* are the most abundant, contradictory results have been published, perhaps reflecting differences in the infection model and the *S. pneumoniae* capsule serotype tested. In an R6 strain background, inactivation of *walK* has no effect on virulence both after intraperitoneal challenge and murine respiratory tract infection model, yet an intranasal challenge of mice with a *walK* mutant of two other serotypes (2 and 6B) was asymptomatic and led to lower recovery of CFUs from lungs and blood compared to the parental strains.[11,30,64] At the molecular level, it was shown that WalKR controlled transcription of genes playing a major role in *S. pneumoniae* virulence: *pspA* and *piaBCDA*, respectively encoding a surface virulence factor involved in the evasion of complement during infection and an ABC transporter involved in iron uptake and required for full virulence.[21,49,54,65-67] This regulation may in part explain the decreased virulence of a *walK* mutant strain, but more work remains to be done to clarify the impact of the WalKR system in *S. pneumoniae* virulence.

In group A *Streptococcus*, a *walR* mutant has been shown to be impaired in its capacity to grow in human blood and to kill mice after subcutaneous infection.[15] Molecular mechanisms responsible for this effect on virulence have yet to be determined.

*S. mutans* is the major agent involved in dental caries, producing sucrose-metabolizing enzymes that are critical in cariogenesis. These enzymes include three glucosyltransferases (encoded by *gtfBCD*) and one fructosyltransferase (encoded by *ftf*) that catalyze the cleavage of sucrose to synthesize glucan and fructan polymers which are both involved in *S. mutans* virulence as they promote its capacity to attach to the tooth surface.[68] It has been shown that expression of *gtfBCD*, *ftf* and *gbpB*, encoding a glucan binding protein, are upregulated by the WalKR system and this has been correlated to a positive control of *S. mutans* sucrose-mediated adherence by WalKR.[14] Biofilm formation by *S. mutans* requires WalK and expression of *atlA*, encoding an autolysin, was also shown to be WalKR dependent.[14,53]

Similarly, in *S. aureus*, it has been shown that the WalKR system promotes biofilm formation, an important factor influencing infection propagation.[19] Additionally, transcriptional WalKR-dependent activation has been shown for two genes potentially involved in *S. aureus* virulence (S. Dubrac, unpublished results). These genes encode SdrD, a sialoprotein-binding protein and EbpS, an elastin-binding protein and are thought to act through interactions with the extracellular host matrix suggesting that the WalKR TCS could be directly involved in *S. aureus* pathogenesis.[10] Furthermore, a *S. aureus isaA sceD* mutant is attenuated for virulence, while SceD is essential for nasal colonization in cotton rats.[52] Both of these genes are controlled by WalKR, providing further evidence for an important role of this system in controlling virulence of *S. aureus*.[19]

While more work is required in order to elucidate the impact of the WalKR system in vivo, the available data strongly suggest that fine-tuning of WalKR activity during the course of infection could be crucial for bacterial propagation and persistence in the host.

## The WalKR TCS as a Target for Antimicrobial Therapy: A Bacterial Achille's Heel?

The emergence of multiple drug-resistant bacterial infections is an escalating problem facing public health. Until recently, antimicrobial molecules targeted relatively few bacterial cellular functions, including cell wall integrity or nucleic acid and protein synthesis. As a consequence, acquired resistance to one antibiotic often leads to a multi-resistance pattern. Given the urgent need to find novel antibiotic targets, the WalKR TCS has appeared as a promising candidate since it is essential for bacterial survival, very well conserved and specific to low G + C% Gram-positive bacteria, many of which are known to develop wide drug-resistance patterns. Inhibitors directed against WalKR would be expected to have a bactericidal effect and to be active against a broad range of major Gram-positive pathogens. A study published in 2001 reports the use of imidazole and zerumbone derivatives as efficient inhibitors of *B. subtilis* WalK autophosphorylation in vitro, as well as growth inhibition of *B. subtilis* by these compounds.[69] But this study did not conclusively demonstrate a direct relationship between cell death and WalKR inhibition rather than some other uncharacterized and nonspecific molecular effect. Such nonspecific effects have been particularly well described for imidazole derivatives that cause disruption of cytoplasmic membranes of bacterial and mammalian cells.[70]

A more promising study based on structure-based virtual screening (SBVS) of chemical molecules potentially targeting the kinase activity of the WalK kinase of *S. epidermidis* was recently published.[18] Among the tested drugs the authors found that several are active either on planktonic cultures or biofilms of *S. epidermidis* but also against other bacteria such as *S. aureus*, *S. pyogenes* and *S. mutans*. Since these drugs appear to be inactive against mammalian erythrocytes, they are considered as lead compounds for developing new reagents against staphylococcal infections. While the two preceding reports relied on the selection of molecules that inhibit WalK autophosphorylation, another drug discovery system has been developed, targeting the dimerization of the WalK kinase, a step essential for its activity. This elegant strategy used a fusion between WalK and the DNA binding domain of IclR, an *E. coli* dimeric repressor and *egfp* as a reporter gene, allowing high throughput screening of molecules targeting dimerization of the hybrid protein.[71] Unfortunately, this system did not allow the characterization of promising new chemical compounds, probably because the screening was done randomly with a library not well adapted for this purpose. But the combination of the use of this system with previous structure-based predictions[18] could be very efficient in identifying new therapeutic molecules.

As mentioned above, until now, the WalKR activating signal has not been characterized and this topic is the subject of intense scrutiny. Beyond the importance of characterizing this signal to understand why the WalKR system is so vital for cell viability, the chemical structure of the signal will likely provide indications about the type of molecules that could specifically bind the receptor domain of the WalK kinase without activating its autophosphorylation. Another unexplored possibility would consist in finding molecules able to constitutively activate this inducible system, which could also lead to a growth defect during host infection. It is clear that the coming years will continue to witness growing interest in this fascinating system.

### *Acknowledgements*

Work in the authors' laboratory was supported by research funds from the European Commission (Grants BACELL Health, LSHG-CT-2004-503468, StaphDynamics, LHSM-CT-2006-019064 and BaSysBio, LSHG-CT-2006-037469), the Centre National de la Recherche Scientifique (CNRS URA 2172) and the Institut Pasteur (Grand Programme Horizontal N° 9).

## References


1. Inouye M, Dutta R, eds. Histidine kinases in signal transduction. San Diego: Academic Press; 2003.
2. Hoch JA, Silhavy TJ. Two-component signal transduction. Washington, DC: ASM Press; 1995.
3. Ausmees N, Jacobs-Wagner C. Spatial and temporal control of differentiation and cell cycle progression in Caulobacter crescentus. Annu Rev Microbiol 2003; 57:225-247.
4. Skerker JM, Prasol MS, Perchuk BS et al. Two-component signal transduction pathways regulating growth and cell cycle progression in a bacterium: a system-level analysis. PLoS Biol 2005; 3:e334.
5. Zahrt TC, Deretic V. An essential two-component signal transduction system in Mycobacterium tuberculosis. J Bacteriol 2000; 182:3832-3838.
6. Fol M, Chauhan A, Nair NK et al. Modulation of Mycobacterium tuberculosis proliferation by MtrA, an essential two-component response regulator. Mol Microbiol 2006; 60:643-657.
7. Fukuchi K, Kasahara Y, Asai K et al. The essential two-component regulatory system encoded by yycF and yycG modulates expression of the ftsAZ operon in Bacillus subtilis. Microbiology 2000; 146:1573-1583.
8. Fabret C, Hoch JA. A two-component signal transduction system essential for growth of Bacillus subtilis: implications for anti-infective therapy. J Bacteriol 1998; 180:6375-6383.
9. Martin PK, Li T, Sun D et al. Role in cell permeability of an essential two-component system in Staphylococcus aureus. J Bacteriol 1999; 181:3666-3673.
10. Dubrac S, Msadek T. Identification of genes controlled by the essential YycG/YycF two-component system of Staphylococcus aureus. J Bacteriol 2004; 186:1175-1181.
11. Throup JP, Koretke KK, Bryant AP et al. A genomic analysis of two-component signal transduction in Streptococcus pneumoniae. Mol Microbiol 2000; 35:566-576.
12. Lange R, Wagner C, de Saizieu A et al. Domain organization and molecular characterization of 13 two-component systems identified by genome sequencing of Streptococcus pneumoniae. Gene 1999; 237:223-234.
13. Echenique JR, Trombe MC. Competence repression under oxygen limitation through the two-component MicAB signal-transducing system in Streptococcus pneumoniae and involvement of the PAS domain of MicB. J Bacteriol 2001; 183:4599-4608.
14. Senadheera MD, Guggenheim B, Spatafora GA et al. A VicRK signal transduction system in Streptococcus mutans affects gtfBCD, gbpB and ftf expression, biofilm formation and genetic competence development. J Bacteriol 2005; 187:4064-4076.
15. Liu M, Hanks TS, Zhang J et al. Defects in ex vivo and in vivo growth and sensitivity to osmotic stress of group A Streptococcus caused by interruption of response regulator gene vicR. Microbiology 2006; 152:967-978.
16. Kallipolitis BH, Ingmer H. Listeria monocytogenes response regulators important for stress tolerance and pathogenesis. FEMS Microbiol Lett 2001; 204:111-115.
17. Hancock LE, Perego M. Systematic inactivation and phenotypic characterization of two-component signal transduction systems of Enterococcus faecalis V583. J Bacteriol 2004; 186:7951-7958.
18. Qin Z, Zhang J, Xu B et al. Structure-based discovery of inhibitors of the YycG histidine kinase: new chemical leads to combat Staphylococcus epidermidis infections. BMC Microbiol 2006; 6:96.
19. Dubrac S, Boneca IG, Poupel O et al. New insights into the WalK/WalR (YycG/YycF) essential signal transduction pathway reveal a major role in controlling cell wall metabolism and biofilm formation in Staphylococcus aureus. J Bacteriol 2007; 189:8257-8269.
20. Bisicchia P, Noone D, Lioliou E et al. The essential YycFG two-component system controls cell wall metabolism in Bacillus subtilis. Mol Microbiol 2007; 65:180-200.
21. Ng WL, Robertson GT, Kazmierczak KM et al. Constitutive expression of PcsB suppresses the requirement for the essential VicR (YycF) response regulator in Streptococcus pneumoniae R6. Mol Microbiol 2003; 50:1647-1663.
22. Ng WL, Kazmierczak KM, Winkler ME. Defective cell wall synthesis in Streptococcus pneumoniae R6 depleted for the essential PcsB putative murein hydrolase or the VicR (YycF) response regulator. Mol Microbiol 2004; 53:1161-1175.
23. Dubrac S, Bisicchia P, Devine K et al. A matter of life and death: cell wall homeostasis and the WalKR (YycGF) regulon. (Submitted for publication) 2008;
24. Szurmant H, Nelson K, Kim EJ et al. YycH regulates the activity of the essential YycFG two-component system in Bacillus subtilis. J Bacteriol 2005; 187:5419-5426.
25. Szurmant H, Mohan MA, Imus PM et al. YycH and YycI interact to regulate the essential YycFG two-component system in Bacillus subtilis. J Bacteriol 2007; 189:3280-3289.

26. Szurmant H, Zhao H, Mohan MA et al. The crystal structure of YycH involved in the regulation of the essential YycFG two-component system in Bacillus subtilis reveals a novel tertiary structure. Protein Sci 2006; 15:929-934.
27. Santelli E, Liddington RC, Mohan MA et al. The crystal structure of Bacillus subtilis YycI reveals a common fold for two members of an unusual class of sensor histidine kinase regulatory proteins. J Bacteriol 2007; 189:3290-3295.
28. Daiyasu H, Osaka K, Ishino Y et al. Expansion of the zinc metallo-hydrolase family of the beta-lactamase fold. FEBS Lett 2001; 503:1-6.
29. Ng WL, Winkler ME. Singular structures and operon organizations of essential two-component systems in species of Streptococcus. Microbiology 2004; 150:3096-3098.
30. Wagner C, Saizieu Ad A, Schonfeld HJ et al. Genetic analysis and functional characterization of the Streptococcus pneumoniae vic operon. Infect Immun 2002; 70:6121-6128.
31. Senadheera MD, Lee AW, Hung DC et al. The Streptococcus mutans vicX Gene Product Modulates gtfB/C Expression, Biofilm Formation, Genetic Competence and Oxidative Stress Tolerance. J Bacteriol 2007; 189:1451-1458.
32. Noone D, Howell A, Collery R et al. YkdA and YvtA, HtrA-like serine proteases in Bacillus subtilis, engage in negative autoregulation and reciprocal cross-regulation of ykdA and yvtA gene expression. J Bacteriol 2001; 183:654-663.
33. Clausen T, Southan C, Ehrmann M. The HtrA family of proteases: implications for protein composition and cell fate. Mol Cell 2002; 10:443-455.
34. Stack HM, Sleator RD, Bowers M et al. Role for HtrA in stress induction and virulence potential in Listeria monocytogenes. Appl Environ Microbiol 2005; 71:4241-4247.
35. Taylor BL, Zhulin IB. PAS domains: internal sensors of oxygen, redox potential and light. Microbiol Mol Biol Rev 1999; 63:479-506.
36. Ulrich LE, Zhulin IB. MiST: a microbial signal transduction database. Nucleic Acids Res 2007; 35:D386-390.
37. Clausen VA, Bae W, Throup J et al. Biochemical characterization of the first essential two-component signal transduction system from Staphylococcus aureus and Streptococcus pneumoniae. J Mol Microbiol Biotechnol 2003; 5:252-260.
38. Depardieu F, Courvalin P, Msadek T. A six amino acid deletion, partially overlapping the VanSB G2 ATP-binding motif, leads to constitutive glycopeptide resistance in VanB-type Enterococcus faecium. Mol Microbiol 2003; 50:1069-1083.
39. Alves R, Savageau MA. Comparative analysis of prototype two-component systems with either bifunctional or monofunctional sensors: differences in molecular structure and physiological function. Mol Microbiol 2003; 48:25-51.
40. Howell A, Dubrac S, Andersen KK et al. Genes controlled by the essential YycG/YycF two-component system of Bacillus subtilis revealed through a novel hybrid regulator approach. Mol Microbiol 2003; 49:1639-1655.
41. O'Connell-Motherway M, van Sinderen D, Morel-Deville F et al. Six putative two-component regulatory systems isolated from Lactococcus lactis subsp. cremoris MG1363. Microbiology 2000; 146:935-947.
42. Mascher T, Helmann JD, Unden G. Stimulus perception in bacterial signal-transducing histidine kinases. Microbiol Mol Biol Rev 2006; 70:910-938.
43. Mizuno T, Tanaka I. Structure of the DNA-binding domain of the OmpR family of response regulators. Molecular Microbiology 1997; 24:665-667.
44. Martinez-Hackert E, Stock AM. Structural relationships in the OmpR family of winged-helix transcription factors. J Mol Biol 1997; 269:301-312.
45. Martinez-Hackert E, Stock AM. The DNA-binding domain of OmpR: crystal structures of a winged helix transcription factor. Structure 1997; 5:109-124.
46. Blanco AG, Sola M, Gomis-Ruth FX et al. Tandem DNA recognition by PhoB, a two-component signal transduction transcriptional activator. Structure 2002; 10:701-713.
47. Makino K, Amemura M, Kawamoto T et al. DNA binding of PhoB and its interaction with RNA polymerase. J Mol Biol 1996; 259:15-26.
48. Trinh CH, Liu Y, Phillips SE et al. Structure of the response regulator VicR DNA-binding domain. Acta Crystallogr D Biol Crystallogr 2007; 63:266-269.
49. Ng WL, Tsui HC, Winkler ME. Regulation of the pspA virulence factor and essential pcsB murein biosynthetic genes by the phosphorylated VicR (YycF) response regulator in Streptococcus pneumoniae. J Bacteriol 2005; 187:7444-7459.
50. Howell A, Dubrac S, Noone D et al. Interactions between the YycFG and PhoPR two-component systems in Bacillus subtilis: the PhoR kinase phosphorylates the noncognate YycF response regulator upon phosphate limitation. Mol Microbiol 2006; 59:1199-1215.

51. Lee SF, Delaney GD, Elkhateeb M. A two-component covRS regulatory system regulates expression of fructosyltransferase and a novel extracellular carbohydrate in Streptococcus mutans. Infect Immun 2004; 72:3968-3973.
52. Stapleton MR, Horsburgh MJ, Hayhurst EJ et al. Characterization of IsaA and SceD, two putative lytic transglycosylases of Staphylococcus aureus. J Bacteriol 2007; 189:7316-7325.
53. Ahn SJ, Burne RA. Effects of oxygen on biofilm formation and the AtlA autolysin of Streptococcus mutans. J Bacteriol 2007; 189:6293-6302.
54. Mohedano ML, Overweg K, de la Fuente A et al. Evidence that the essential response regulator YycF in Streptococcus pneumoniae modulates expression of fatty acid biosynthesis genes and alters membrane composition. J Bacteriol 2005; 187:2357-2367.
55. Aguilar PS, Hernandez-Arriaga AM, Cybulski LE et al. Molecular basis of thermosensing: a two-component signal transduction thermometer in Bacillus subtilis. EMBO J 2001; 20:1681-1691.
56. Deng DM, Liu MJ, ten Cate JM et al. The VicRK system of Streptococcus mutans responds to oxidative stress. J Dent Res 2007; 86:606-610.
57. Ramadurai L, Jayaswal RK. Molecular cloning, sequencing and expression of lytM, a unique autolytic gene of Staphylococcus aureus. J Bacteriol 1997; 179:3625-3631.
58. Sakata N, Mukai T. Production profile of the soluble lytic transglycosylase homologue in Staphylococcus aureus during bacterial proliferation. FEMS Immunol Med Microbiol 2007; 49:288-295.
59. Shemesh M, Tam A, Feldman M et al. Differential expression profiles of Streptococcus mutans ftf, gtf and vicR genes in the presence of dietary carbohydrates at early and late exponential growth phases. Carbohydr Res 2006; 341:2090-2097.
60. Martin PK, Bao Y, Boyer E et al. Novel locus required for expression of high-level macrolide-lincosamide-streptogramin B resistance in Staphylococcus aureus. J Bacteriol 2002; 184:5810-5813.
61. Friedman L, Alder JD, Silverman JA. Genetic changes that correlate with reduced susceptibility to daptomycin in Staphylococcus aureus. Antimicrob Agents Chemother 2006; 50:2137-2145.
62. Jansen A, Turck M, Szekat C et al. Role of insertion elements and yycFG in the development of decreased susceptibility to vancomycin in Staphylococcus aureus. Int J Med Microbiol 2007; 297:205-215.
63. Mwangi MM, Wu SW, Zhou Y et al. Tracking the in vivo evolution of multidrug resistance in Staphylococcus aureus by whole-genome sequencing. Proc Natl Acad Sci USA 2007; 104:9451-9456.
64. Kadioglu A, Echenique J, Manco S et al. The MicAB two-component signaling system is involved in virulence of Streptococcus pneumoniae. Infect Immun 2003; 71:6676-6679.
65. Brown JS, Gilliland SM, Holden DW. A Streptococcus pneumoniae pathogenicity island encoding an ABC transporter involved in iron uptake and virulence. Mol Microbiol 2001; 40:572-585.
66. Ren B, McCrory MA, Pass C et al. The virulence function of Streptococcus pneumoniae surface protein A involves inhibition of complement activation and impairment of complement receptor-mediated protection. J Immunol 2004; 173:7506-7512.
67. Ren B, Szalai AJ, Hollingshead SK et al. Effects of PspA and antibodies to PspA on activation and deposition of complement on the pneumococcal surface. Infect Immun 2004; 72:114-122.
68. Munro C, Michalek SM, Macrina FL. Cariogenicity of Streptococcus mutans V403 glucosyltransferase and fructosyltransferase mutants constructed by allelic exchange. Infect Immun 1991; 59:2316-2323.
69. Yamamoto K, Kitayama T, Minagawa S et al. Antibacterial agents that inhibit histidine protein kinase YycG of Bacillus subtilis. Biosci Biotechnol Biochem 2001; 65:2306-2310.
70. Hilliard JJ, Goldschmidt RM, Licata L et al. Multiple mechanisms of action for inhibitors of histidine protein kinases from bacterial two-component systems. Antimicrob Agents Chemother 1999; 43:1693-1699.
71. Furuta E, Yamamoto K, Tatebe D et al. Targeting protein homodimerization: a novel drug discovery system. FEBS Lett 2005; 579:2065-2070.
72. Thompson JD, Higgins DG, Gibson TJ. CLUSTAL W: improving the sensitivity of progressive multiple sequence alignment through sequence weighting, position-specific gap penalties and weight matrix choice. Nucleic Acids Res 1994; 22:4673-4680.

# CHAPTER 16

# Inhibitors Targeting Two-Component Signal Transduction

**Takafumi Watanabe, Ario Okada, Yasuhiro Gotoh and Ryutaro Utsumi***

## Abstract

A two-component signal transduction system (TCS) is an attractive target for antibacterial agents. In this chapter, we review the TCS inhibitors developed during the past decade and introduce novel drug discovery systems to isolate the inhibitors of the YycG/YycF system, an essential TCS for bacterial growth, in an effort to develop a new class of antibacterial agents.

## Introduction

In bacteria, signal transduction in response to a wide variety of environmental stimuli is mediated by pairs of proteins that communicate with each other through a two-component signal transduction system (TCS, Fig. 1) involving protein phosphorylation. TCS consists of a histidine kinase (HK) and a response regulator (RR), which play global roles in bacterial growth as well as drug-resistance, virulence, biofilm formation, quorum sensing and regulation of receptors of plant hormones such as ethylene and cytokinine.[1] In particular, TCSs are attractive as targets for antimicrobials for the following reasons: i) Although many HK and RR genes are coded on the bacterial genome, few are found in lower eukaryotes. For example, only one HK and two RRs are present in the complete genome sequence of *Saccharomyces cerevisiae* and none has been found from mammalian genomes. The HK/RR signal transduction system is distinct from serine/threonine and tyrosine phosphorylation in higher eukaryotes. ii) HKs and RRs possess a high degree of homology around their active sites. HKs are characterized by a conserved C-terminal catalytic domain containing a histidine residue, which is the site of autophosphorylation and the N, G1, F and G2 boxes, which presumably form a nucleotide-binding surface.[1,2] RRs contain a conserved N-terminal domain that includes the aspartate residue that is phosphorylated, a pair of aspartate residues preceding this site and a lysine residue, all of which contribute to the acidic pocket for the phosphorylation site.[2] Such a high degree of structural homology in the catalytic domain of HKs and in the receiver domain of RRs suggest that multiple TCSs within a single bacterium could be inhibited simultaneously by a single inhibitor, thereby lowering the frequency of the appearance of drug-resistance strains. iii) Some TCSs are essential for bacterial viability (Table 1) such as CckA/CtrA in *Caulobacter crescentus* (Chapter 8), MtrB/MtrA in *Mycobacterium tuberculosis*,[3] HP166/HP165 in *Helicobacter pylori*[4] and YycG/YycF in *Bacillus subtilis*,[5] *Enterococcus faecalis*,[6] *Streptococcus pneumoniae*[7,8] and *Staphylococcus aureus*.[9] YycG/YycF is considered a novel target for antibacterial agents against multidrug-resistant bacteria, including methicillin-resistant *S. aureus* (MRSA) and vancomycin-resistant *E. faecalis* (VRE)(Chapter 15).

*Corresponding Author: Ryutaro Utsumi—Department of Bioscience, Graduate School of Agriculture, Kinki University, 3327-204 Nakamachi, Nara 631-8505, Japan. Email: utsumi@nara.kindai.ac.jp

*Bacterial Signal Transduction: Networks and Drug Targets*, edited by Ryutaro Utsumi.

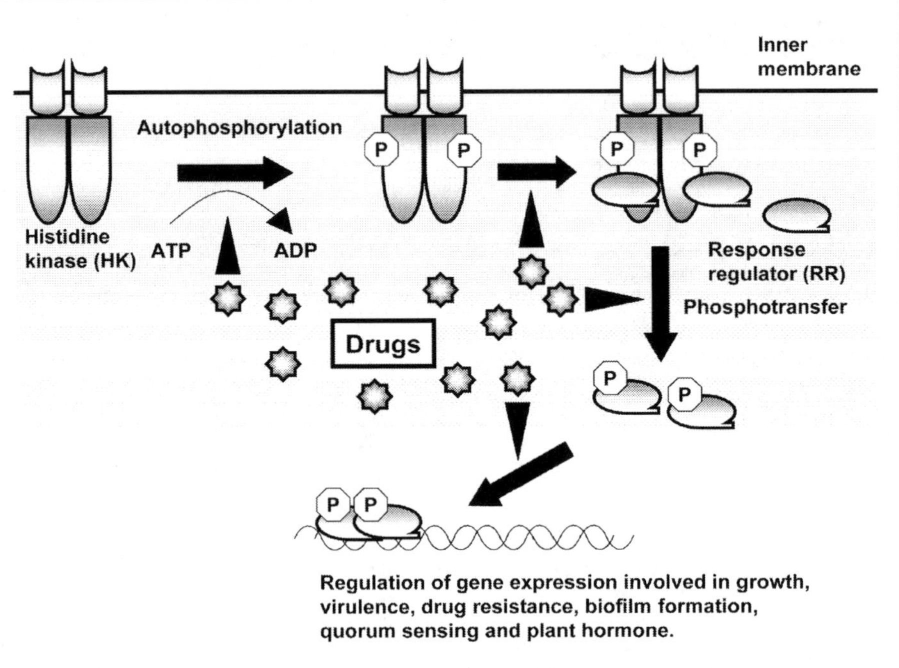


Figure 1. TCS for drug discovery. ► indicates the drug-target sites.

Here, we present an overview of the inhibitors of HK and introduce novel drug discovery systems to isolate the inhibitors of the YycG/YycF system, an essential TCS for bacterial growth, in an attempt to develop a new class of antibacterial agents.

## HK Inhibitors[10-12]

The first synthetic TCS inhibitors were reported in a search for compounds that modulate the AlgR2/AlgR1, regulating the production of alginate, a virulence factor of *Pseudomonas aeruginosa* expressed in the lungs of cystic fibrosis patients during infection.[13] Isothiazolone and imidazolium salt inhibited *algD* promoter activation linked to a *xylE* reporter gene in a whole cell assay employ-

***Table 1. Essential TCS for cell growth[6,31-36*]***

| Strains | Genome Size (Mb) | TCS* | | Essential TCS |
|---|---|---|---|---|
| | | HK | RR | |
| *Caulobacter crescentus CB15* | 4.0 | 34 | 44 | CckA-CtrA |
| *Mycobacterium tuberculosis* H37Rv | 4.4 | 15 | 15 | MtrB-MtrA |
| *Helicobacter pylori* 26695 | 1.7 | 4 | 7 | HP165-HP166 |
| *Bacillus subtilis* 168 | 4.2 | 37 | 34 | YycG-YycF |
| *Enterococcus faecalis* V583 | 3.2 | 17 | 18 | YycG-YycF |
| *Streptococcus pneumoniae* R6 | 2.0 | 13 | 14 | YycG-YycF |
| *Staphylococcus aureus* Mu50 | 2.9 | 17 | 17 | YycG-YycF |

*Number of TCSs on chromosomal DNA.

ing a *P. aeruginosa* strain ($IC_{50}$, 1-2 μM). Neither compound inhibited *P. aeruginosa* cell growth. Isothiazolones (Fig. 2A) had a significant effect on the autophosphorylation of AlgR2 (HK) above a threshold concentration of 2 μM, but they did not inhibit the binding of AlgR1 (RR) to the *algD* promoter. On the other hand, imidazolium salt and analogue (Fig. 2B) inhibited the kinase activities of CheA, NRII and KinA. A series of the structurally related imidazole compounds were also reported as inhibitors of four HKs, i.e., EnvZ, PhoQ, BvgS and EvgS.[14] In this case, the prototypical members of the series inhibited the growth of bacteria including oxacillin-resistant *S. aureus*, penicillin-resistant *S. pneumoniae* and VRE at MIC of 0.39-3.12 μg/ml.

Recently, Gilmour et al[15] identified a novel thienopyridine (TEP) compound (Fig. 2C) in a high-throughput screen of compound libraries. In this screen, a coupled assay containing the HpkA (HK) and DrrA (RR) of *Thermotoga maritima* was used. TEP inhibits HpkA ($IC_{50}$, 5.5 μM) competitively with respect to ATP but does not comparably inhibit mammalian serine/threonine kinases and does not inhibit the growth of bacterial or mammalian cells. TEP could serve as a starting compound for a new class of HK inhibitors with antibacterial activity.

Previous work reported five chemotypes (cyclohexene, closantel, benzimidazole, trityl and bisphenol) (Fig. 2D, E, F, G and H) that possess inhibitory activity against KinA (an HK with an $IC_{50}$ value ranging from 2-20 μM) and that result in antibacterial activity against gram-positive bacteria (MIC, 2 μg/ml).[16,17] These compounds also had an appreciable effect on the cell membrane integrity or caused hemolysis of equine erythrocytes.[16]

A microbial extract screening program to identify inhibitors of HK had also been performed and the streptopyrroles (Fig. 2I) were isolated. These inhibited NRII, a HK with an $IC_{50}$ of 20 μM, and exhibited antimicrobial activity against a wide range of bacteria and fungi (MICs, 0.78-12.5 μg/ml).[18]

While these compounds inhibited both bacterial growth and HK, no absolute proof has been presented showing that growth inhibition was the direct consequence of HK inhibition. Since

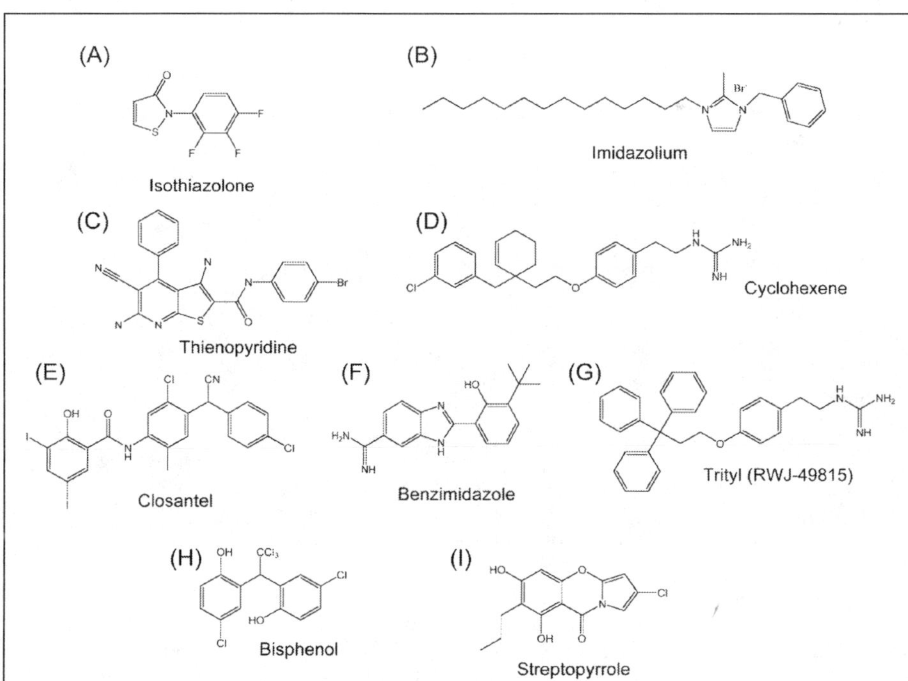


Figure 2. Histidine kinase inhibitor.

they are inhibitors against non-essential HKs, including KinA, AlgR2, VanS, CheA, NRII, EnvZ, PhoQ, BvgS and EvgS, it has not been established whether the inhibition of HK accounts for the bactericidal activity.

On the other hand, inhibitors targeting YycG (HK), essential for the growth of *B. subtilis*, were investigated to find imidazole derivatives (Fig. 4A, NH125) with $IC_{50}$ of 6.6-40 μM and the zerumbone derivative NH0891.[19,20] Imidazole derivatives had antibacterial activity in drug-resistant *S. aureus*, *E. faecalis* and *S. pneumoniae* as well as *B. subtilis* with MICs of 0.39-6.25 μg/ml. These results show a good correlation between the growth inhibition of gram-positive bacteria and the inhibition of YycG, indicating that the bacterial signal transduction YycG/YycF is a promising target for antibacterial agents.[5,12] In the following section, we present novel drug discovery systems (differential growth assay and homodimerization assay) by which the inhibitors targeting YycG/YycF are selectively screened with natural products and a synthetic compound library to develop a new class of antibacterial agents.

## Inhibitors Targeting an Essential TCS, YycG/YycF

### *Differential Growth Assay*[21,22]

Hypersensitive strains have been used for the discovery of new drugs targeting specific cellular pathways.[23,24] The specific hypersensitivities displayed by temperature- sensitive (ts) mutants indicate that use of these mutants in whole cell screening provides a rapid method to develop target-specific screens for the identification of novel compounds.

By a differential growth assay using a temperature-sensitive *yycF* mutant (CNM2000) of *B. subtilis*, which is supersensitive to inhibitors of HK,[21] inhibitors of YycG were isolated by screening samples of acetone extracts from 4000 microbes. A total of 11 samples showed greater activity against CNM2000 than against strain 168 (wild type). Seven of those samples significantly inhibited the autophosphorylation activity of YycG. Starting with the most potent extract from

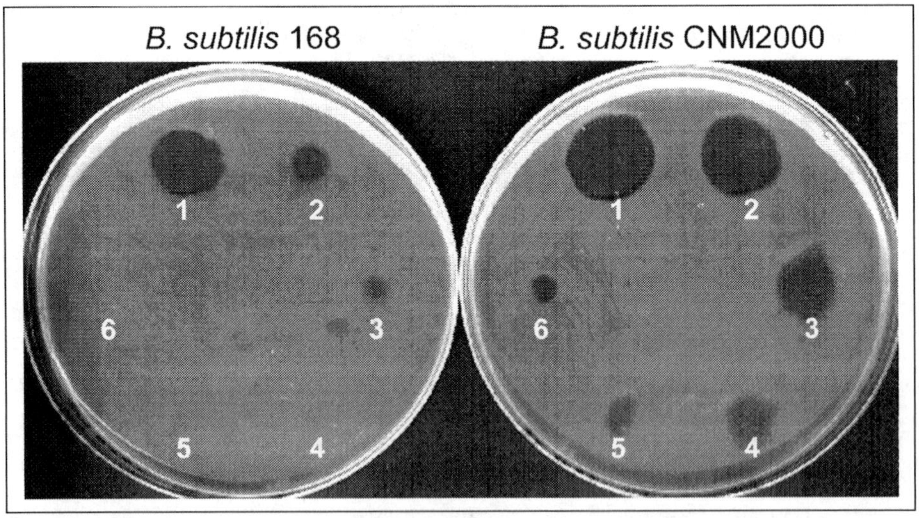


Figure 3. Differential growth assay using HK inhibitor aranorosinol B. One microliter of aranorosinol B at six concentrations ( μg/μ l) was spotted on trypticase soy agar plates containing 168 and CNM2000: 10 (1), 5(2), 2.5(3), 1.25(4), 0.625 (5) and 0.313 (6). Reproduced with permission from Watanabe T, Hashimoto Y, Yamamoto K, et al. Characterization of Inhibitors of the Essential Histidine Kinase, YycG in *Bacillus subtilis* and *Staphylococcus aureus*. J Antibiot 2003 (56):1045-1052.

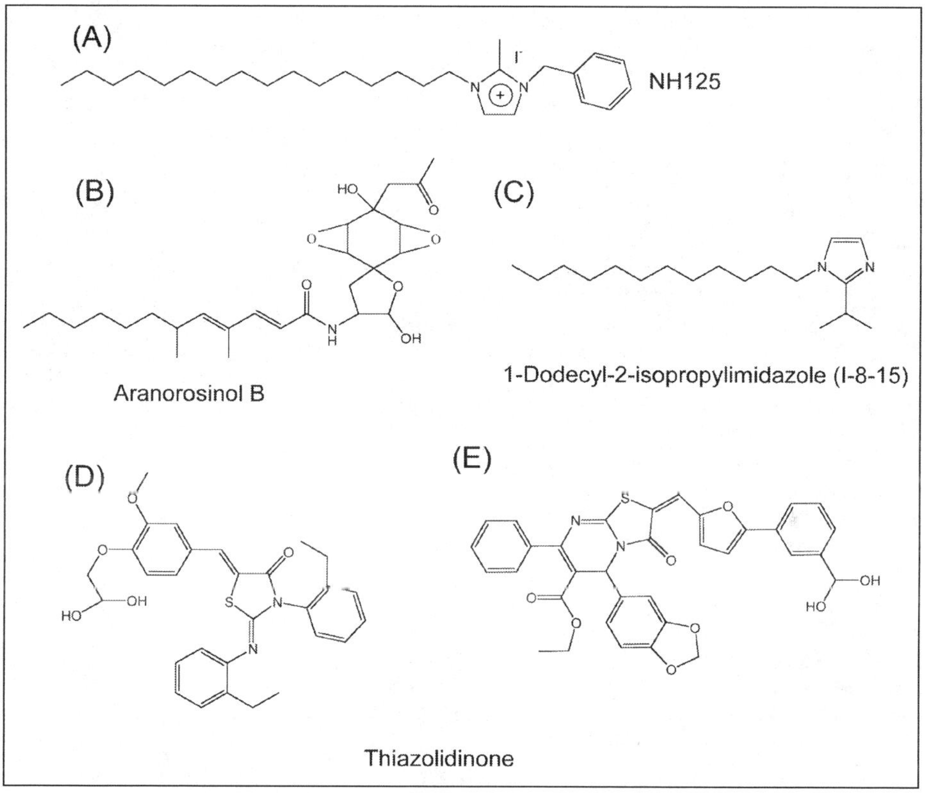


Figure 4. Inhibitors of YycG.

the microbe WF140196, identified as *Gymnascella dankaliensis* (*Pseudoarachniotus roseus*), an active compound was purified that was more potent against CNM2000 than against strain 168 (Fig. 3). This compound was identified as aranorosinol B (Fig. 4B).[22] The purified aranorosinol B inhibited autophosphorylation of YycG from both *B. subtilis* and *S. aureus* as well as 13 other HKs from *Escherichia coli* ($IC_{50}$, 70-700 μM). Aranorosinol B did not show antibacterial activity against gram-negative bacteria including *E. coli* and *P. aeruginosa*, while MICs against *B. subtilis* and *S. aureus* were 31.25 and 15.62 μg/ml, respectively.

## *Homodimerization Assay*[21,25]

Cytoplasmic or truncated forms of HKs have been known to dimerize in vitro.[26] Solution state NMR[27] indeed revealed dimerization of the homodimeric core domain in EnvZ (HK), which belongs to the same Pho subfamily of HKs as YycG. Dimerization of YycG is considered an essential step in the autophosphorylation of this enzyme.[28] To study the function of YycF in the essential process, we characterized a YycF (H215P) mutation that caused temperature-sensitive growth in *B. subtilis*.[28] The response regulators YycF and YycF (H215P) were analyzed using circular dichroism spectroscopy, whose Tm values were 56.0 and 45.9°C, respectively, suggesting that the protein structure of YycF (H215P) is significantly sensitive to an increase in temperature. Furthermore, using the gel mobility shift assay and DNase I footprinting, we investigated the YycF (H215P) binding to the YycF box of the *ftsAZ* operon of *B. subtilis*. The replacement of the histidine 215 with proline resulted in a decrease of the DNA-binding ability of YycF in vitro. In vivo, using *E. coli* two-hybrid and homodimerization assays, it was clarified that H215 of YycF plays a crucial role in homodimerization of the protein.[28] Thus, the essential genes involved in the growth of *B. subtilis* appear to be regulated by the homodimerization of YycF. These results suggest that YycG and YycF dimerization is an excellent target for the discovery of novel antibiotics. To identify such drugs, we have developed a high-throughput genetic system for targeting the homodimerization of HK and RR, which are based on the dimerization properties of the IclR repressor of *E. coli*, using GFP as the reporter gene (Fig. 5).

To identify the inhibitors against YycG and YycF, we screened a chemical library using *E. coli* cultures, JM109/pFI028 and JM109/pFI014, respectively. Through this screening process, we identified two compounds named I-8-15 (Fig. 4C)[25] and D897 (to be submitted) that significantly increased the fluorescent intensity of JM109/pFI028 and JM109/pFI014, respectively. I-8-15 indeed inhibited autophosphorylation of YycG with an $IC_{50}$ of 76.5 μM and also significantly inhibited the growth of MRSA and VRE with MICs at 25 and 50 μg/ml, respectively. D897 did not affect the autophosphorylation of YycG but did inhibit the growth of MRSA and VRE with MICs of 100 μg/ml as well as the DNA binding of YycF to the promoters of *tagA* and *yocH* in *B. subtilis* (to be submitted). When aranorosinol B, which was isolated using the differential growth assay, was added in the culture of JM109/pFI028, no increase in fluorescent intensity was observed.[25] I-8-15 acted on YycG in a different manner from that of aranorosinol B and exerted its bactericidal activity by inhibiting the autophosphorylation of YycG. Consequently, by using both a differential growth assay and a high-throughput genetic system, it is possible to isolate various types of TCS inhibitors with different modes of action (Fig. 6).

## Structure-Based Virtual Screening

Using structure-based virtual screening from a small molecular lead-compound library, the inhibitors (thiazolidinone derivatives, Fig. 4D, 4E) of YycG of *Staphylococcus epidermidis* were discovered with $IC_{50}$ of 6.5-29 μM.[29] They displayed bactericidal effects on biofilm cells of *S. epidermidis*, indicating that they can serve as potential agents against *S. epidermidis* biofilms.[30]

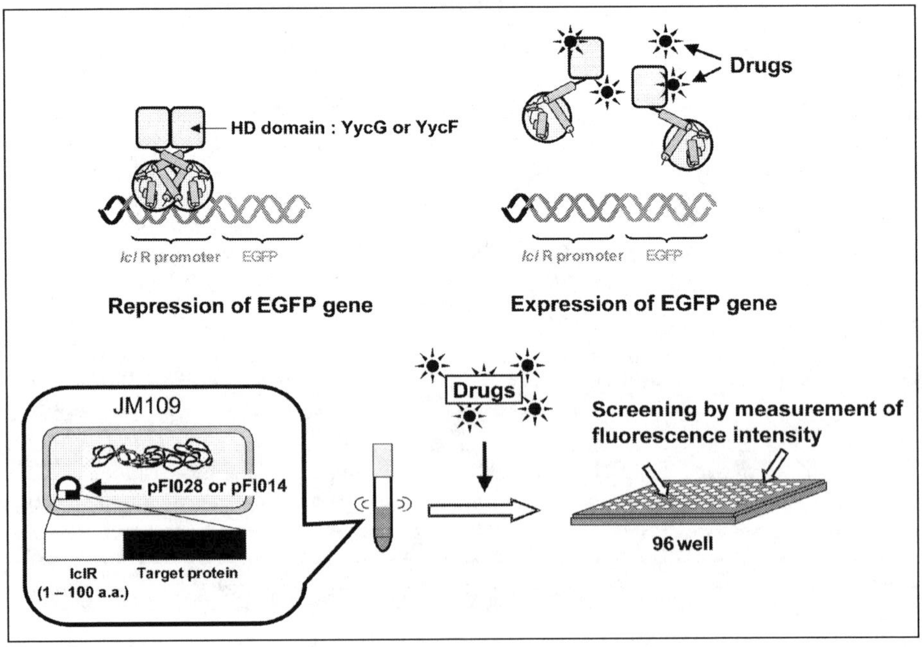


Figure 5. Homodimerization assay system (HD system).

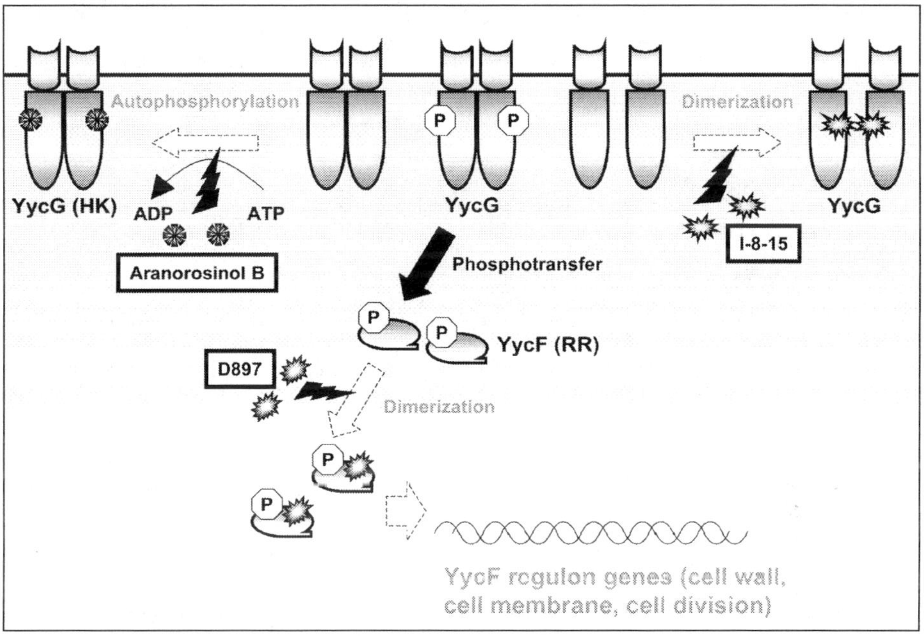


Figure 6. A new class of antibacterial agents against YycG/YycF.

Importantly, these compounds did not affect the stability of mammalian cells or hemolytic activities. The structure-based drug discovery system can also be used in developing potential TCS inhibitors.

## *Acknowledgements*

This work was supported by the Research and Development Program for New Bio-industry Initiatives (2006-2010) of the Bio-oriented Technology Research Advanced Institution (BRAIN), Japan and the Academic Frontier Project for Private Universities through a matching fund subsidy from the Ministry of Education, Culture, Sports, Science and Technology (2004-2008), Japan. It was also supported in part by the Nara Prefecture Collaboration of Regional Entities for the Advancement of Technological Excellence, JST (2005-2009), Japan.

## References

1. Inouye M, Dutta R, eds. Histidine kinases in signal transduction: San Diego: Academic Press, 2003.
2. Parkinson JS, Kofoid EC. Communication module in bacterial signaling proteins. Annu Rev Genet 1992; 26:71-121.
3. T Deretic V. An essential two-component signal transduction system in Mycobacterium tuberculosis. J Bacteriol 2000; 182:3832-3838.
4. Beir D, Frank R. Molecular characterization of two-component systems of Helicobacter pylori. J Bacteriol 2000; 182:2068-2076.
5. Fabret C, Hoch JA. A two-component signal transduction system essential for growth of Bacillus subtilis: implications for anti-infective therapy. J Bacteriol 1998; 180:6375-6383.
6. Hancock L, Perego M. Two-component signal transduction in Enterococcus faecalis. J Bacteriol 2002; 184:5819-5825.
7. Lange R, Wagner C, de Saizieu A et al. Domain organization and molecular characterization of 13 two-component systems identified by genome sequencing of Streptococcus pneumoniae. Gene 1999; 237:223-234.
8. Throup JP, Koretke KK, Bryant AP et al. A genomic analysis of two-component signal transduction in Streptococcus pneumoniae. Mol Microbiol 2000; 35:566-576.

9. Martin PK, Li T, Sun D et al. Role in cell permeability of an essential two-component system in Staphylococcus aureus. J Bacteriol 1999; 181:3666-3673.
10. Macielag MJ, Goldschmidt R. Inhibitors of bacterial two-component signaling systems. Exp Opin Invest Drugs 2000; 9:2351-2369.
11. Matsushita M, Janda KD. Histidine kinases as targets for new antimicrobial agents. Bioorg Med Chem 2002; 10:855-867.
12. Barrette JF, Hoch JA. Two-component signal transduction as a target for microbial anti-infective therapy. Antimicrob Agents Chemother 1998; 42:1529-1536.
13. Roychoudhurry S, Zielinski NA, Ninfa AJ et al. Inhibitor of two-component signal transduction systems: Inhibition of alginate gene activation in Pseudomonaas aeruginosa. Proc Natl Acad Sci USA 1993; 90:965-969.
14. Yamamoto K, Kitayama T, Ishida N et al. Identification and characterization of a potent antibacterial agent, NH125 against drug-resistant bacteria. Biosci Biotechnol Biochem 2000; 64:912-923.
15. Gilmour R, Foster JE, Sheng Q et al. New class of competitive inhibitor of bacterial histidine kinases. J Bacteriol 2005; 187:8196-8200.
16. Hilliard JJ, Goldschmidt RM, Licata L et al. Multiple mechanisms of Action for inhibitors of histidine protein kinases from bacterial two-component systems. Antimicrob Agents Chemother 1999; 43:1693-1699.
17. Barrett JF, Goldschmidt RM, Lawrence LE et al. Antibacterial agents that inhibit two-component signal transduction systems. Proc Natl Acad Sci USA 1998; 95:5317-5322.
18. Trew SJ, Wrigley SK, Pairet L et al. Novel Streptopyrroles from Streptomyces rimousus with bacterial protein histidine kinase inhibitory and antimicrobial activities. J Antibiotics 2000; 53:1-11.
19. Yamamoto K, Kitayama T, Minagawa S et al. Antibacterial agents that inhibit histidine protein kinase YycG of Bacillus subtilis. Biosci Biotechnol Biochem 2001; 65:2306-2310.
20. Kitayama T, Iwabuchi R, Minagawa S et al. Unprecedented olefin-dependent histidine-kinase inhibitory of zerumbone ring-opening material. Bioorgn Medicin Chem. Lett 2004; 14:5943-5946.
21. Okada A, Gotoh Y, Watanabe T et al. Targeting two-component signal transduction:a novel drug discovery system. Methods Enzymol 2007; 422:386-395.
22. Watanabe T, Hashimoto Y, Yamamoto K et al. Isolation and characterization of inhibitors of the essential histidine kinase, YycG in Bacillus subtilis and Staphylococcus aureus. J Antibiot 2003; 56:1045-1052.
23. Kamogashira T, Takegata S. A screening method for cell wall inhibitors using a D-cycloserine hypersensitive mutant. J Antibiot 1988; 41:803-806.
24. Numata K, Yamamoto H, Hatori M et al. Isolation of an aminoglycoside hypersensitive mutant and its application in screening. J Antibiot 1986; 39:994-1000.
25. Furuta E, Yamamoto K, Tatebe D et al. Targeting protein homodimerization: A novel drug discovery system. FEBS Lett 2005; 579:2065-2070.
26. Hidaka Y, Park H, Inouye M. Demonstration of dimer formation of the cytoplasmic domain of a transmembrane osmosensor protein, EnvZ of Escherichia coli using N-histidine tag affinity chromatography. FEBS Lett 1997; 400:238-243.
27. Tomomori C, Tanaka T, Dutta R et al. Solution structure of the homodimerization core domain of Escherichia coli histidine kinase EnvZ. Nat Struc Biol 1999; 6:729-734.
28. Watanabe T, Hashimoto Y, Umemoto Y et al. Molecular characterization of the essential response regulator protein YycF in Bacillus subtilis. J Mol Microbiol Biotechnol 2003; 6:155-163.
29. Qin Z, Zhang J, Xu B et al. Structure-based discovery of inhibitors of the YycG histidine kinase: new chemical leads to combat Staphylococcus epidermidis infections. BMC Microbiol 2006; 6:1-18.
30. Qin Z, Lee B, Zhang J et al. Antimicrobial activities of YycG histidine kinase inhibitors against Staphylococcus epidermidis biofilms. FEMS Microbiol Lett in press.
31. Nierman WC, Feldblyum TV, Laub MT et al. Complete genome sequence of Caulobacter crescentus. Proc Natl Acad Sci USA 2001; 98:4136-4141.
32. Cole ST, Brosch R, Parkhill J et al. Deciphering the biology of Mycobacterium tuberculosis from the complete genome sequence. Nature 1998; 393:537-544.
33. Tomb JF, White O, Kerlavage AR et al. The complete genome sequence of the gastric pathogen Helicobacter pylori. Nature 2000; 388:539-547.
34. Kunst F, Ogasawara N, Moszer I et al. The complete genome sequence of the gram-positive bacterium Bacillus subtilis. Nature 1997; 390:249-256.
35. Hoskins J, Alborn WE Jr, Arnold J et al. Genome of the bacterium Streptococcus pneumoniae strain R6. J Bacteriol 2001; 183:5709-5717.
36. Kuroda M, Ohta T, Uchiyama I et al. Whole genome sequencing of methicillin-resistant Staphylococcus aureus. Lancet 2001; 357:1225-1240.

# Index

## D



## E



## F



## G

## H



## I



## L



## M



## N



## O



## P

## Q



## R

## V



## Y